CATALYTIC OXIDATIONS WITH HYDROGEN PEROXIDE AS OXIDANT

Catalysis by Metal Complexes

VOLUME 9

The titles published in this series are listed at the end of this volume.

CATALYTIC OXIDATIONS WITH HYDROGEN PEROXIDE AS OXIDANT

Edited by

GIORGIO STRUKUL

Department of Chemistry,
University of Venezia, Italy

KLUWER ACADEMIC PUBLISHERS

DORDRECHT / BOSTON / LONDON

Library of Congress Cataloging-in-Publication Data

Catalytic oxidations with hydrogen peroxide as oxidant / Giorgio Strukul (editor).
p. cm. -- (Catalysis by metal complexes ; v. 9)
ISBN 0-7923-1771-8 (hb : acid free paper)
1. Oxidation. 2. Hydrogen peroxide. 3. Catalysis. I. Strukul, Giorgio. II. Series.
QD63.O9C38 1992
541.3'93--dc20 92-13459

ISBN 0-7923-1771-8

Published by Kluwer Academic Publishers,
P.O. Box 17, 3300 AA Dordrecht, The Netherlands.

Kluwer Academic Publishers incorporates
the publishing programmes of
D. Reidel, Martinus Nijhoff, Dr W. Junk and MTP Press.

Sold and distributed in the U.S.A. and Canada
by Kluwer Academic Publishers,
101 Philip Drive, Norwell, MA 02061, U.S.A.

In all other countries, sold and distributed
by Kluwer Academic Publishers Group,
P.O. Box 322, 3300 AH Dordrecht, The Netherlands.

Printed on acid-free paper

Printed in the Netherlands

TABLE OF CONTENTS

Foreword

Hydrogen peroxide is a chemical that is becoming increasingly fashionable as an oxidant both in industry and in academia and whose production is expected to increase significantly in the next few years. This growth in popularity is largely due to the fact that at the present time the environmental impact of chemical productions is under severe scrutiny and chemical Companies are facing increasing restrictions (and therefore inreasing costs) to reduce environmental pollution. From this point of view hydrogen peroxide is unique as an oxidant, its by-product being only water.

To date this chemical is largely employed as a non selective oxidant in operations like the bleaching of paper, cellulose and textiles, or in the formulation of detergents and only to a minimal extent in the manufacture of organic chemicals. It seems predictable that because of the advantages related to the above environmental considerations, this amount will increase in the future.

Unfortunately, over the years, hydrogen peroxide has not witnessed the rapid blooming of applications of e.g. the oxidations with alkylhydroperoxides catalyzed by transition metal species, the number of catalytic systems capable of delivering selectively an oxygen atom coming from hydrogen peroxide to an organic molecule being still quite limited. In reviewing metal-catalyzed epoxidation of olefins some years ago, R. Sheldon wrote: "Indeed it is significant that commercially operated epoxidation processes utilizing hydrogen peroxide are conspicuous by their absence". Although this sentence is ten years old and despite the increased interest toward hydrogen peroxide, the situation has not substantially changed.

Being aware of this limitation, this book has been organized to cover the different aspects of the chemistry of hydrogen peroxide, including those that are not strictly related to transition metal catalysis, in this respect being partly at odd with the general title of this Series. The various chapters in which the book is divided have been written critically by the authors (all well-known scientists in their respective fields) with the general aim to stimulate new ideas and emphasizing those aspects that are likely to lead to new developments in organic synthesis in the coming future. I thank all of them for accepting my invitation to review their respective arguments with enthusiasm and for their substantial contributions to this book.

Finally, I wish to express my sincere gratitude to the two Editors of this Series: Brian James for his long lasting friendship and scientific advice and mostly Renato Ugo for the initial idea of making a book on hydrogen peroxide and for convincing me to undertake the task.

Giorgio Strukul
Venezia 1991

List of Authors

Dr. Valeria CONTE
Centro di Studio sui Meccanismi di Reazioni Organiche del CNR,
Dipartimento di Chimica Organica, via Marzolo 1,
I-35131 Padova, ITALY

Professor Ruggero CURCI
Dipartimento di Chimica, Università di Bari, via Amendola 173,
I-70126 Bari, ITALY

Professor Fulvio DI FURIA
Centro di Studio sui Meccanismi di Reazioni Organiche del CNR,
Dipartimento di Chimica Organica, via Marzolo 1,
I-35131 Padova, ITALY

Professor John O. EDWARDS
Department of Chemistry, Brown University,
Providence, RI 02912, U.S.A.

Dr. Gustaaf GOOR
Department of Organic and Biological Chemistry Research, Degussa AG,
D-6450 Hanau, F.R.G.

Professor Craig L. HILL
Department of Chemistry, Emory University,
Atlanta, GA 30322, U.S.A.

Dr. Bernard MEUNIER
Laboratoire de Chimie de Coordination du CNRS,
205 route de Narbonne,
F-31077 Toulouse cedex, FRANCE

Professor Giorgio STRUKUL
Dipartimento di Chimica, Università di Venezia, Dorsoduro 2137,
I-30123 Venezia, ITALY

CHAPTER 1

Introduction and Activation Principles

Giorgio STRUKUL

Dipartimento di Chimica, Università di Venezia, Dorsoduro 2137, 30123 Venezia, Italy

1. Introduction and Historical Background

Catalytic oxidation is an area of particular significance for homogeneous catalysis, in fact it is the field where the largest contribution of homogeneous catalysis to industrial organic chemistry are found. In this field, even for the production of bulk chemicals, homogeneous catalysts appear to compete effectively with heterogeneous ones. This observation is self evident if data taken from the current literature [1] on the world capacity of some of the largest industrial productions involving catalytic oxidation are considered (Table 1). A classification by size clearly indicates that most of the reported processes are practised in the homogeneous phase. These include for example the oxidation of *p* -xylene to terephthalic acid, the production of propylene oxide from propylene, or the oxidation of acetaldehyde and light paraffins to acetic acid. In addition, in the case of acetic acid another fair 2.2 x 10^6 tons/year came in the same year 1986 from the rapidly expanding process involving the carbonylation of methanol catalyzed by Rh complexes, bringing the total contribution of homogeneous catalysis to acetic acid to 4.5 x 10^6 tons/year.

With the notable exception of propylene oxide, all these chemicals are produced with processes that utilize molecular oxygen as the primary oxidant. Particularly, some of the largerst productions like e.g. those of terephthalic acid, phenol or acetic acid are all homolytic in nature.

The chemistry that constitutes the basis for these technologies has been known for several decades and involves initially the autooxidation of the alkanes from which hydroperoxides are produced cyclically.
In Scheme 1 the representative case of cyclohexane is shown as an example. The dangerous building up of hydroperoxides in these systems is prevented by their subsequent decomposition from the metal catalyst (typically Co(II) salts)

G. Strukul (ed.), Catalytic Oxidations with Hydrogen Peroxide as Oxidant, 1–11.

Table 1: Catalytic oxidation processes for the production of bulk chemicals[a].

product	reactant	year	10^6 t/y	oxidant	process
terephthalic acid[b]	*p*-xylene	1987	10.1	O_2	homogeneous, homolytic
ethylene oxide	ethylene	1984	8.5	O_2	heterogeneous, heterolytic
phenol	toluene, cyclohexane benzene/propylene	1987	4.2[c]	O_2	homogeneous, homolytic
formaldehyde	methanol	1984	3.9	O_2	heterogeneous, heterolytic
acetaldehyde	ethylene	1986	3.0	O_2	homogeneous, heterolytic
vinyl acetate	ethylene/HOAc	1986	2.6[c]	O_2	heterogeneous, heterolytic
acetic acid	acetaldehyde light paraffins	1986	2.3[c]	O_2	homogeneous, homolytic
propylene oxide	propylene	1985	1.3	t-BuOOH PhEtOOH	homogeneous[d], heterolytic

[a] Estimated world capacity; data taken from ref. 1. [b] Includes dimethyl terephthalate. [c] Only oxidation processes are considered. [d] Includes Shell heterogeneous process based on TiO_2/SiO_2.

Scheme 1:

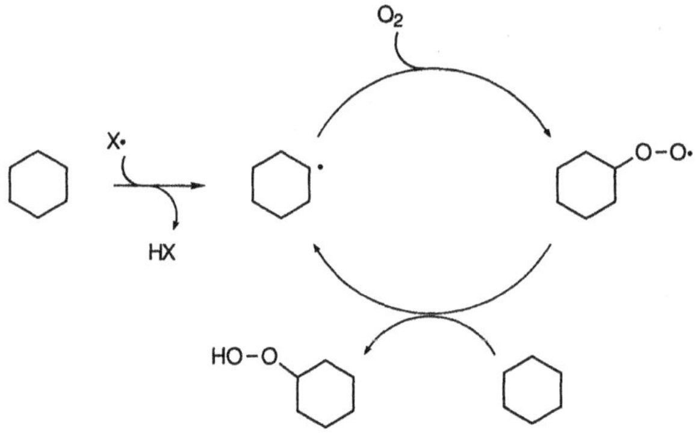

Co^{II} + CyOOH ⟶ Co^{III} + CyO• + HO^-

Co^{III} + CyOOH ⟶ Co^{II} + CyOO• + H^+

CyO• + Cy-H ⟶ Cy-OH + Cy•

2 CyOO• ⟶ O_2 + Cy-OH + Cy=O

that produces oxy and peroxy radicals through a pathway that is known as the Haber-Weiss mechanism [2]. Product formation takes place from hydrogen abstraction by the alkoxy radical and the virtually non catalyzed decomposition of the peroxyradical. As is clear from Scheme 1, while the role of the metal is very important in sustaining the radical chain reaction (i.e. for the reaction rate), it is relatively marginal for the selectivity of these systems which ranges typically around 40 - 65%. Energetic radicals such as RO• are rather indiscriminate in abstraction of hydrogen from carbon, especially at high temperatures. Moreover, the selectivity of the overall process is affected by the thermodynamics of oxidation reactions involving O_2, the stability of the various intermediates to oxidation being:

$$\text{alkane} > \text{alcohol} > \text{aldehyde, ketone} > \text{carboxylic acid}$$

with $CO_2 + H_2O$ as the thermodynamic sink.

These observations are widely known and are largely responsible for the practical need of operating the production plants at low conversion per pass and their being equipped with costly recycling facilities. Despite the many disadvantages related to the relatively low yields and extensive product separation, the fact that these processes are still industrially practised indicates clearly that no real alternatives have been found to date. Indeed, this situation is certainly due to the poor ligand character of simple, saturated (and to a lesser extent aromatic) hydrocarbons and hence the lack of catalysts capable to promote a heterolytic, non-radical activation of these substrates and their subsequent functionalyzation. As will be seen in other chapters of this book this situation is gradually improving, since for example the direct hydroxylation of phenol with H_2O_2 catalyzed by Ti silicalites is becoming an industrial reality.

Historically, the need for more selective metal systems capable of direct activation of dioxygen and transfer to suitable substrates prompted inorganic chemists to explore during the Sixties the reactivity of group VIII transition metals [3] following the early discovery by Vaska [4] that the well known Ir(I) complex $(PPh_3)_2Ir(CO)Cl$ was able to bind dioxygen reversibly. The general hope at that time was that analogously to other simple gas molecules like H_2 and CO, even the activation of O_2 with these metals would result in the catalytic oxidation of unsaturated hydrocarbons. This scientific effort led to the synthesis of a wide variety of stable dioxygen complexes of group VIII transition metals and the discovery of a very rich chemistry in reactions with a

Scheme 2:

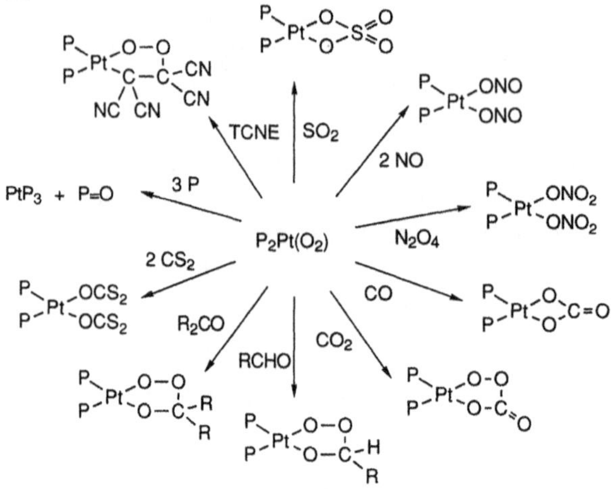

variety of substrates. The prototype complex of this class is the well known $(PPh_3)_2Pt(O_2)$ and a summary of its oxygen transfer reactions is shown in Scheme 2 [3c]. However, despite the expectations, none of the reactions known for these metal-dioxygen complexes resulted in the development of catalytic processes of some synthetic utility. It is now clear that these metal complexes have to be converted into hydroperoxy or alkylperoxy species in order to be able to perform oxygen transfer (see chapter 6).

In the same period industrial organic chemists were studying the reactivity of early transition metals toward what can be considered as a partially reduced form of dioxygen, i.e. organic hydroperoxides, that organic chemists had widely studied and learned to prepare in a stable form in the previous two decades. This approach proved much more rewarding industrially, since it led researchers at Atlantic Richfield and Halcon to the development and subsequent joint commercialization of the Oxirane process for the production of propylene oxide from propylene and alkylhydroperoxides catalyzed by Mo(VI) species [5] (see Scheme 3 for reaction mechanism [6]). This process was subsequently modified by Shell with the use of a heterogeneous catalysts based on TiO_2 dispersed on silica [7] and stands among the largest productions involving catalytic oxidation (Table 1).

Scheme 3:

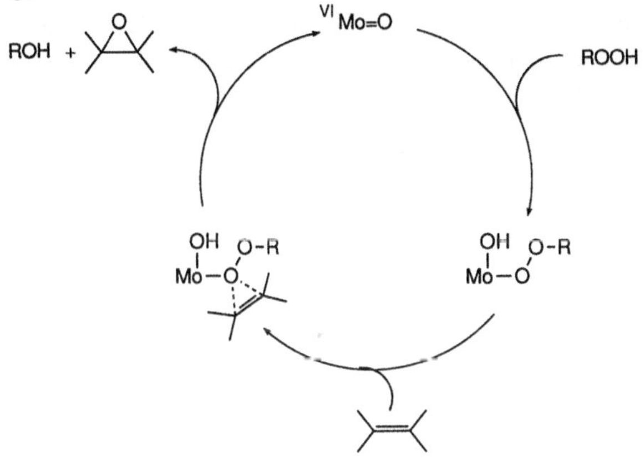

Other metal centers like Ti(IV), V(V), W(VI) [7a] behave similarly and in the past 20 years this chemistry has generated a wealth of interesting applications even in the synthesis of fine chemicals, the most famous one being probably the enantioselective epoxidation of allylic alcohols with *t* -BuOOH catalyzed by the so-called "Sharpless reagent" i.e. $Ti(OPr^i)_4$ modified with diethyltartrate [8].

Now the demand for catalytic oxidation processes even in fine chemistry seems to be rapidly increasing and this point has been keenly elucidated by Sheldon in a few occasions [9]. Low-salt or byproduct-free technologies are becoming more and more fashionable under the push of severe environmental constraints. For instance, in the fine chemical industry the use of dichromate or permanganate as oxidants is still quite common, although technologies based on these and similar oxidants will rapidly become prohibitive because of the increasing costs for chemical companies to clean up their effluents before disposing outside the production plants. In this respect hydrogen peroxide appears to be particularly appealing since it is the clean oxidant *par excellence*, its reduction product being only water.

2. Hydrogen Peroxide as Oxidant: An Overview

Hydrogen peroxide has a number of advantages compared to other common low cost oxidants.

The first one is that some of these oxidants like for example sodium perborate, potassium hydroperoxysulfate or even many organic peroxy acids are prepared from hydrogen peroxide itself. This implies that their use while leading to the ultimate consumption of hydrogen peroxide, on the other hand it requires further chemical operations and plant facilities necessary to the recovery of the reduction products and their subsequent conversion into the active form.

A second advantage comes from the active oxygen content. In Table 2 the active oxygen content of some common oxidants is reported.

Table 2: Single oxygen atoms donors.

Donor	% Active oxygen	Product
H_2O_2	47.0[a]	H_2O
O_3	33.3	O_2
t-BuOOH	17.8	*t*-BuOH
NaClO	21.6	NaCl
NaBrO	13.4	NaBr
HNO_3	25.4	NO_x
$KHSO_5$	10.5	$KHSO_4$
$NaIO_4$	7.2[b]	$NaIO_3$
PhIO	7.3	PhI

[a] Calculated on 100% H_2O_2. [b] Assuming only one oxygen atom is utilized.

This is calculated as the ratio between the weight of oxygen that can be transferred to a suitable substrate and the molecular weight of the oxidant. As is clear from Table 2, H_2O_2 largely exceeds all the others with the exception of ozone which suffers from being highly noxious and requiring costly equipments for its generation [10]. This observation has the obvious consequence that the price per kilo in the case of hydrogen peroxide corresponds to 470 g of active oxygen, while in the case of e.g. potassium hydroperoxysulfate this figure drops down to 105 g.

A third advantage comes from another economic consideration related to Table 2: the lower amount of effluents that is a consequence of the higher oxygen content. This is illustrated with the example below (Table 3) where the oxidation of propylene to propylene oxide with suitable oxidants is compared. If a theoretical yield of 100% is assumed in all cases, H_2O_2 produces a far lower amount of effluent (H_2O!) compared to the others. Although in practice

100% yields are rarely obtained, the potential economic advantages intrinsic in the use of hydrogen peroxide are quite evident.

Table 3: Amounts of oxidant required and product formed[a] to produce 1 kilo of propylene in the reaction below:

$$C_3H_6(0.724Kg) + \text{oxidant} \rightarrow C_3H_6O(1\ kg) + \text{product}$$

Oxidant (kilos)	Product (kilos)
H_2O_2 (0.586)[b]	H_2O (0.310)
t-BuOOH (1.551)	*t*-BuOH (1.276)
NaClO (1.283)	NaCl (1.007)
$KHSO_5$ (2.622)	$KHSO_4$ (2.346)
PhIO (3.790)	PhI (3.514)

[a] Assuming 100% yield. [b] 100% H_2O_2.

The estimated world capacity for hydrogen peroxide in 1987 was 881 x 10^3 tons/year [11]. Of these about 95% came from the antraquinone process (see chapter 2), the remaining being due mostly to the autooxidation of isopropanol which is practised only in the USSR [11, 12]. Because of the environmental safety of hydrogen peroxide, it is predictable that the increasing demand for this chemical will require in the near future simpler, more economical production technologies, the direct combination of hydrogen and oxygen being of ultimate conceivable simplicity. This is a long standing idea and patents on this subject have already appeared in the past [13]. These processes are based either on the use of soluble complexes or on heterogeneous Pd metal catalysts working in an organic solvent. Recently (1988), Du Pont has patented a new process [14] where H_2 and O_2 are directly combined under pressure on a heterogeneous Pd metal catalyst dispersed on "adsorbent" carbon and operating in an acidic aqueous suspension. Although at the moment this process seems to be limited by the small plant capacity (3300 tons/year), it is predictable that this technology will gain significance in the near future.

Industrially, hydrogen peroxide is used mainly as a non selective oxidant, i.e. for paper, textile and cellulose bleaching, water purification and, particularly in Europe, in the manufacture of perborate and percarbonate used in detergents. A modest, approximate 10% of the total world production is estimated to be employed in the manufacture of organic chemicals. These include mainly glycerol from allyllic alcohol, epoxidized soybean oil used as plasticizer, organic peroxides used as free radical initiators in polymerizations, lauryldimethylamine oxide used in rinsing agents for dishwashers [11].

The paucity of practical applications of hydrogen peroxide in the synthesis of organic chemicals contrasts sharply with the number of advantages listed above related to its use.

Indeed, the chemistry of hydrogen peroxide in selective oxidation has not parallelled the tremendous growth of applications of alkylhydroperoxides, despite the similar nature.

Like *t* -BuOOH or PhEtOOH, H_2O_2 is relatively stable and requires to be converted in a more active form in order to be effective as oxidant, this is generally accomplished either with organic or with inorganic (transition metals) promoters. However, unlike *t* -BuOOH or PhEtOOH two major drawbacks have thwarted the use of hydrogen peroxide in reactions with organic chemicals.

The first one is related to the unavoidable presence of water that is both the reduction product and the solvent of commercial hydrogen peroxide. This point is better clarified with an example. Both Bayer-Degussa [15] and Ugine-Kuhlmann [16] have patented processes where propylene is converted to propylene oxide using peracids (formed in situ from H_2O_2) as actual oxidants. In both cases the peracid aqueous solution causes both problems in contact between the oxidant and the hydrophobic olefin and oxirane ring opening with subsequent lowering of yields.

Technically, the problem was solved in one case (Bayer-Degussa) by producing the oxidant (perpropionic acid) from propionic acid and H_2O_2 in aqueous H_2SO_4. The peracid is separated by countercurrent extraction with benzene followed by dehydration via azeotropic removal of water. The water free benzene solution is then transferred to a separate unit and contacted with propylene to produce the epoxide in an uncatalyzed oxygen transfer reaction. Propionic acid has to be recovered and sent back to the first unit to regenerate the peracid.

In the Ugine-Kuhlmann process the same result was accomplished by removing water by continuous distillation during the reaction and using metaboric acid $(HBO)_n$ to absorb residual water. This is converted in situ to boric acid which catalyzes the epoxidation reaction, but must be also partly removed in a continuous operation to be dehydrated in an external step to regenerate the metaboric acid.

The second major disadvantage with respect to the above hydroperoxides is the lower stability of aqueous hydrogen peroxide with respect to radical decomposition. For example, as said above Ti(IV) species are excellent catalysts for the stereoselective epoxidation of olefins with *t* -BuOOH [8], while on the other hand, species of Ti(IV) decompose catalytically H_2O_2 (see chapter 4).

Some excellent reviews dedicated to the subject of catalytic oxidations [3c, 7a, 17], including some books [3d, 18], have appeared over the years. Specifically, on hydrogen peroxide three, rather general books have been published [19], although the most recent one is over ten years old.

Because of the increasing interest in the use of hydrogen peroxide as oxidant for specific organic transformations it seems appropriate to review the argument of transition metal catalyzed oxidations. However the number of metal systems that are capable to promote applications of this oxidant is still rather limited, and a broadening of their scope or simply of their number is certainly required in the future to meet the increasing demand for clean oxidation technologies. Therefore, in reviewing this subject where some purely organic systems have also been included, stress has been placed on those innovation aspects that are likely to stimulate new ideas and lead to developments in the future.

3. Principles of Activation of Hydrogen Peroxide

In this introductory chapter, it seems appropriate to survey briefly the general principles through which activation of hydrogen peroxide in the presence of transition metals can be accomplished. As stated above, due to the relative stability of aqueous solutions of hydrogen peroxide, the general aim is to increase its reactivity by converting H_2O_2 into a different, more active species. This is accomplished by three general methods:

1. By reacting H_2O_2 with metals capable to generate metal peroxy or hydroperoxy species, where the peroxo bond remains intact. These are formed for instance with Ti(IV), V(V), Mo(VI), W(VI) (d^0 metal centers) or with Pd(II) and Pt(II) (d^8 metal centers) and the result is an increase of either the electrophilic or the nucleophilic character of the peroxygens with respect to the original oxidant.

2. By using H_2O_2 as a mooxygen donor to produce highly reactive metal-oxo species. This behavior is typical of biomimetic systems based for example

on Fe(II), Ru(II), Mn(II) and the study of the reactivity and oxidation properties of these systems has been one of the fields of most intense scientific endeavor during the past 10 -15 years. However in these studies, the most widely successful (and used) monooxygen donors are iodosyl arenes or hypochlorite or even tert-butylhydroperoxide and hydroperoxy-sulfate, hydrogen peroxide being employed in only a limited number of cases. Recently, a new class of metal-oxo oxidants, polyoxometallates which can be regenerated from H_2O_2, has rapidly emerged as one of the most promising sectors of this chemistry.

3 Through radical decomposition of H_2O_2 with one-electron redox couples like Fe(II)/ Fe(III) or Ti(III)/Ti(IV). These are the basis for the well known Fenton and Udenfriend systems and a way of generating the highly reactive HO• radical. Although these systems are not very selective they do have some interesting synthetic applications.

According to this classification this book has been divided into different chapters where the relevant chemistry has been described. Moreover, since some of these principles pertaining to metals are shared also with some purely organic systems a chapter devoted to this topic has been included. Finally (or better initially) to address the current interests of industry in this field, the next chapter deals with the industrial aspects related to the manufacture of hydrogen peroxide and its industrial use for the production of organic chemicals.

References

1. K. Weissermel and H.-J. Arpe, '*Industrielle Organische Chemie*', third edition, VCH, Weinheim 1988.
2. F. Haber and J. J. Weiss, *Proc. Roy. Soc. London, Ser. A*, **147**, 332 (1934).
3. For leading reviews see: (a) L. Vaska, *Chem. Rev.*, **9**, 175 (1976); (b) V. J. Choy and C. J. O'Connor, *Coord. Chem. Rev.*, **9**, 145 (1972/73); (c) J. E. Lyons, *Asp. Hom. Catal.*, vol 3, Ed. R. Ugo, Reidel, Dordrecht 1977, p. 1.; (d) R. A. Sheldon and J. K. Kochi, '*Metal Catalyzed Oxidations of Organic Compounds*', Academic, New York 1981, chapter 4; and references therein
4. L. Vaska, *Science*, **140**, 809 (1963).
5. (a) R. Landau, *Hydrocarbon Process.*, **46**, 141 (1967); (b) R. Landau, G. A. Sullivan and D. Brown, *CHEMTECH*, 602 (1979); (c) J. Kollar, *US Patent*, 3 350 422 (1967) and 3 351 635 (1967), both to Halcon Int.
6. A. O. Chong and K. B. Sharpless, *J. Org. Chem.*, **42**, 1587 (1977).
7. (a) R. A. Sheldon, *Asp. Hom. Catal.*, vol 4, Ed. R. Ugo, Reidel, Dordrecht 1981, p. 3; (b) H. P. Wulff, *US Patent*, 3 923 843 (1975), *Brit. Patent*, 1 249 079 (1975), both to Shell Oil.

8. See: M. G. Finn and K. B. Sharpless in '*Asymmetric Synthesis* ', Ed. J. D. Morrison, Academic, New York 1986, vol. 5, p. 247 and references therein.
9. (a) R. A. Sheldon, *Stud. Surf. Sci. & Catal.*, vol. 55, Eds. G. Centi and F. Trifirò, Elsevier, Amsterdam 1991, p. 1; (b) R. A. Sheldon, *CHEMTECH* , 566 (1991).
10. H. Ledon, *Stud. Surf. Sci & Catal.*, vol. 66, Ed. L. Simandi, Elsevier 1991, p. 603.
11. W. Büchner, R. Schliebs, G. Winter and K. H. Büchel '*Industrial Inorganic Chemistry*', VCH, Weinheim 1989, p. 21.
12. G. Goor, W. Kunkel and O. Weiberg in '*Ullmann's Encyclopedia of Industrial Chemistry*', VCH, Weinheim 1989, vol. A13, p. 443.
13. (a) G. W. Hooper, Brit. Patent, 1 056 126 (1967) to ICI; (b) P. M. Dyer and F. Moseley, *US Patent* , 4 128 627 (1978) to Air Products; (b) L. Kim and G. W. Schoenthal, *US Patent* , 4 007 256 (1977) to Shell Oil; (d) R. C. Michaelson, *US Patent* , 4 347 231 (1982) to FMC Corp.
14. L. W. Gosser, *US Patent*, 4 681 751 (1988) to Du Pont.
15. G. Prescher, G. Schreyer, O. Weiberg, R. Wirthwein, H. Waldmann, H. Seifert, W. Schwerdtel and WN Swodenk, *Ger. Patent* , 2 519 297 (1976) and 2 519 298 (1976) to Baeyer-Degussa.
16. (a) J.-P. Schirmann, M. Pralus and S. Y. Delavarenne, *Ger. Patent* , 2 803 757 (1977) to PCUK; (b) J.-P. Schirmann and S. Y. Delavarenne, *Ger. Patent* , 2 803 791 (1977) and 2 752 626 (1978) to PCUK
17. See for example: (a) J. A. Connor and E. A. V. Ebsworth, *Adv. Inorg. Chem. Radiochem.*, **6**, 279 (1964); (b) G. Henrici-Olivè and S. Olivè, *Angew. Chem. Int. Ed. Engl.*, **13**, 29 (1974); (c) R. A. Sheldon and J. K. Kochi, *Adv. Catal.*, **25**, 272 (1976); (d) Catalytica Associates, Inc., "*Selective Catalytic Oxidation of Hydrocarbons, A Critical Analysis* ", Multiclient Study No. 1077, Santa Clara, California 1979; (e) R. A. Sheldon in '*The Chemistry of Funtional Groups: Peroxides* ', Ed S. Patai, Wiley, New York 1983, p. 161; (f) H. Mimoun in '*Comprehensive Coordination Chemistry* ', Eds. G. Wilkinson, R.D. Gillard and J. McCleverty, Pergamon, Oxford 1987, chapter 61.3, p. 317; (g) K. A. Jørgensen, *Chem. Rev.*, **89**, 431 (1989).
18. (a) R. L. Augustine and D. J. Trecker (Eds.), '*Oxidation* ' vol. I and II, Dekker, New York 1969; (b) K. B. Wiberg (Ed), '*Oxidation in Organic Chemistry* ' part A, Academic, New York 1965; (c) W. S. Trahanovsky (Ed), '*Oxidation in Organic Chemistry* ' part B, Academic, New York 1973 and part C, Academic, New York 1978;
19. (a) W. C. Schumb, C. N.Satterfield and R. L. Wentworth, "*Hydrogen Peroxide* ", Reinhold, New York 1955; (b) W. Weigert (Ed), "*Wasserstoffperoxid und Seine Derivate* ", Hüthig, Heidelberg 1978; (c) J. P. Schirmann and S. Y. Delavarenne, "*Hydrogen Peroxide in Organic Chemistry* ", Informations Chimie, Paris 1979.

CHAPTER 2

Hydrogen Peroxide: Manufacture and Industrial Use for Production of Organic Chemicals

Gustaaf GOOR

Department of Organic and Biological Chemistry Research
Degussa AG, 6450 Hanau, F. R. G.

1. Introduction

In 1818 J. L. Thenard discovered hydrogen peroxide by reacting barium peroxide with nitric acid [1]. Based on this reaction commercial production of hydrogen peroxide began around 1880 [2]. The very diluted hydrogen peroxide produced by the barium peroxide process found only limited use due to high production costs, low hydrogen peroxide content and poor stability.

The disadvantages of the barium peroxide process were overcome by the electrochemical process which was based on the electrolysis of sulfuric acid solution and subsequent hydrolysis of the formed peroxy disulfuric acid to give hydrogen peroxide and sulfuric acid [3]. The electrochemical process allowed production of pure and stable concentrated hydrogen peroxide solutions. The first hydrogen peroxide plant based on sulfuric acid electrolysis went on stream in 1908 at the Österreichische Chemische Werke, Weißenstein in Austria. This Weißenstein process was modified by using ammonium sulfate instead of sulfuric acid.

In 1901 Manchot observed that autoxidizable compounds like hydroquinones or hydrazobenzenes react quantitatively under alkaline conditions with oxygen to form peroxides [4]. In 1932 Walton and Filson proposed to produce hydrogen peroxide by alternating oxidation and reduction of hydrazobenzenes [5]. Subsequently, Pfleiderer developed a process for the alkaline autoxidation of hydrazobenzene in which sodium peroxide is obtained and sodium amalgam is used to reduce azobenzene [6]. A commercial plant, based on this technology, was operated by Kymmene AB in Kuusankoski (Finland).

The major technical drawbacks of the azobenzene process, i.e. hydrogenation of azobenzene with sodium amalgam and oxidation of hydrazobenzene in alkaline solution, were solved by Riedl in using polynuclear

G. Strukul (ed.), Catalytic Oxidations with Hydrogen Peroxide as Oxidant, 13–43.

hydroquinones. Based on the work done by Riedl and Pfleiderer, BASF developed between 1935 and 1945 the anthraquinone process (so-called AO-process) in a pilot plant with a monthly production of 30 tons. Construction of two large plants, at Heidebreck and at Waldenberg, each having a capacity of 2000 t (calculated as 100 %) was started. Both plants were partially completed when construction was stopped at the end of World War II. In 1953, E.I. du Pont de Nemours started up the first hydrogen peroxide plant using the anthraquinone process. Several other companies began producing hydrogen peroxide by this process and production capacity increased greatly. In 1988 world capacity stood at 1,095,000 tons to H_2O_2 100 % [7].

Another autoxidation process, based on oxidation of 2-propanol, was developed and used by Shell in its 15,000 t Norco facility between 1957 and 1980.

2. Manufacture of Hydrogen Peroxide by AO-Process

2.1 Chemistry of AO-Process

The AO-process involves the hydrogenation of a 2-alkyl-9,10-anthraquinone with hydrogen to the corresponding hydroquinone and oxidation of the hydroquinone with oxygen, usually air, to yield hydrogen peroxide and the starting 2-alkyl-9,10-anthraquinone (Scheme 1). Hydrogen peroxide is extracted with water, whereas the quinone is recycled to the hydrogenator.

Scheme 1:

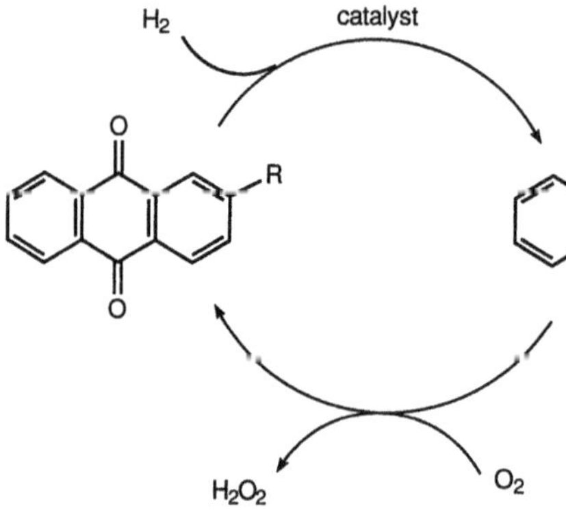

The anthraquinone, commonly called the working compound, is dissolved in a solvent or solvent mixture for hydrogenation, oxidation and extraction.

This anthraquinone solvent mixture is called the working solution. During the cyclic process both quinone and hydroquinone, formed in the hydrogenator, must remain dissolved. Because quinone and hydroquinone have different solubility, quinones dissolve well in nonpolar aromatic solvents (quinone solvents) and hydroquinones dissolve well in polar solvents (hydroquinone solvents), solvent mixtures are mostly used. Examples of quinone solvents include benzene [8], t-butylbenzene [9], t-butyltoluene [10], trimethyl benzene [11], methylnaphthalene [12] and polyalkylated benzenes [13]. Examples of hydroquinone solvents include esters of phosphonic [14] or phosphoric acid [15], nonyl alcohols [12], esters of methyl cyclohexanol [16], N,N-dialkylamides of carboxylic acids [17, 18], tetraalkyl urea compounds [19-21] and N-alkyl-2-pyrrolidones [22].

The solvent or solvent mixture should be chemically stable during hydrogenation and oxidation, poorly soluble in water and aqueous hydrogen peroxide solution and non toxic and should have a high flash point and high distribution coefficient of hydrogen peroxide between water and the working solution. Furthermore, it should have good solubility for quinone and hydroquinone as the maximum capacity of a working solution, defined as the maximum amount of hydrogen peroxide that can be produced per process cycle, is limited by the solubility of the working compound in the oxidized and the reduced state.

Besides using more powerful solvents or solvent mixtures, maximum capacity of a working solution can also be varied by modifying the substituent on the anthraquinone skeleton. Examples of working compounds proposed in the patent literature are 2-ethylanthraquinone [23], 2-t-butylanthraquinone [24], mixed 2-amylanthraquinones [25] and 2-neopentylanthraquinone [26]. Improved solubility is also claimed for mixtures of alkylanthraquinones [27]. Although in most cases the substituent of monoalkylated anthraquinones is located at the 2-position, polyalkylated anthraquinones, where the alkyl groups contain a total of three to fifteen carbon atoms, were proposed [28].

The working compound should be chemically stable during the process. The formation of degradation products and their ability to be regenerated to active quinones also influences the choice of the working compound. Apart from hydroquinone formation, a number of secondary reactions occur in the hydrogenator; particularly hydrogenation of the aromatic rings, preferably the one that is not substituted by an alkyl group, is important. 2-Alkyl- 5,6,7,8-tetrahydro-9,10-dihydroxyanthracene is formed (eq. 1), which on oxidation with oxygen quantitatively yields hydrogen peroxide and 2-alkyl-5,6,7,8-

tetrahydroanthraquinone (commonly called tetra) (eq. 2).

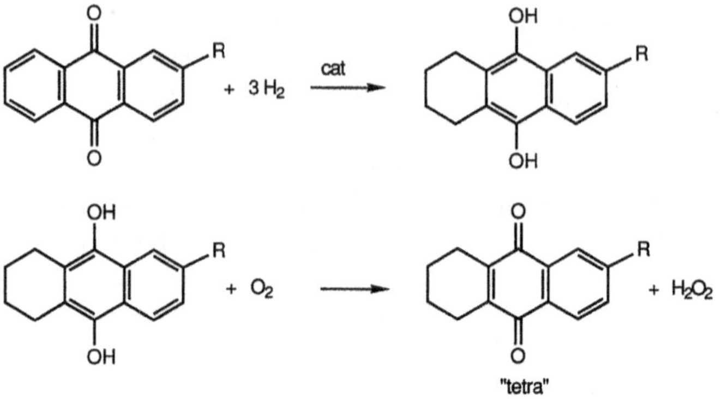

(1)

(2)

Whereas the tetra is more easily hydrogenated than the corresponding anthraquinone, the resulting hydroquinone of the tetra is much more difficult to oxidize than the readily oxidizable anthrahydroquinone.

The formation of the tetra during the process cycle depends on the process conditions. To overcome the problems which arise from the poor oxidation rate of the hydroquinone of the tetra, actions were proposed to keep the content of the tetra in the working solution low by either minimizing its formation or by dehydrogenation into the anthraquinone.

Tetra formation is suppressed by using selective catalysts [29], special solvents [30] or working compounds [28] and mild hydrogenation conditions [31]. For permanent dehydrogenation of the tetra-compound, use of olefins in the presence of activated alumina has been recommended [32]. Tautomerisation of 2-alkyl-9,10-dihydroxyanthracene yields the 2-alkyl-10-hydroxy-9-anthrone (oxanthrone) [33], [34] which can be further reduced to 2-alkyl-9-anthrone (anthrone)(eq. 3).

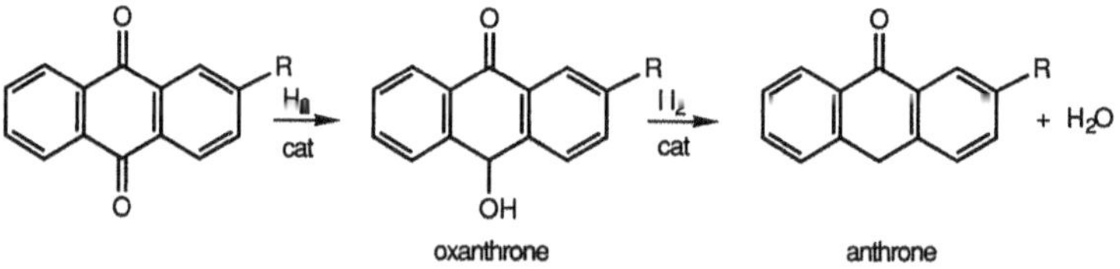

(3)

If no precautions are taken to minimize the tetra content, an equilibrium will result in which the hydroquinone fed to the oxidizer consists exclusively of the hydroquinone of the tetra. This status of the working solution is commonly called all-tetra-system. If both anthraquinone and tetracompound are hydrogenated, hydrogen transfer occurs (eq. 4):

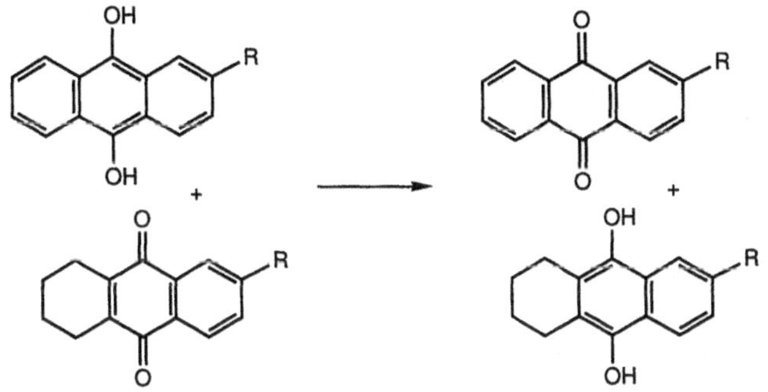

(4)

During the oxidation of the tetra hydroquinone, small amounts of a tetrahydroanthraquinone epoxide are formed in a side reaction. This epoxide does not participate in hydrogen peroxide formation and has to be considered as a loss of the working compound. Therefore, steps were suggested to regenerate tetra from the epoxide, e.g. treatment of the working solution with activated alumina [35] (Scheme 2)

Scheme 2:

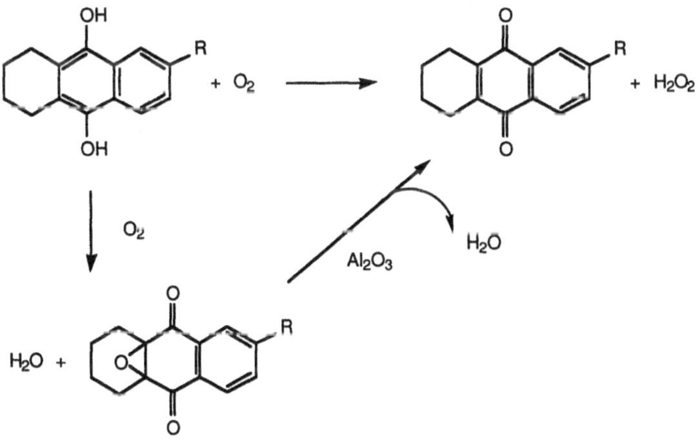

On high tetra level, further ring reduction generates the 1,2,3,4,5,6,7,8-octahydroanthrahydroquinone. Oxidation of the octa hydroquinone also yields hydrogen peroxide; but the reaction is extremely slow, so that octa represents a loss of active quinone.

2.2 Description of AO-Process

As already mentioned in section 1, all operating AO-plants are based on the Riedl-Pfleiderer process. A schematic flow sheet is depicted in fig. 1 [36]. The working solution, stored in the storage tank or hydrogenator feed tank (a), is fed into the hydrogenator, (b) where it is hydrogenated in the presence of a catalyst. The hydrogenator is operated at a slightly elevated pressure and at a temperature below 100°C. In most of the patent examples, the depth of hydrogenation , i.e. conversion of quinone to hydroquinone, is kept below 60% to minimize by-product formation although conversions above 80% have been claimed [37].

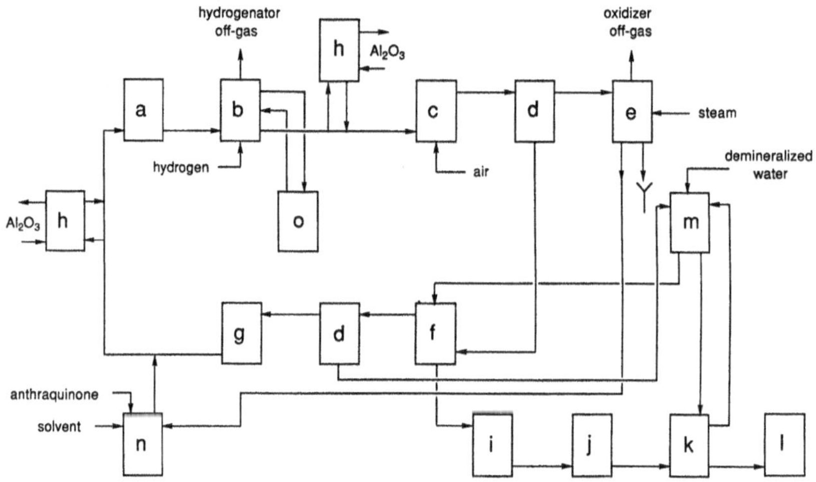

Fig. 1: Flow sheet of AO-process: (a) working solution storage tank or hydrogenator feed tank; (b) hydrogenator including catalyst filtration; (c) oxidizer; (d) separator; (e) activated carbon adsorber; (f) extractor; (g) drier; (h) regeneration and purification of working solution; (i) crude product purification; (j) crude product storage tank; (k) concentrator; (l) hydrogen peroxide storage tank; (m) demineralized water feed tank; (n) working solution make-up tank; (o) catalyst regeneration.

Riedl and Pfleiderer used Raney nickel as a catalyst. Raney nickel is not only pyrophoric, but it is also deactivated by hydrogen peroxide and shows excessive tetra formation. Small quantities of residual hydrogen peroxide in the working solution leaving the drier were catalytically destroyed. To improve the selectivity, i.e. the ratio of hydroquinone formation to tetra formation, the catalyst was pretreated, e.g. with ammonium formate [38], nitriles [39] or amines [40].

Today, most AO-plants prefer palladium catalysts because of their higher selectivity and simpler handling. Palladium catalyst can be used as palladium black [41] or supported on a carrier for slurry [42] or fixed-bed operation [43]. For fixed-bed catalysts, it was shown that almost no by-products were formed for supports having a specific surface area of less than 5 m^2g^{-1} and a pore volume below 0.03 cm^3g^{-1} [43]. Pretreatment of alumina support with alkali is reported to lower quinone consumption [44]. Activity and selectivity of palladium supported catalysts is improved by using α-alumina, [45] whereas addition of metals like zirconium, thorium, cerium or titanium results in extremely stable palladium/silica catalysts [46].

Several hydrogenator types have been patented, e.g. tubular [41, 47], draft-tube [48] and fixed-bed reactors [49, 50]. Dispersing hydrogen within the working solution prior to contacting the hydrogenation catalyst has reportedly raised the reaction rate of a fixed-bed hydrogenator [51].

Hydrogenated working solution fed into the oxidizer (c) cannot contain any remaining catalyst: both palladium and Raney nickel catalyze hydrogen peroxide decomposition. Carbon filters used with palladium black catalyst have been patented [52]. In the oxidizer, a catalyst-free working solution is oxidized with an oxygen-containing gas, mostly air. The reaction is not catalyzed (autoxidation) and the overall rate depends on temperature, oxygen partial pressure, and phase-boundary surface area. Several oxidizers have been proposed in the literature: liquid and gas may flow either cocurrently or countercurrently in a single- or multiple-column system with or without packings [53-55]. Low loss of solvent and minimum by-product formation is reached in a countercurrent, packed column by limiting the contact time of solution with air [56]. After oxidation, the air and the working solution are separated (d). The air, saturated with solvent, is purified with active carbon, (e) which periodically is regenerated, e.g. with steam, to recover the solvent.

Hydrogen peroxide is recovered from the oxidized working solution by water extraction (f). Beside sieve-tray extraction columns, used in the BASF-process, a number of other extractors, e.g. packed columns, pulsed packed columns, spray columns and mechanical extractors have been suggested. Emulsification of the working solution in the aqueous phase can be avoided by using water which has been adjusted to an acid pH value [57]. Polyethylene, polypropylene, polytetrafluoroethylene or polyvinylchloride improve coalescence of the working solution droplets [58]. Free water leaving the extractor with the working solution is separated (d) and the water content in the working solution is adjusted in the drier (g) to prevent agglomeration of the suspended catalyst. It was proposed to operate the extractor at a relatively low temperature, followed by separation of the free water and then to increase the working solution temperature by 2.8 - 28°C before it was recycled to the hydrogenator [59]. Also, use of oxidizer off-gas is reported to reduce the relative moisture in the working solution[60].

During the cyclic reduction and oxidation of the working solution, degradation products are formed from both the working compound and the solvent. These products must be removed or regenerated. A large variety of processes have been proposed to purify the working solution and to regenerate the active working compound from the quinone degradation products, either by using oxidized or hydrogenated working solution. Examples include treatment with sodium aluminum silicates [61], activated alumina or magnesium oxides [62], calcium, magnesium or zinc silicates [63], aqueous solution of sodium or potassium hydroxide [64] and sulfuric acid [65]. Treatment of hydrogenated working solution with alkaline substances converts epoxide into the corresponding tetra compound [35]. Although alumina and sodium aluminum silicate lose activity on regenerating the degraded working solution, their original activity can be restored [66, 67].

The crude aqueous hydrogen peroxide from the extractor has a H_2O_2 concentration of 15 - 30 wt % and is contaminated. Numerous methods are claimed to lower the amount of dissolved organic compounds such as treatment with hydrocarbons [68, 69], polyethylene [70], anion exchange resins [71], activated carbon [72] and adsorber resins [73, 74]. Use of anionic exchange resins containing chelating agents [75] and use of reverse osmosis with membranes made of polyamides, polypiperazineamides, polysulfones or polyacrylonitriles [76] have been suggested to remove inorganic impurities from aqueous hydrogen peroxide. The purified crude product is further purified and concentrated in a concentration unit (k) to the commercial grades of 50 - 70 wt % [77 - 80], stabilized [81 - 82] and stored (l). Water vapor, produced during concentration, is condensed and recycled into the water storage tank.

Continuous AO-operation is further sustained by additional activities. For example, to maintain activity of hydrogenation catalyst, part of it is separated from the liquid and regenerated (o) by calcination at an elevated temperature to remove organic material [83]. To compensate for quinone and solvent losses, the working solution is periodically made up (n) with quinone and solvents. Make-up exclusively with tetra compound is reported [84].

3. Properties of Hydrogen Peroxide

Hydrogen peroxide is a clear, colorless liquid which is completely miscible with water. Concentrated aqueous solutions are considerably more soluble in organic solvents than water. Some physical constants of the main commercially available grades are recorded in Table 1.

Table 1: Physical properties of commercial H_2O_2 solutions.

Property	H_2O_2 strength (wt%) 35	50	70
Density at 20°C (g/cm^3)	1.1312	1.1953	1.2886
Viscosity at 20°C (mPa s)	1.11	1.17	1.23
Freezing point (°C)	-33	-52.2	-40.3
Boiling point (°C)	107.9	113.8	125.5

Numerous other physical data appear in the literature [85].
The decomposition of hydrogen peroxide

$$2\,H_2O_2 \rightarrow 2\,H_2O + O_2 \qquad (5)$$

is extremely important in handling hydrogen peroxide because the generation of gas and heat (-98,3 kJ mol^{-1}) may cause safety problems. Decomposition depends on temperature, pH and presence of impurities, e.g. metal ions or metal oxides and hydroxides. Substitution reaction of hydrogen peroxide with organic reagents leads to a variety of peroxygen compounds, e.g. with carboxylic acids percarboxylic acids are formed. Furthermore, a number of compounds, such as sodium carbonate and urea, form peroxyhydrates with hydrogen peroxide. Although most of the applications of hydrogen peroxide are based on its strong oxidizing property, it can also behave as a reducing agent. Reduction of ceric sulphate and potassium permanganate are used for the volumetric determination of hydrogen peroxide.

4. Preparation of Percarboxylic Acids

Percarboxylic acids are conveniently prepared by reaction of hydrogen peroxide with carboxylic acids, acid chlorides or anhydrides [86]. Especially the lower aliphatic percarboxylic acids, i.e. performic, peracetic and perpropionic acid, have reached technical importance for oxidation reactions like epoxidation and hydroxylation of olefins. Because of their hazardous nature, these lower aliphatic percarboxylic acids are never isolated as pure substances, but used either as preformed peracid, dissolved in an inert solvent, or by the in-situ process. The stability of percarboxylic acids increases with increasing molecular weight and several aromatic percarboxylic acids like *m*-chlorperbenzoic acid or monoperphthalic acid are so stable that they can be isolated in the pure form.

4.1 Reaction of Carboxylic Acids with Hydrogen Peroxide

The reaction between hydrogen peroxide and lower aliphatic carboxylic acids is reversible and a strong acid catalyst is required (eq. 6) to ensure a sufficient reaction rate.The equilibrium mixture contains percarboxylic acid, the corresponding acid, hydrogen peroxide, water and the catalyst. As a catalyst, mostly sulfuric acid or strong acid cation exchange resins [87] are used. The preparation of performic and trifluoroperacetic acid does not require any catalyst since the starting acids are strong enough to catalyze the reaction. The equilibrium composition depends mainly upon the molar ratio of the reactants and the initial hydrogen peroxide concentration. The time needed to reach equilibrium is influenced by the temperature and the catalyst concentration. Of the lower aliphatic equilibrium percarboxylic acids, equilibrium peracetic acid is the most important.

$$RC(O)OH + H_2O_2 \overset{H^+}{\rightleftharpoons} RC(O)OOH + H_2O \qquad (6)$$

Starting from the equilibrium-mixture solutions of percarboxylic acids free of carboxylic acid, hydrogen peroxide, catalyst and water can be prepared. This is done by combining the equilibrium reaction with either a continous distillation or an extraction step.

Distilled aqueous solutions of peracetic, perpropionic and perbutyric acid are prepared in excellent yield by introducing a mixture of the corresponding acid and hydrogen peroxide in a ratio H_2O_2 : acid of 10 : 1 to the bottom of a distillation column containing sulfuric acid as the catalyst and distilling off a solution of percarboxylic acid in water under reduced pressure [88, 89]. During the distillation an equimolar mixture of hydrogen peroxide and carboxylic acid is continously charged to the reactor. The water content of the distillate is adjusted in a way that both dilution- and reaction-water are removed. Solutions containing up to 60 wt % of peracetic acid in water can be obtained. Because distilled peracetic acid contains no sulfuric acid, it is used in applications where the presence of sulfuric acid may initiate side reactions.

To prepare solutions of percarboxylic acids substantially free of water, distilled aqueous percarboxylic acids are extracted with inert organic solvents, and subsequently the remaining water in the extract is azeotropically removed. Preferred solvents are butylacetate [90], tributyl phosphate [91] and chlorinated or aromatic hydrocarbons [92]. The disadvantage of high energy consumption, required for preparation of distilled percarboxylic acid, is overcome by first reacting hydrogen peroxide and the organic acid in strong sulfuric acid medium and then countercurrently extracting the reaction solution with benzene [93]. The extract is azeotropically dehydrated and a solution of approximately 20 wt % percarboxylic acid in benzene, which contains unreacted hydrogen peroxide, results. To lower the hydrogen peroxide content, the benzene extract is subjected to an extraction with water, involving at least three stages [94]. Use of propylene dichloride as an inert solvent for extraction of perpropionic acid is also recommended [95].

An efficient and low-energy-consuming continuous process for preparing organic solutions of percarboxylic acids, substantially free of mineral acid, consists (Figure 2) of reacting an organic acid and hydrogen peroxide in a strong acidic medium, distilling off water, percarboxylic acid and carboxylic acid, adsorbing percarboxylic and carboxylic acid in tributyl phosphate and final desorbing with carboxylic acid esters [96]. A solution consisting of approx. 25 wt % percarboxylic acid, 5 wt % carboxylic acid and 2 wt % water is obtained; for reactions which are sensitive to hydrolysis, e.g. preparation of epoxy silanes, this solution can be further azeotropically dehydrated. A significant advantage of this process is that it gives some freedom as to the choice of carboxylic acid and solvent which may ease separation after reaction.

The direct acid-catalyzed reaction of long-chain aliphatic carboxylic acids with hydrogen peroxide proceeds at an extremely slow rate because of poor solubility of the organic acid in the reaction medium. However, taking advantage of the fact that concentrated sulfuric acid is a good solvent for aliphatic acids, long-chain aliphatic percarboxylic acids are prepared in excellent yields by using an excess of sulfuric acid and hydrogen peroxide [97]. Dipercarboxylic acids are similarly made [98]. Some of these dipercarboxylic acids, i.e. α,ω-C_8-C_{13} dipercarboxylic acids, are considered low temperature bleaching agents [99].

Because of its strong oxidation-sulfonation properties, the sulfuric acid method fails to prepare aromatic percarboxylic acids. Replacement of sulfuric acid by methanesulfonic acid provides aromatic percarboxylic acids in high yields [100].

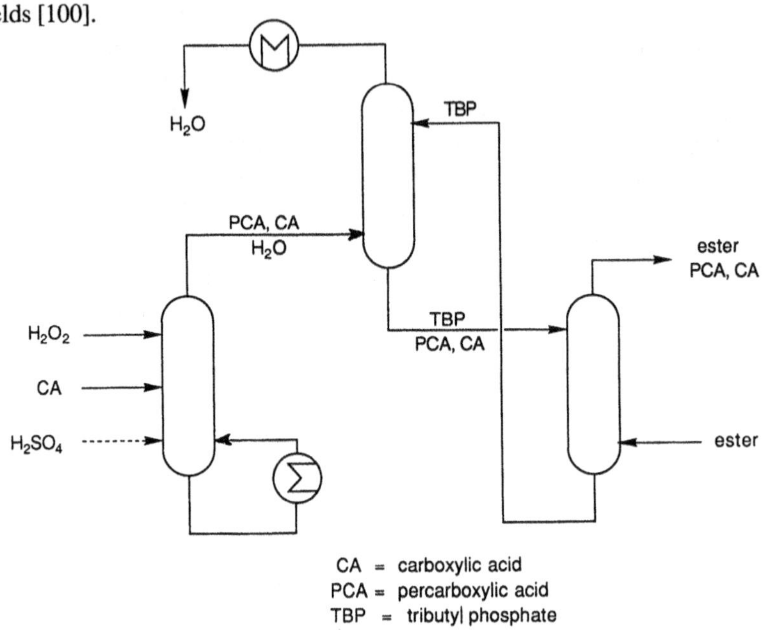

Fig. 2: Preparation of percarboxylic acids in esters.

4.2 Reaction of Acid Anhydrides with Hydrogen Peroxide

Solutions of percarboxylic acid in carboxylic acid can be made by reacting an excess of the corresponding carboxylic acid anhydride with hydrogen peroxide at ambient temperature in the presence of a strong catalyst (eq.7).

$$(RCO)_2O + H_2O_2 \xrightarrow{H^+} RCO_3H + RCO_2H \qquad (7)$$

Because of the possibility of forming hazardous diacyl peroxides, this method is not recommended to prepare lower aliphatic percarboxylic acids.

4.3 Reactions of Acid Chlorides with Hydrogen Peroxide

Percarboxylic acids can also be prepared by reacting a carboxylic acid chloride with hydrogen peroxide in an alkaline medium (eq.8).

$$RCOCl + H_2O_2 \xrightarrow{B^-} RCO_3H \qquad (8)$$

This method is very often used for preparation of aromatic percarboxylic acids [101], but again possible formation of hazardous diacyl peroxides has to be considered.

5. Epoxidation

By reacting percarboxylic acids or hydrogen peroxide with olefinic double bonds, epoxides are formed (eq.9). Epoxides are extremely important intermediates for the chemical industry because of the many reactions they undergo. Most interesting of these is the addition of active hydrogen compounds like water, alcohols, organic acids, ammonia, amines and hydrogen sulfide. The reaction products find application as surfactants, detergents, antistatic agents, corrosion inhibitors, additives to lubricating oils, textiles and cosmetics. Epoxidized soybean oil is used mainly as a plasticizer and stabilizer for polyvinylchloride.

(9)

Most epoxidations of industrial interest are performed with lower aliphatic percarboxylic acids, especially performic and peracetic acid [102, 103]. Cleavage of the epoxide ring with formation of hydroxy esters, which in the past invariably occured, is nowadays almost prevented by proper choice of the reaction conditions. The rate of epoxidation largely depends on the temperature and the molecular structure of the olefin as well, i. e. electron-donating substituents at the carbon carbon double bond enhance the reaction rate, whereas electron-withdrawing substituents have the opposite effect [104]. The yield and purity of the epoxide obtained are further influenced by the stability of the epoxide against other components in the reaction mixture. Electron-withdrawing substituents stabilize the epoxide ring. Cleavage of the epoxide ring decreases in the order sulfuric acid > formic acid > acetic acid > propionic acid.

5.1 Epoxidation with Performic Acid

Epoxidation with performic acid is mostly carried out in situ. In general hydrogen peroxide is added in portions to a mixture of olefin and formic acid while the solution is stirred. In doing this only small amounts of performic acid are formed which react immediately with the double bond. A further advantage is that build-up of substantial quantities of detonable performic acid is prevented. To minimize ring opening use of less than one mole of formic acid for each double bond is recommended [105]. Also, employment of an inert solvent helps to reduce by-product formation [105]. The formic acid used in the reaction is normally not recovered: during the reaction, partial decomposition of formic acid occurs, the remaining goes to waste.

Olefins which are epoxidized in this way include soybean oil, linseed oil, natural rubber [106] and long-chain olefins with terminal [107] or internal [108] double bonds. Continuous epoxidation of fatty acids or esters in counter-current flow with hydrogen peroxide and formic acid in a cascade reactor having more than three stages, each stage comprising a stirred reactor and a phase separator, is reported [109].

5.2 Epoxidation with Peracetic Acid

To epoxidize oils and fats, in-situ peracetic acid can be used in the same way as in-situ performic acid. However, in-situ peracetic acid requires the presence of a strong acid which can cause severe loss of epoxide by ring-opening if epoxidation is carried out homogeneously. Therefore, use of inert solvents such as heptane and toluene is suggested [110]. Again, less than one mole of acetic acid for each epoxidizable bond is used.

Commercial equilibrium peracetic acid contains a small amount of sulfuric acid, up to 1 wt %, which can initiate side reactions of the epoxide. This is eliminated by adding anhydrous sodium acetate to neutralize the sulfuric acid [111]. For preparation of epoxides, sensitive to ring-opening by acetic acid, addition of large amounts of acid-neutralizing compounds during epoxidation is recommended [112]; alternatively, epoxidation at a constant pH value of 3 - 6 in the presence of acid-neutralizing compounds is reported [113]. A continuous process for countercurrent epoxidation of terminal and/or internal olefins with equilibrium peracetic acid, generated in the presence of strongly acidic cation exchange resins, is mentioned in the literature [114]. After the epoxidation, the aqueous phase is recycled to the peracetic acid generator.

Distilled peracetic acid can be used in the same way as equilibrium peracetic acid, yet with the advantage that no buffering agent is required. Distilled peracetic acid is a versatile peracid, and it can be used to epoxidize almost all olefins, except propylene. Even from allyl alcohol, glycidol is obtained in high yield under the proper reaction conditions [115]. Although the concentrated glycidol contains less than 0.1 wt % acetic acid, it does not have satisfactory storage stability [116].

5.3 Epoxidation with a Solution of a Percarboxylic Acids in an Inert Solvent

Epoxidations with aqueous percarboxylic acid suffer from two major disadvantages: *i.* side reactions between epoxide and water, which can lower the yield and purity of the epoxide obtained, *ii.* recovery of the carboxylic acid. Several continuous processes for epoxidation of olefins with solutions of percarboxylic acids in inert solvents, which give excellent yields of epoxide with minimal by-product formation, are described. Some examples are listed in Table 2.

Table 2: Epoxidation of olefins with a solution of a percarboxylic acid in an inert solvent.

Olefin	Percarboxylic acid	Solvent	Reference
propylene	$C_2H_5CO_3H$	benzene	117, 118
propylene	$C_2H_5CO_3H$	1,2-dichloroethane	119
propylene	$C_2H_5CO_3H$	1,2-dichloropropane	120
allyl alcohol	CH_3CO_3H	acetone	121
allyl chloride	$C_2H_5CO_3H$	1,2-dichloropropane	122
α-olefins	$C_2H_5CO_3H$	benzene	123
olefins	$C_2H_5CO_3H$	benzene	124
olefins	$C_2H_5CO_3H$	1,2-dichloropropane	125
polybutadienes	$C_2H_5CO_3H$	benzene	126

Particularly, preparation of propylene oxide gained high interest: excess propylene is reacted at 400 kPa with a solution of perpropionic acid in benzene at temperatures below 90°C. The reaction mixture is separated by distillation: non converted propylene is returned to the epoxidation reactor, whereas propionic acid and benzene are recycled to the perpropionic acid generation step; overall yield of propylene oxide, based on hydrogen peroxide charged to the peracid unit is 94.9% [118]. A peracetic acid-based propylene oxide plant is operated by Daicel in Japan [127]. At this facility, peracetic acid is produced by air oxidation of acetaldehyde . Generation of peracetic acid, dissolved in ethyl acetate, from acetaldehyde and use of this peracid solution for epoxidations was first industrially applied by Union Carbide [128, 129].

5.4 Catalytic Epoxidations with Hydrogen Peroxide

Water-soluble olefins are readily epoxidized with hydrogen peroxide in the presence of salts or oxides of W, Mo or Se [130, 131]. For example, reaction of allyl alcohol with hydrogen peroxide catalyzed by 1.5 - 2 g $NaHWO_4$ per mole of hydrogen peroxide at 40 - 45°C affords glycidol in yields of 81 - 83 %, based on hydrogen peroxide [132] (eq. 10).

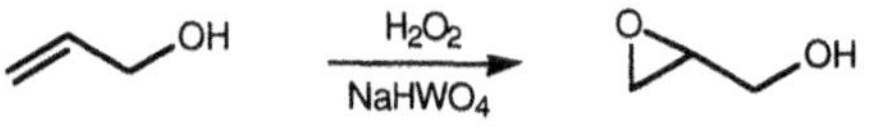

(10)

Distillation of the reaction mixture gives glycidol with a purity of approximately 99.5 %. The $NaHWO_4$ catalyst can be recovered by air oxidation in a fluid bed [133, 134]. Also, the tungstate catalyzed epoxidation of maleic acid (eq. 11) at a pH of 4 - 5 and subsequent hydrolysis of the intermediate 2,3-epoxy-succinic acid to yield D,L-tartaric acid, including recovery of the catalyst and of unreacted maleic acid from the aqueous reaction mixture, was described [135, 136].

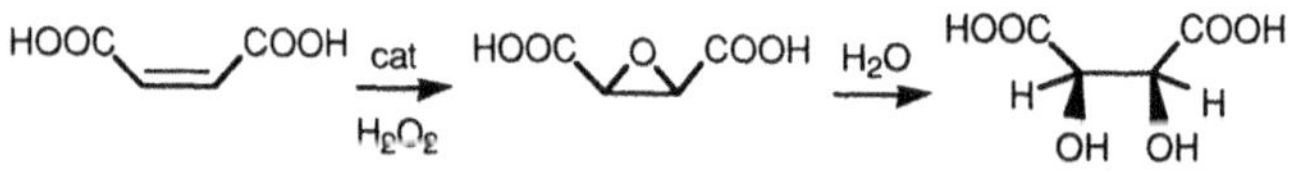

(11)

In the last two decades, several catalysts for epoxidation of olefins with hydrogen peroxide have been developed: As compounds [137], Sn compounds with carboxylic acids as co-solvents [138], W, Mo or V compounds with continuous water removal [139], As, Sb or Bi compounds with V, Mo, W or Ti compounds as co-catalysts [140], B oxides in polyether solvents [141], Te bounded macromolecular styrene-divinylbenzene copolymers [142], Ti containing zeolites [143] and silylated Ti containing zeolites [144]. Propylene can be epoxidized in 40 min at 40°C and at 600 kPa with 34 wt % hydrogen peroxide in 92% yield and a propylene oxide selectivity of 97% [144].

Although most of the catalytic epoxidations are carried out in a polar solvent to homogenize the reaction mixture, they can also be conducted with excellent yields in an aqueous organic two-phase system in the presence of a tungstate/phosphate/tetraalkylammonium salt catalyst [145]. Good catalysts must be stable, highly reactive and selective, and be easily recycled.

6. Hydroxylation of Olefinic Double Bonds

Hydroxylation of olefinic double bonds is the addition of two hydroxyl groups, one to each carbon of the unsaturated bond, which leads to the formation of vic-diols. Long chain aliphatic vic-diols and their derivatives have properties which make them useful as modifiers in protective coatings, plastics, lubricants, waxes, emulsifiers and textile finishing agents.

vic-Diols are prepared by hydrolysis of epoxides. Mostly, because of better yields, the intermediate epoxide is not isolated. A convenient and efficient method is to dissolve the olefin in an excess of formic acid and to add slowly hydrogen peroxide to the mixture [146]. Because of the high acidity of the formic acid, the epoxide is rapidly converted into an hydroxy formate which is readily hydrolyzed (eq. 12). Many attempts are reported to lower formic acid and hydrogen peroxide consumption [147 - 149]. Also, use of peracetic acid dissolved in acetic acid is described [150].

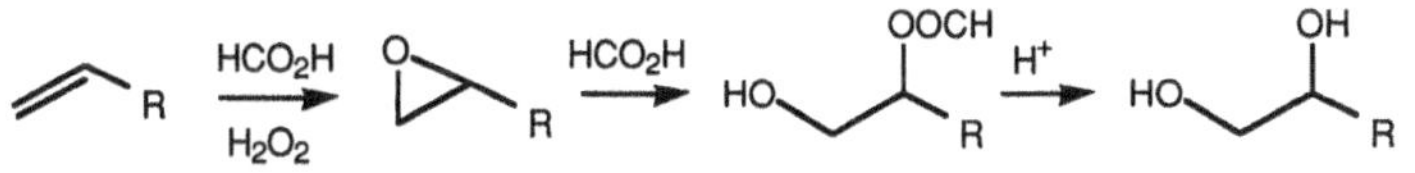

(12)

1,2-Pentanediol, an important intermediate for production of fungicides [151], can be continuously prepared by reacting 1-pentene with a solution of perpropionic acid in benzene and subsequently hydrolyzing the epoxide in benzene solution [152]. The yield, based on 1-pentene, of 92% is much higher than that with performic acid [149].

Compounds prepared by reacting tungstate with phosphoric acid and hydrogen peroxide, and neutralizing the reaction product with a quaternary ammonium salt appear to be effective catalysts for the direct hydroxylation of olefins with hydrogen peroxide [153].

7. Nuclear Hydroxylation of Aromatic Hydrocarbons

Hydroxylation of benzene with hydrogen peroxide to phenol in the presence of oxidizable metal ions has been comprehensively reviewed [154].

Besides phenol also catechol and hydroquinone in varying ratios but in low yield, were formed. In general nuclear hydroxylation of aromatic hydrocarbons is characterized by low yield and selectivity. By reacting polynuclear aromatic hydrocarbons with peracetic acid, mainly ring cleavage, yielding dicarboxylic acids, was observed [155] (eq. 13).

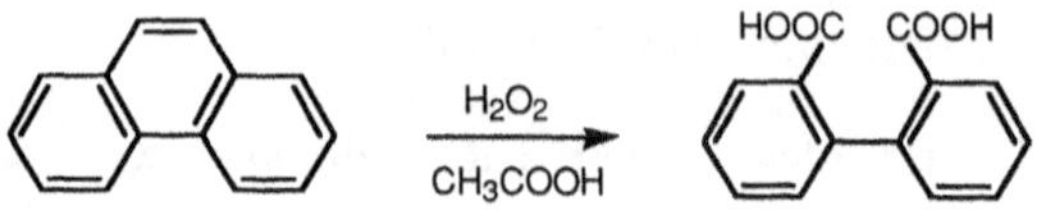

(13)

Hydroxylation of phenol with hydrogen peroxide to catechol and hydroquinone was extensively studied (eq. 14).

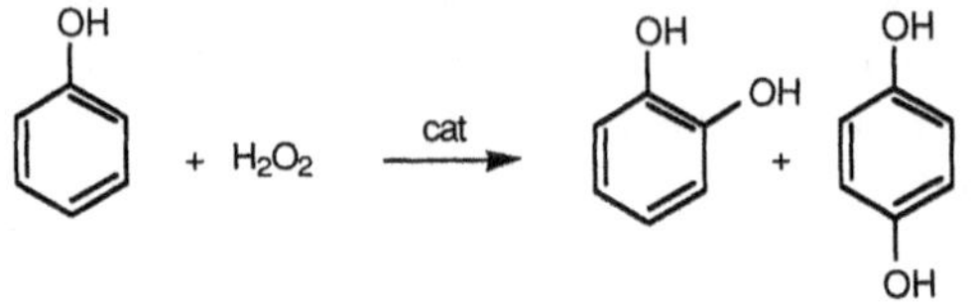

(14)

Catechol is used as an intermediate for production of agrochemicals, pharmaceuticals, and fragrances and flavours, whereas hydroquinone is mainly consumed in photographic industry.

In Europe three different processes for production of catechol and hydroquinone, all using hydrogen peroxide, are commercially exploited by Rhône-Poulenc [156], Brichima [157] and Enichem [158]. Rhône-Poulenc uses a mixture of strong acids like $HClO_4$ and H_3PO_4 as the catalyst [159]. The process developed by Brichima is a free-radical chain process, based on combined iron and cobalt salts [160]. In the Enichem process, phenol is hydroxylated over Ti containing zeolites [161]. What these processes all have in common is that the conversion is limited by using an excess of phenol. In doing this, tar formation mainly due to further oxidation and coupling reactions of the products formed can be minimized. Comparison of the main features of the three processes (Table 3) shows that the Enichem process can operate at a higher phenol conversion, resulting in high selectivities on both hydrogen peroxide and phenol and keeping tar formation at an acceptable level [162].

Table 3: Phenol hydroxylation with hydrogen peroxide.

Features	Rhône-Poulenc	Brichima	Enichem
Catalyst	$HClO_4$, H_3PO_4	Fe^{2+}/Co^{2+}	TS-1
Phenol conversion (%)	5	10	25
H_2O_2 yield (%)	70	50	70
Phenol selectivity (%)	90	80	90
Catechol / hydroquinone	1.4	2.3	1
Tars (%)	10	20	12

Other reaction systems for hydroxylation of phenol were recently suggested including hydrogen peroxide in the presence of a shape-selective zeolite catalyst such as HZSM-5 giving exclusively hydroquinone [163], hydrogen peroxide in the presence of a perfluorosulfonated polymer partly loaded with Ti or V [164], hydrogen peroxide with sulfur dioxide or selenium dioxide [165], anhydrous solutions of hydrogen peroxide in organic solvents that do not form a low-boiling azeotrope with hydrogen peroxide and sulfur dioxide [166] and percarboxylic acids [167].

8. Baeyer-Villiger-Oxidation of Ketones

Ketones can be oxidized to esters in the so-called Baeyer-Villiger reaction with percarboxylic acids (eq. 15).

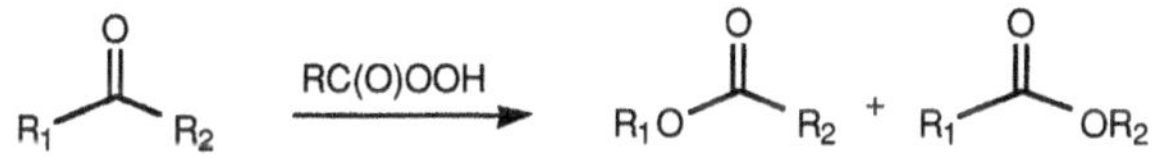

(15)

This reaction has been the subject of a number of reviews. Ketones used include aliphatic, alicyclic and aromatic ketones, α,β-unsaturated ketones, α-diketones and α-keto acids [168 -170]. Also, oxidations with hydrogen peroxide in alkaline medium [171] or in the presence of catalysts [172] are described. This H_2O_2 route, however, is generally inferior to the percarboxylic acid reaction.

Oxidation of cyclic ketones with percarboxylic acids gives lactones in good yields. ε-Caprolactone is commercially produced by Baeyer-Villiger reaction of cyclohexanone (eq. 16). Low molecular weight polyesters, made from ε-caprolactone, are interesting intermediates for production of

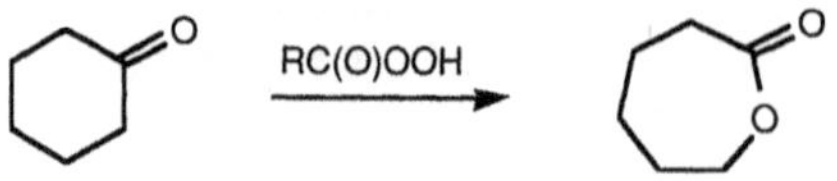

(16)

high performance polyurethane elastomers and coatings [173]. Oxidation of cyclohexanone is carried out using either peracetic acid in an anhydrous medium such as ethyl acetate [174] or distilled aqueous peracetic acid [175]. In both cases an excess of cyclohexanone is used and yield of ε-caprolactone based on peracetic acid is better than 90%. Strong acids cannot be present in the percarboxylic acid, as ε-caprolactone easily undergoes ring opening and polymerisation.

9. Oxidation of Nitrogen Compounds

9.1 Amine Oxides

Aliphatic tertiary amines are readily oxidized with aqueous hydrogen peroxide. For example, lauryldimethylamine reacting with 35 wt % hydrogen peroxide at 60-75°C affords lauryldimethylamine oxide in 99% yield [176]. (eq. 17)

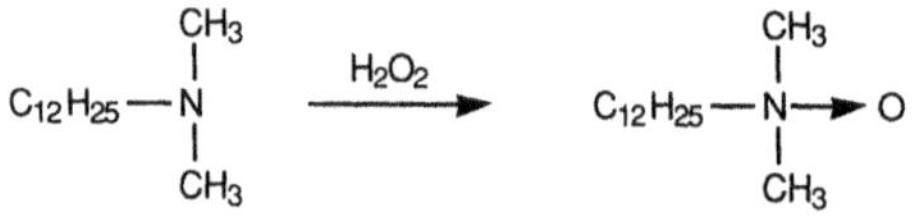

(17)

Lauryldimethylamine oxide is primarily used in light-duty liquid dishwashing detergents. Less reactive tertiary amines, especially the heterocyclic amines, require percarboxylic acids [177]. (eq. 18)

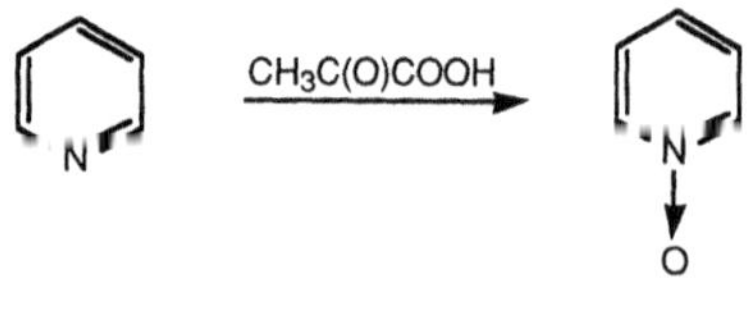

(18)

9.2 Hydrazine

Hydrazine hydrate is commercially produced by oxidation of ammonia with chlorine or hypochlorite (Raschig process; Baeyer process). Large quantities of sodium chloride are formed as a by-product.

A new process, developed by Produits Chimiques Ugine Kuhlmann, uses hydrogen peroxide as the oxidant in the presence of a ketone and either a nitrile [178] or an amide and disodium hydrogen phosphate as a co-catalyst [179]. Hydrolysis of the intermediate ketazine yields hydrazine. Use of methyl ethyl ketone is favored.

The mechanism is not completely understood since both hydrogen peroxide and ammonia do not react. Probably, periminoacids, formed as intermediates by perhydrolysis of the nitrile or the amide present in the reaction mixture act as oxygen transferring agents. Ammonia reacts with methyl ethyl ketone (Scheme 3) to form a ketimine which is oxidized to an oxaziridine. The oxaziridine oxidizes ammonia and the hydrazone obtained is converted with excess ketone to the ketazine. The methyl ethyl ketazine, which is insoluble in the reaction mixture, is removed by decantation. The ketazine is distilled and subsequently hydrolyzed under pressure to give aqueous hydrazine and methyl ethyl ketone, which is then recycled [180]. The aqueous phase, remaining after ketazine decantation, is concentrated to remove reaction water and then recycled to the reactor. The peroxide process which eliminates salts as by-products is commercially used by Produits Chimiques Ugine Kuhlmann and by Mitsubishi Gas Chemical Co. [181].

Scheme 3:

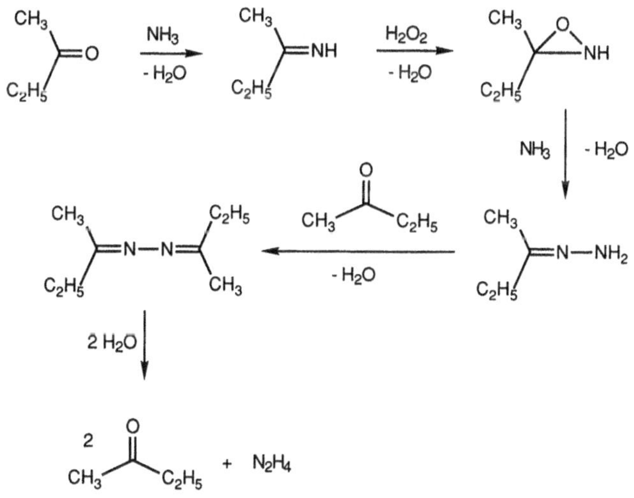

9.3 ε-Caprolactam

The technical interest in ε-caprolactam, monomer for nylon-6, has initiated extensive work on cyclohexanone oxime formation by oxidation with hydrogen peroxide.

Oxidation of primary amines with hydrogen peroxide catalyzed by tungstate gives oximes in high yield [182, 183]. (eq. 19)

$$RCH_2{-}NH_2 \xrightarrow[\text{cat}]{H_2O_2} RCH_2{-}NHOH \longrightarrow RCH{=}NOH \qquad \textbf{(19)}$$

Thus, reaction of cyclohexylamine with hydrogen peroxide in the presence of sodium tungstate at 10 - 20°C affords a 98% yield of cyclohexanone oxime [183]. In the absence of a catalyst, oxidation of primary amines leads to a mixture of oxime, nitroso-, bisnitroso- and nitro-compounds.

Another process (eq.20) uses the reaction of hydrogen peroxide and excess ammonia with cyclohexanone in the presence of catalysts, e.g. sodium tungstate, at 20-30°C; yields of cyclohexanone oxime up to 97% are obtained [184, 185].

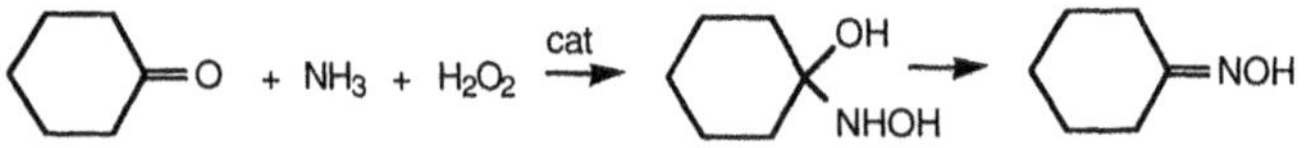

(20)

1-Hydroxy cyclohexyl hydroxylamine is formed as intermediate. Hawkins showed that the uncatalyzed reaction of cyclohexanone with hydrogen peroxide and ammonia gives either 1-hydroperoxycyclohexylamine or 1,1'-peroxydicyclohexylamine, depending on the ketone to hydrogen peroxide ratio [186] (eq.21).

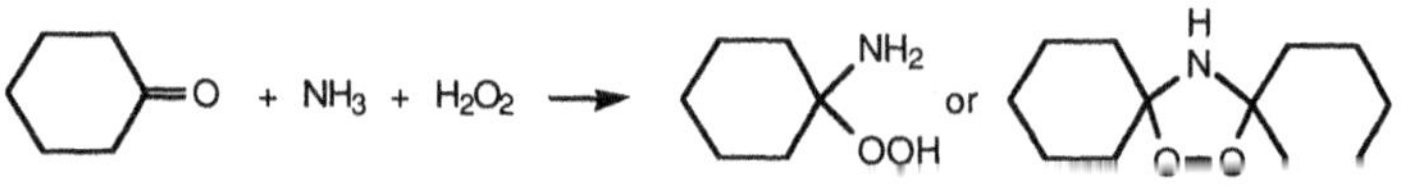

(21)

Base catalyzed rearrangement of 1,1'-peroxydicyclohexylamine in the presence of lithium compounds affords ε-caprolactam and cyclohexanone [187]. By vapor-phase pyroloysis of 1,1'-peroxycyclohexylamine, 11-cyanoundecanoic acid is obtained, [188] which can be either hydrogenated to

12-aminododecanoic acid, the nylon-12 monomer, or hydrolyzed to 1,12-dodecanedioic acid. A commercial plant for production of 12-aminododecanoic is operated by UBE Industries Ltd. [189]. Recently, a process for preparing cyclohexanone oxime by reacting cyclohexanone with ammonia and hydrogen peroxide in the liquid phase, which is catalysed by a Ti-containing zeolite, was reported [190]. This process is being further developed in a pilot plant [191].

1,1'-Peroxydicyclopentylamine, useful as a precursor for sebacic acid and 10-aminodecanoic acid, is obtained in good yield by reacting cyclopentanone with ammonia and hydrogen peroxide [192].

9.4 Reaction with Hydrogen Cyanide

Reacting aqueous hydrogen peroxide with hydrogen cyanide in the presence of copper and iron salts at ambient temperature yields almost quantitatively cyanogen [193] (eq. 22).

$$2\,HCN + H_2O_2 \xrightarrow[Cu^{2+}]{Fe^{3+}} (CN)_2 + 2\,H_2O \qquad \textbf{(22)}$$

The reoxidation of copper(I)-ions is so fast that no precipitation of copper(I) cyanide occurs. Use of copper(I) sulphate is favored. With copper(I) chloride small amounts of cyanogen chloride are formed. Under proper reaction conditions, no hydrolysis of cyanogen is observed.

10. Oxidation of Sulfur Compounds

Thioethers are readily converted with aqueous hydrogen peroxide into sulfoxides [194]. Solvents like acetone, alcohol [194] and acetic acid [195] have been used although acetone and ketones in general cannot be recommended because of possible formation of ketone peroxides. In order to avoid sulfone formation, the temperature should be kept below 50°C and stoichiometrical quantities of hydrogen peroxide should be applied.

Oxidation of sulfides with hydrogen peroxide at higher temperature provides sulfones [196]. This reaction is promoted by Mo compounds [197], Zr salts [198] and acids [199]. Peracids are used in preparing sulfoxides [200]. Sulfides can also be oxidized to sulphones with alkaline hydrogen peroxide [201]. Further oxidation reactions of sulfur compounds include conversions of mercaptans to disulphides with alkaline hydrogen peroxide [202], preparation of sulfonic acids by reacting mercaptans or disulfides with percarboxylic acids [203], formation of methyl isothiocyanate from sodium N-methyl dithiocarbamate with slightly alkaline hydrogen peroxide [204] and oxidation of thiourea to formamidine sulfinic acid [205] (eq. 23, 24).

$$CH_3\text{-}NH\text{-}C(S)\text{-}SNa \xrightarrow[OH^-]{H_2O_2} CH_3\text{-}N{=}C{=}S \qquad (23)$$

$$NH_2\text{-}C(S)\text{-}NH_2 \xrightarrow{H_2O_2} NH{=}C(SO_2H)\text{-}NH_2 \qquad (24)$$

11. Miscellaneous

Reaction of olefins with hydrogen peroxide and either hydrogen halide or elemental halogen gives halohydrins [206] (eq. 25).

$$2\,C{=}C + H_2O_2 + X_2 \rightarrow 2\,X\text{-}C\text{-}C\text{-}OH \qquad (25)$$

An excellent example of the oxidation of halide ions is demonstrated in the preparation of tetrabromobisphenol A in 95% yield by brominating bisphenol A with elemental bromine and hydrogen peroxide [207].

Water soluble, open-chain and cyclic 1,2-diols are oxidatively cleaved to carboxylic acids with aqueous hydrogen peroxide in the presence of tungstate and phosphate ions under acidic conditions [208]. A mixture of cis and trans 1,2-cyclohexanediol affords adipic acid in 94% yield. Also, direct oxidative cleavage of olefins into carboxylic acids with hydrogen peroxide catalyzed by acids or heteropolyacids of Mo and W is reported [209-211]. Refluxing cyclohexene with hydrogen peroxide and catalytic amounts of $H_3PMo_{12}O_{40}$ gives adipic acid in 92% yield [210].

Oxidation of primary and secondary alcohols to aldehydes and ketones is normally carried out employing strong inorganic oxidizing agents. These oxidations suffer mostly from poor selectivity: the aldehydes and ketones obtained are readily further oxidized to the corresponding carboxylic acids. This side reaction is completely suppressed by using a synthetic zeolite containing titanium and hydrogen peroxide [212]. For example, benzyl alcohol is oxidized in 90 % yield to benzaldehyde; benzoic acid is not formed.

12. Safety Aspects

Hydrogen peroxide decomposes to form water and oxygen with the generation of heat

$$H_2O_2\,(liq) \rightarrow H_2O\,(liq) + O_2\,(g) \qquad \Delta H = -93.8\ kJ/mol$$

The rate of decomposition increases with temperature by a factor of approximately 2.2 times for each 10°C rise over the range of 20 - 100°C [213]. Decomposition can be further accelerated by the presence of impurities or by contact with incompatible surfaces. Therefore, to safely store and handle hydrogen peroxide, equipment should be made of appropriate materials, be adequately vented and be maintained free of contaminants [214].

Explosivity of hydrogen peroxide/water vapor mixtures has been studied over a wide range of pressure [215 - 219]. At atmospheric pressure, the lower explosive limit occurs at 26 mol% [215] which corresponds to the equilibrium vapor concentration over a boiling solution of approximately 75 wt% H_2O_2. At reduced pressure, the minimum hydrogen peroxide concentration in the vapor, which can explode, increases.

It is apparently impossible to obtain a propagating detonation in pure 90 wt % hydrogen peroxide at room temperature [220, 221]. The sensitivity of 86 wt % and 90.7 wt % to detonation shock, initiated by a tetryl booster, depends on container diameter and temperature [222]. The hazard of an explosion, however, increases rapidly when hydrogen peroxide is mixed with organic material. Data have been published on the explosive limits for several ternary mixtures containing hydrogen peroxide, water and organic compounds [220, 223 -226]. The explosive effect varies somewhat with the nature of organic compound, but more strongly with the concentration of hydrogen peroxide and with the ratio of hydrogen peroxide and organic compound. The most sensitive compositions are those that contain stoichiometric quantities of hydrogen peroxide, required to oxidize completely the organic compound to water and carbon dioxide. This critical composition can also result from mixtures of organic and diluted hydrogen peroxide if excess unreacted hydrogen peroxide is concentrated. It should be noted that a reaction between hydrogen peroxide and organic compound may take place yielding intermediates which have completely different explosive properties. Thus, mixing of formic acid and hydrogen peroxide produces on standing performic acid and the resulting mixture is much more sensitive than the original one [224]. Also the presence of sulphuric acid can raise the hazardous potential of an hydrogen peroxide organic system. Therefore, unknown systems should not be tested on a large scale operation without having already defined the safety limits on a smaller scale version.

References

1 L. J. Thenard, *Ann. Chim. Phys.*, **8**, 306 (1818).
2. W. C. Schumb, C. N. Satterfield and R. L. Wentworth, '*Hydrogen Peroxide* ', Reinhold Publ. Co., New York 1955, p. 18.
3. H. Berthelot, *Compt. Rend.*, **86**, 71 (1878).
4. W. Manchot, *Liebigs Ann. Chem.*, **314**, 177 (1901); *ibid.*, **316**, 331 (1901).
5. J. H. Walton and G. W. Filson, *J. Am. Chem. Soc.*, **54**, 3228 (1932).
6. BASF, DRP 649234, 1934 and DRP 6583767 , 1935 (G. Pfleiderer).
7. *European Chemical News* , May 30, 1988, p. 9.
8. I.G. Farbenindustrie, GB 508081, 1939.
9. Degussa, DE 1945750, 1969 (G. Käbisch and R. Trübe).
10. Edogawa, DE 1112051, 1958 (K. Hiratsuka et al.).
11. Laporte Chemicals Ltd., DE 953790, 1953 (C.W. Lefeuvre).
12. E.I. du Pont de Nemours & Co., DE 888840, 1950 (C.R.Harris and J.W. Sprauer).
13. Degussa, DE 1261838, 1963 (G. Käbisch).
14. Becco, US 2537516, 1950 (L.H. Dawsey, C.K. Muehlhausser and R.R. Umhoefer).
15. Becco, US 2537655, 1950 (L.H. Dawsey, C.K. Muehlhausser and R.R. Umhoefer).
16. Laporte Chemicals Ltd., DE 933088, 1953 (C.W. Lefeuvre).
17. E.I. du Pont de Nemours & Co, BE 819676, 1974.
18. Air Liquide, EP 287421, 1988 (C. Pralus).
19. Degussa, DE 2018686, 1970 (G. Giesselmann, G. Schreyer and W. Weigert).
20. Degussa, US 4349526, 1982 (G. Goor and W. Kunkel).
21. Eka Nobel AB, EP 284580, 1988 (E. Bengtsson and U.M. Andersson).
22. FMC, EP 95822, 1983 (W. Ranbom).
23. I.G. Farbenindustrie, US 2158525, 1939 (H.J. Riedl and G. Pfleiderer).
24. E.I. du Pont de Nemours & Co., US 2689169, 1954 (W.S. Hinegardner).
25. Edogawa, DE 1085861, 1957 (K. Hiratsuka et al.).
26. BASF, DE 3732015, 1987 (M. Eggersdorfer, J. Henkelmann and W. Grosch).
27. Degussa, DE 1063581, 1957 (L.R. Darbee and D.F. Kreuz).
28. Degussa, DE 1051257, 1957 (G. Käbisch and E. Richter).
29. Laporte Chemicals Ltd., GB 741444, 1953 (W.R. Holmes et al.).
30. Solvay & Cie., DE 1041928, 1956 (M. Gonze and E. Leblon).
31. E.I. du Pont de Nemours & Co., US 2673140, 1954 (J.W. Sprauer).
32. Edogawa, FR 1319025, 1962.
33. W.M. Weigert, H. Delle and G. Käbisch, *Chem. Ztg.*, **99**, 101 (1975).
34. Edogawa, DE 2141234, 1971 (H. Shin et al.).
35. Degussa, DE 1273499, 1964 (G. Schreyer O. Weiberg and G. Giesselmann).
36. '*Ullmann's Encyclopedia of Industrial Chemistry* ', 5th Edition, Vol. A 13, 450.
37. Oxysynthese, GB 1119820.
38. BASF, DE 801840, 1949 (G. Pfleiderer, H. J. Riedl and W. Deuschel).
39. FMC, US 2720532, 1955 (R. R. Umhoefer).
40. FMC, US 2730533, 1956 (R. R. Umhoefer).
41. Degussa, DE 1542089, 1965 (G. Käbisch and H.Herzog).
42. Laporte Chemicals Ltd., GB 718306, 1953 (C.A.Morgan)

43. Degussa, DE 1079604, 1959 (D. H. Porter).
44. E.I. du Pont de Nemours & Co., US 4061598, 1977 (K.M. Makar).
45. E.I. du Pont de Nemours & Co., US 4800075, 1989 (C.L. Jenkins).
46. E.I. du Pont de Nemours & Co., US 4521531, 1985 (J.S. Coates).
47. Degussa, US 4428923, 1984 (W. Kunkel, J. Kemnade and D. Schneider).
48. Laporte Chemicals Ltd., GB 718307, 1952 (J.A. Williams and C.W. Lefeuvre).
49. FMC, US 3009782, 1961 (D. H. Porter).
50. EKA AB, EP 102934, 1983 (T. Berglin and W. Herrmann)
51. FMC, US 4428922, 1984 (Q.G. Hopkins).
52. Degussa, GB 1096260, 1966.
53. *Br. Chem. Eng.*, **2**, 88 (1959).
54. Degussa, DE 2003268, 1970 (M. Liebert, H. Delle and G. Käbisch).
55. EKA AB, DE 2419534, 1974 (B. G. Franzen).
56. Allied Chemical Corp., US 2902347, 1959 (J.N. Cosby, R.L. Harris and F.H. Rissmiller).
57. Kali Chemie, US 3126257, 1964 (H.Kunowski and G.Fix)
58. Laporte Chemicals Ltd. DE 1159398, 1959 (W.R.Holmes and J. Mc Kenzie).
59. PPG, DE 977130, 1955 (J.R. Cox and T.E. Brown).
60. EKA AB, US 4503028, 1985 (B.G. Franzen and W. Herrmann).
61. PPG, US 3055838, 1962 (J.W. Moore).
62. E.I. du Pont de Nemours & Co., US 2739875, 1956 (J.W. Sprauer and T.V. Williams).
63. Laporte Chemicals Ltd., GB 928784, 1959 (B.K. Howe).
64. Laporte Chemicals Ltd., GB 991413, 1962 (B.K. Howe).
65. E.I. du Pont de Nemours & Co., CA 649850, 1962 (W.S. Hinegardner and H.R. Eisenhauer).
66. Degussa, DE 2036553, 1970 (G. Käbisch and W.Kunkel).
67. Oxysynthese, FR 1340901, 1962 (M. Charret).
68. E.I. du Pont de Nemours & Co., DE 1036225, 1956 (J.W. Sprauer).
69. Laporte Chemicals Ltd., GB 841323, 1957 (J.G. Lait and H.B. Hopkins).
70. Columbia-Southern Chemical Corp., GB 795433, 1956.
71. PPG, GB 924625, 1959.
72. E.I. du Pont de Nemours & Co., US 3016290, 1962 (A.A. Elston and W.S. Hinegardner).
73. Oxysynthese, FR 1539843, 1967 (P. Thirion).
74. Mitsubishi Gas Chemical, US 4792403, 1988 (S. Togo, Y. Sugihara and T. Ikehe).
75. Tokai Denka Kogyo K.K., DE 3826720, 1988 (S. Watanabe).
76. E.I. du Pont de Nemours & Co., US 4879043, 1989 (J.H. Boughton et. al.).
77. Laporte Chemicals Ltd. US 3073755, 1963 (R.H. Banfield and A. Hildon).
78. Columbia-Southern Chemical Corp., US 2960448, 1960 (W.S. Straub and C.J. Sindlinger).
79. Degussa, US 3060105, 1962 (J. Müller).
80. FMC, US 3152052, 1964 (T.M. Jenney, D.H. Porter and R.H. Rice).
81. Laporte Chemicals Ltd. GB 893069, 1959 (P.V. Rutledge and D.S. Bell).
82. Columbia-Southern Chemical Corp., US 2904517, 1959 (P.H. Baker).
83. BE 900626, 1985 (M. Bollen).
84. Produits Chimiques Ugine Kuhlmann, US 4508696, 1985 (M. Coingt).
85. '*Gmelins Handbuch der Anorganischen Chemie* ', 8 Auflage, Sauerstoff, Lieferung 7, System-Nummer 3, Verlag Chemie, 1966.

86. D. Swern in '*Organic Peroxides* ', Ed. D. Swern, vol. 1, Wiley Interscience, New York 1970, p. 313.
87. Degussa, DE 2018713, 1970. (P. Hoffmann, O. Weiberg and W. Weigert).
88. Degussa, DE 1165576, 1960 (O. Weiberg).
89. Degussa, DE 1170926, 1962 (O. Weiberg).
90. Degussa, DE 2141155, 1971 (P. Kleinschmit, G. Schreyer, O. Weiberg and W. Weigert).
91. Degussa, DE 2145604, 1971 (G.Schreyer and O.Weiberg)
92. Degussa, DE 2141156, 1971 (M. Krüger, H. Mönch, G. Schreyer and O. Weiberg).
93. Degussa, US 4101570, 1978 (M. Krüger, G. Schreyer and O. Weiberg).
94. Bayer and Degussa, US 4059619, 1977 (G. Prescher, G. Schreyer, O. Weiberg, R. Wirthwein, H. Waldmann, H. Seifert, W. Schwerdtel and W. Swodenk).
95. Propylox, DE 2807344, 1978 (A. Hildon, T.D. Manly and A. J. Jaggers).
96. Degussa, DE 3720562, 1987 (G. Böhme, W. Hofen, G. Prescher and R. Siegmeier).
97. W.E. Parker, C. Ricciuti, L. L. Logg and D. Swern, *J. Am. Chem. Soc.*, **77**, 4037 (1955).
98. W.E. Parker, L.P. Witnauer and D. Swern, *J. Am. Chem. Soc.*, **79**, 1929 (1957).
99. P. Kuzel, T. Lieser and M. Dankowski, *Seifen-Öle-Fette-Wachse* , 111, 448 (1985).
100. L.S. Silbert, E. Siegel and D. Swern, *J. Org. Chem.*, **27**, 1336 (1962).
101. FMC, US 3462480, 1969 (D.G. Mc Kellar and R. von Falkenstein).
102. J. Gordon, *Hydr. Process. & Petr. Refiner* , **41**, 141 (1962); **41**, 185 (1962).
103. *Chem. Week* , 92, April 6, 1963, p. 55.
104. D. Swern, *J. Am. Chem. Soc.*, **69**, 1692 (1947).
105. Rohm & Haas, US 2485160, 1949 (W.D. Niederhauser et al.).
106. Rubber Research Institute of Malaysia, GB 2179355, 1987 (C. Seong-Fong).
107. Degussa, DE 3002826, 1980 (G. Käbisch, R. Trübe, H. Wittmann, S. Raupach, H. Malitius and M. Lindemann)
108. Henkel and Degussa, DE 3002861, 1980 (M. Lindemann, H. Weiss, D. Schweizer, G. Käbisch, H. Malitius, S. Raupach, R. Trübe and H. Wittmann).
109. Henkel, DE 3320219, 1983 (G. Dieckelmann, K.Eckwert, L. Jeromin, E. Peukert and U. Steinberner).
110. Ashland Oil and Refining Co., US 3360531, 1971 (W.H. French).
111. D. Swern in '*Organic Peroxides* ', Ed. D. Swern, vol. 2, Wiley-Interscience, New York 1971, p. 436.
112. Peroxid-Chemie, EP 55387, 1981 (W. Edl and G. Sienel).
113. Degussa, DE 3723843, 1987 (R. Siegmeier, R. Wirthwein and U. Brandt).
114. Henkel, US 4647678, 1987 (K. Eckwert, L. Jeromin, A. Meffert, E. Peukert and D. Gutsche).
115. Degussa, DE 2123930, 1971 (A. Kleemann, G. Schreyer and O. Weiberg).
116. A. Kleemann and R. Wagner, '*Glycidol* ', Hüthig Verlag, Heidelberg 1981, p. 32.
117. Degussa and Bayer, DE 2 519 297, 1975 (G. Prescher, G. Schreyer, O. Weiberg, R. Wirthwein , H. Waldmann, H. Seifert, W. Schwerdtel and W. Swodenk).
118. Degussa and Bayer, DE 2519298, 1975 (G. Prescher, G. Schreyer, O. Weiberg, R. Wirthwein , H. Waldmann, H. Seifert, W. Schwerdtel and W. Swodenk).
119. Atochem, EP 61393, 1982 (J. C. Lecoq, M. Pralus and J. P. Schirmann).
120. Propylox, DE 2747761, 1977 (A. Hildon and P. Greenhalgh).
121. FMC, US 3954815, 1976 (W.C. Fisher, S.M. Linder, R.L. Pelley and H.P. Liao).
122. Interox Chemicals, DE 2602776, 1976 (A. Hildon and P. Greenhalgh).
123. Bayer, EP 56932, 1982 (G. Rauleder and H. Waldmann).

124. Degussa, DE 3528005, 1985 (A. Grund, G. Prescher, G. Böhme, W. Hofen and H. Petsch).
125. Solvay, EP 19322, 1980 (R. Walraevens and L. Lerot).
126. Degussa, DE 3528007, 1985 (R. Siegmeier, A. Grund, G. Prescher and U. Brandt).
127. *Hydr. Process.*, 225 (1981).
128. Union Carbide Corp., US 2804473, 1957 (B. Phillips, F.C. Frostick and P.S. Starcher).
129. *Chem. Eng.*, July 25, 1960, p. 68.
130. G. Payne and P. Williams, *J. Org. Chem.*, **24**, 54 (1959).
131. Z. Raciszewski, *J. Am. Chem. Soc.*, **82**, 1267 (1960).
132. Degussa, DE 2408948, 1974 (W. Heim, A. Kleemann, H. Kolb and G. Schreyer).
133. Degussa, DE 2454630, 1974 (H. Friedrich, W. Heim, A. Kleemann, H. Kolb and G. Schreyer).
134. Degussa, DE 2454572, 1974 (H. Friedrich, W. Heim, A. Kleemann, H. Kolb and G. Schreyer).
135. Degussa, DE 2508228, 1975 (G. Prescher and G. Schreyer).
136. Degussa, DE 2543332, 1975 (G. Prescher and G. Schreyer).
137. Union Carbide Corp., DE 2607768, 1976 (C. H. McMullen).
138. Mitsubishi Gas Chemical, EP 74259, 1982 (Y. Kuriyama, M. Kakuda and S. Nitok)
139. Produits Chimiques Ugine Kuhlmann, DE 2752626, 1977 (J. P. Schirmann and S. Y. Delavarenne)
140. Produits Chimiques Ugine Kuhlmann, DE 2605041, 1976 (M. Pralus, J. P. Schirmann and S. Y. Delavarenne).
141. Produits Chimiques Ugine Kuhlmann, DE 2916834, 1979 (J. C. Lecoq, J. P. Schirmann and M. Pralus).
142. Halcon, US 4558026, 1985 (W. F. Brill).
143. Anic, EP 100119, 1983 (C. Neri, B. Anfossi, A. Esposito and F. Buonomo).
144. Eniricerche, EP 230949, 1987 (M. G. Clerici and U. Romano).
145. Instituto Guido Donegani, GB 2055821, 1980 (C. Venturello, E. Alneri and G. Lana).
146. D. Swern, N. Billen and J. T. Scanlan, *J. Am. Chem. Soc.*, **68**, 1504 (1946).
147. Degussa, EP 25940, 1980 (G. Käbisch, H. Malitius, S. Raupach, R. Trübe and H. Wittmann).
148. Degussa, DE 2937840, 1979 (G. Käbisch, H. Malitius, S. Raupach, R. Trübe and H. Wittmann).
149. Ciba-Geigy, EP 141775, 1984 (H. Issler and R. Maul).
150. Chevron, US 3991126, 1976 (R. Bacskai).
151. Janssen Pharmaceutica, US 4079062, 1978 (G.Van Reet, J. Heeres and L. Wals).
152. Degussa, DE 3442937, 1984 (R. Siegmeier, G. Prescher, H. Maurer and G. Hering).
153. Montedison, US 4562299, 1983 (C. Venturello and M. Gambaro).
154. G. Sosnovsky and D.J. Rawlinson in '*Organic Peroxides* ', Ed. D. Swern, Vol. 2, Wiley Interscience, New York 1971, p. 277.
155. C.A. Buehler and H. Hart, *J. Am. Chem. Soc.*, **85**, 2177 (1963).
156. J. Varagnat, *Ind. Eng. Chem. Prod. Res. Dev.*, **15**, 212 (1976).
157. P. Maggioni and F. Minisci, *Chim. Ind. (Milan)*, **59**, 239 (1977).
158. U. Romano, A. Esposito, F. Maspero, C. Neri and M. G. Clerici, *Stud. Surf. Sci & Catal.*, vol. 55, Elsevier , Amsterdam, 1990, p. 33.
159. Rhone-Poulenc, DE 2064497, 1970 (F. Bourdin, M. Constantini, M. Jouffret and G. Lartigau).
160. Brichima, DE 2341743, 1973 (P. Maggioni).

161. Anic, GB 2116974, 1985 (A. Esposito, M. Taramasso, C. Neri and F. Buonomo).
162. B. Notari, *Stud. Surf. Sci. & Catal.*, vol. 37, Elsevier, Amsterdam, 1987, p. 413.
163. Mobil Oil Corp., US 4578521, 1986 (C.D. Chang and S.D. Hellring).
164. FMC, EP 132783, 1984 (R.A. Bull).
165. Degussa, DE 3602180, 1986 (K.H. Drauz, A. Kleemann, G. Prescher and G. Ritter).
166. Degussa, US 4551562, 1985 (K.H. Drauz and A. Kleemann).
167. H. Waldmann, C. Jupe, J. Baumert, H. Seifert and G. Schümmer, *Ger. Chem. Eng.*, **5**, 64 (1982).
168. C.H. Hassall, *Org.React.*, **9**, 73 (1957).
169. P.A.S. Smith in '*Molecular Rearrangements*', Ed. P. de Mayo, vol. 1, Wiley Interscience, New York 1963, p. 568.
170. B. Plesnicar in 'Oxidation in Organic Chemistry', part C, Ed. W. S. Trahanovsky, Academic Press, New York 1978, p. 211.
171. M. Bogdanowicz, T. Ambelang and B. M. Trost, *Tetrahedron Lett.*, 923 (1973).
172. J.D. McClure and P.H. Williams, J. Org. Chem., 27, 627 (1962).
173. R. E. Redfern in '*Speciality Chemicals*', May 1982.
174. Union Carbide Corp., DE 1086686, 1955 (C.B. Phillips and P.S. Starcher).
175. Degussa, DE 1258858, 1965.
176. K. Lindner, *Tenside*, **1**, 112 (1964).
177. H.S. Mosher, L. Turner and A. Carlsmith, *Org. Synth. Coll.*, **4**, 828 (1963).
178. Produits Chimiques Ugine Kuhlmann, US 3869541, 1975 (F. Weiss, J. P. Schirmann and H. Mathais).
179. Produits Chimiques Ugine Kuhlmann, US 3972878, 1976 (J. P. Schirmann, J. Combroux and S. Y. Delavarenne).
180. Produits Chimiques Ugine Kuhlmann, DE 2639010, 1976 (J. P. Schirmann, J. Combroux and S. Y. Delavarenne).
181. *Japan High Tech. Report*, **41**, 3 (1990).
182. Inventa, DE 939808, 1952 (K. Kahr).
183. K. Kahr, *Angew. Chem.*, **72**, 135 (1960).
184. O. L.Lebedev and S. N. Kazarnovski, *Zhur. Obsch. Khim.*, **30**, 1631 (1960) (through *C.A.*, **55**, 1473a (1961)).
185. Toa Gosei, DE 1245371, 1960; DE 1265745, 1967; DE 1274124, 1967; DE 1277252, 1967.
186. E.G. Hawkins, *Angew. Chem.*, **85**, 850 (1973).
187. BP Chemicals Ltd., DE 1814288, 1969; DE 1803872, 1971; DE 1670246, 1971.
188. BP Chemicals, GB 1198422, 1966 (E. Hawkins).
189. *Hydr. Process.*, 132 (1981).
190. Montedipe, US 4745221, 1988 (P. Roffia, M. Padovan, E. Moretti and G. De Alberti).
191. *Chem.Eng.*, July 18, 1988, p. 19.
192. Idemitsu Kosan, JP 6267.067, 1987 (Y. Hironaka) (through *C.A.*, **108**, 38584 (1988)).
193. R. Fahnenstich, W. Heimberger, F. Theissen and W. Weigert, *Chem. Ztg.*, **96**, 388 (1972).
194. D. Tarbell and C. Weaver, *J. Am. Chem. Soc.*, **63**, 2941 (1941).
195. T. Douglas, *J. Am. Chem. Soc.*, **68**, 1072 (1946).
196. M. Lister and L. Sutton, *Trans. Faraday Soc.*, **35**, 495 (1939).
197. General Anilin and Film Corp., US 3005852, 1959 (H.B. Freyermuth, H.S. Schultz and S.R. Buc).
198. Eastman Kodak, FR 2095718, 1972 (N.W. Connon).

199. N. Novitskaya and S. Chernikova, *Neftekhimiya*, **10**, 429 (1970) (through *C.A.*, **73**, 76486 (1970)).
200. C. Cumper. J. Read and A. Vogel, *J. Chem. Soc.*, 5323 (1965).
201. H. Boehme and U. Sitorus, *Chem. Ztg.*, **96**, 37 (1972).
202. J. Hesse and F. Truby, *Chem. Ind. (London)*, **16**, 680 (1965).
203. H. Nawa, W. Brady, M. Koike and L. Reed, *J. Am. Chem. Soc.*, **82**, 896 (1960).
204. Degussa, DE 2105473, 1972 (G. Gießelmann, W. Schwarze and W. Weigert).
205. SKW Trostberg, DE 2736943, 1977 (K. Büchler, L. Mittermeier and W. Pollwein).
206. Degussa, DE 2160613, 1971; DE 2204548, 1972 (W. Weigert, H. Klebe, A. Meffert, G. Käbisch and A. Langenfeld).
207. Degussa, DE 2227439, 1972 (W. Weigert, H. Hein, H. Mechler, E. Meyer-Simon and K. Janzon).
208. C. Venturello and M. Ricci, *J. Org. Chem.*, **51**, 1599 (1986).
209. New Japan Chemical, EP 201719, 1986 (M. Nakazawa and K. Fujitani).
210. Shinippon Rika KK, JP 63-93746, 1988 (T. Fujitani and M. Nakazawa) (through *C.A.*, **109**, 230298 (1988)).
211. T. Oguchi, T. Ura, Y. Ishii and M. Ogawa, Chem. Lett. 1989, 857.
212. Anic, EP 102655, 1983 (A. Esposito, C. Neri and F. Buonomo).
213. Chemical Safety Data Sheet SD-53. Properties and Essential Information for Safe Handling and Use of Hydrogen Peroxide, Manufacturing Chemists Association, Washington D.C., 1969.
214. J. Mackenzie, *Chem. Eng.*, June 1990, 84.
215. C.N. Satterfield, G.M. Kavanagh and H. Resnick, *Ind. Eng. Chem.*, **43**, 2507 (1951).
216. C.N. Satterfield, P.J. Ceccotti and A.H. Feldbrugge, *Ind. Eng. Chem.*, **47**, 1040 (1955).
217. C.N. Satterfield, G.M. Kavanagh and H. Resnick, *J. Chem. Eng. Data* , **4**, 131 (1959).
218. J. M. Monger, H. J. Baumgartner, G.C. Hood and C.E. Sanborn, *J. Chem. Eng. Data* , **9**, 119 (1964).
219. G. A. Campbell and P. V. Rutledge, *Inst. Chem. Eng., Symp. Ser.*, **33**, 37 (1972).
220. E. S. Shanley and F. P. Greenspan, *Ind. Eng. Chem.*, **39**, 1536 (1947).
221. F. Bellinger, H.B. Friedman, W.H. Bauer, J.W. Eastes and W.C. Bull, *Ind. Eng. Chem.*, **38**, 310 (1946).
222. Bureau of Mines, Information Circular 8387 (1968).
223. E. Haeuseler, *Explosivstoffe* , **6/7**, 64 (1953).
224. J.M. Monger, H. Sello and D.C. Lehwalder, *J. Chem. Eng. Data* , **6**, 23 (1961).
225. Bureau of Mines, Report of Investigations 5877 (1961).
226. Shell Chemical Corp., Bull. SC 59-44, '*Concentrated Hydrogen Peroxide: Summary of Research Data on Safety Limitations* '.

CHAPTER 3

Activation of Hydrogen Peroxide by Organic Compounds

Ruggero CURCI * and John O. EDWARDS #

* *Dipartimento di Chimica, Università di Bari,*
Via Amendola, 173; 70126 Bari, Italy

Department of Chemistry, Brown University,
Providence, RI 02912, U.S.A.

1. Introduction

Hydrogen peroxide is a powerful oxidant, as demonstrated by the E° value of 1.76 V measured for the half-reaction $H_2O_2 + 2\,H^+ + 2\,e^- \rightarrow 2\,H_2O$; yet it reacts quite slowly with some substrates of interest such as olefins and aromatic hydrocarbons. In order to take advantage of its power, it may be required to "activate" the peroxide functionality. By activation we mean here improving its reactivity toward a substrate through the formation of an intermediate peroxide, whose rate constant with the given substrate is significantly larger than that of hydrogen peroxide itself.

As reported in other chapters of this book, the activation of hydrogen peroxide by transition metal ions (M) has been much studied [1] and the resultant reactive species have various functionalities, such as in **1** or **2** [1,2].

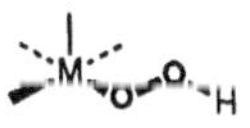

1

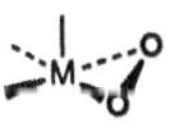

2

However, activation can also occur by interaction of hydrogen peroxide with an organic compound. A classic example is its reaction with acetic anhydride, yielding peroxyacetic acid (Eq. 1), which is capable of reacting with electron pair-donors (R_2S, Br^-, etc.) five powers of ten more rapidly than hydrogen peroxide [3].

G. Strukul (ed.), Catalytic Oxidations with Hydrogen Peroxide as Oxidant, 45–95.

$$(CH_3CO)_2O + H_2O_2 \xrightarrow{H^+} CH_3\text{-}\overset{\overset{\displaystyle O}{\|}}{C}\text{-}OOH + CH_3\text{-}\overset{\overset{\displaystyle O}{\|}}{C}\text{-}OH \qquad (1)$$

Of course, several other methods allow one to generate reactive species starting from hydrogen peroxide. For instance, thermolysis or photolysis [4] of H_2O_2 generate the reactive hydroxyl radical HO•, which is an extremely effective oxidant [5]. One-electron reduction of H_2O_2, as promoted either electrochemically or upon reaction with certain transition metal ions [6], can also generate hydroxyl radical. In these systems, a radical chain mechanism is often set forth, and high selectivities are rare to attain.

In a review of this length, the range and depth of topics must be limited. The activation modes considered here are those which involve mechanisms largely heterolytic, i.e. polar (two-electron) instead of radical. Also, an exhaustive coverage of the literature is not feasible, so that the reader should refer to more detailed surveys in order to probe into the many aspects of hydrogen peroxide chemistry [1-11]. We shall restrict the subject to sketch out the relevant principles that rule H_2O_2 activation by organic compounds and constrain ourselves to present a limited number of pertinent examples. As already mentioned, activation by metal complexes [1], including the important metalloporphyrin reactions [11], will be also omitted since this subject is dealt with in other chapters.

In spite of the above restrictions, it will be seen that transformation of hydrogen peroxide to a more reactive organic derivative is an excellent method of H_2O_2 activation. It finds many practical applications, not least of which is the *in situ* conversion of H_2O_2 to peroxyacetic acid CH_3CO_3H by tetraacetylethylenediamine (TAED) in the bleaching of soiled clothes at lower temperatures [12,13].

2. Activation Mechanisms

2.1 Polar reactions of H_2O_2 and of organic peroxides

In order to understand the mechanisms of rate enhancement through the interaction of H_2O_2 with organic "activators", it is convenient to review first the key features of peroxide polar reactions [14].

Let us consider the energy required to cleave a peroxide bond, first by homolytic processes (Eq. 2) and then by heterolysis.

$$ZO-OY \longrightarrow ZO^{\bullet} + YO^{\bullet} \qquad (2)$$

The amount of energy required to carry out the homolytic cleavage of a generic peroxide ZOOY (Eq. 2) varies from 47 kcal mol^{-1} for HOOH, down to values such as 29 kcal mol^{-1} for dialkyl peroxodicarbonates [15] RO-C(O)-OOR (which decompose slowly at room temperature), and to 23 kcal mol^{-1} for phenylacetylperoxide $[PhCH_2C(O)O]_2$, which undergoes multiple bond cleavage [16]. In the latter, contributing to the ease of homolytic bond scission is an entropy of activation of about 12 cal mol^{-1}K^{-1}. Some experimental values of $\Delta H^{\neq}$ and of $\Delta S^{\neq}$ for peroxide bond homolysis [17] are given in Table 1.

Table 1: Some activation parameters for peroxide bond homolyses.[a]

Peroxide	$\Delta H^{\neq}$ (kcal mol^{-1})	$\Delta S^{\neq}$ (cal mol^{-1}K^{-1})
HO–OH	47	11
t-BuO–OH	41	12
RO–OR [b]	~ 36	~ 14
$Ph\overset{O}{\overset{\Vert}{C}}O-OBu\text{-}t$	35	8
$^{-}O_3SO-OSO_3^{-}$	32	11
$R\overset{O}{\overset{\Vert}{C}}O-O\overset{O}{\overset{\Vert}{C}}R$ [b]	30	6
$RO\overset{O}{\overset{\Vert}{C}}O-O\overset{O}{\overset{\Vert}{C}}OR$ [b]	29	5

[a] Most data are from Ref. 17. [b] In these compounds R represents a simple alkyl group (Et, n-Bu, etc.).

The data in Table 1 can be used to estimate $\Delta H^{\neq}$ for homolysis of other peroxide species, and hence rate constants at various temperatures, provided one makes the reasonable assumption that $\Delta H^{\neq}$ for a mixed peroxide ZOOY is the mean of the activation enthalpies (i.e., bond dissociation energies) of ZOOZ and YOOY. For example, for the process in Eq.3, we employ the enthalpy values of 47 and 30 kcal mol^{-1} for HOOH and RCO-OO-COR , respectively (Table 1). The mean of these two is 38.5 kcal mol^{-1}. Using this for $\Delta H^{\neq}$, and the value of +12 cal mol^{-1}K^{-1} for $\Delta S^{\neq}$, the rate constant for homolysis of the peroxyacid at 25 °C is estimated to be 6×10^{-12} s^{-1}.

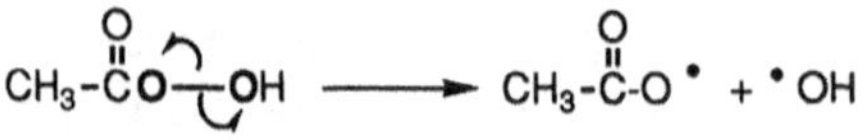

(3)

The very low rate estimated suggests that simple (i.e, uncatalysed) homolysis of peroxyacids in solution is usually well hidden by solvent and trace-metal induced decomposition. On the other hand, heterolysis – i.e. the process wherein the pair of bonding electrons goes to one oxygen of the O-O moiety – is not readily discussed in term of experimental data, largely because this is never seen as a clean first-order process (Eq.4).

(4)

In fact, the usual heterolytic reaction is bimolecular with another species (a nucleophile, Nu:) bringing up a pair of electrons to that oxygen which is losing its grip on the O-O bonding electrons (Eq.5).

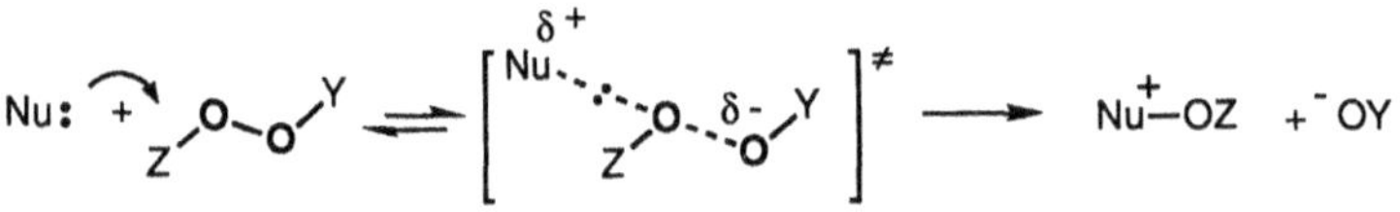

(5)

Because of the endothermic charge development associated to electron-deficient formation in Eq. 4, the unimolecular heterolysis has a very high activation energy; in fact one can estimate [14] that in the gas phase it is disfavoured with respect to homolysis by more than 250 kcal mol^{-1}. Then, it is not surprising that it has not been possible to find compelling evidence for free HO^+, or other simple oxacations RO^+, even in peroxide reactions catalysed by strong acids.

On the other hand, in polar solvents heterolysis by bimolecular steps (possibly aided by structural factors) can be a dominant process. The rates of these heterolytic reactions can have markedly different rate constants, largely depending on the leaving group YO^- and nucleophile nature (Eq. 5). In Table 2, some data are presented concerning reactions of HOOH (bimolecular and

Table 2: Nucleophilic reactivities toward hydrogen peroxide.[a]

Nucleophile	E_n[b]	k_2[c] ($M^{-1}s^{-1}$)	k_3[c] ($M^{-2}s^{-1}$)	Ref.
R_3P:	--	10^2 [d]	$2. \times 10^3$ [d]	19
HS^-	--	0.48	--[f]	20
SO_3^{2-}	2.57	0.20	--[f]	21
$S_2O_3^{2-}$	2.52	2.5×10^{-2}	1.7	22
$(H_2N)_2C{=}S$:	2.18	7×10^{-2}	1.42	23
$PhCH_2SMe$ [e]	--	5×10^{-2} [d]	--[f]	24
I^-	2.06	1.1×10^{-2}	0.32	25
H_2S	--	8×10^{-3}	--	20
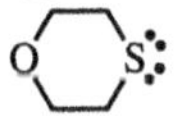	--	2.6×10^{-3}	3.5×10^{-2}	26
CN^-	2.02	1×10^{-3}	--[f]	27
SCN^-	1.83	5.1×10^{-4}	2.5×10^{-2}	28
Et_3N:	--	3.3×10^{-4}	--[f]	29
H_2NNH_2	--	1.8×10^{-5}	--[f]	30
NO_2^-	1.73	3×10^{-7}	--[f]	31
HO^-	1.65	$< 1 \times 10^{-7}$	--[f]	32
Br^-	1.51	3.8×10^{-8}	2.3×10^{-4}	25
Cl^-	1.24	1.8×10^{-9}	8.3×10^{-7}	25
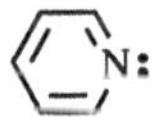	1.20	(slow oxidation at 90 °C)		
H_2O	0.00	(^{18}O exchange rate $< 1.2 \times 10^{-10}$)		32

[a] In water at 25 °C, unless noted otherwise. [b] A nucleophilicity parameter, see Ref. 18. [c] Based on a two-term rate law, i.e. $v = k_2[Nu][HOOH] + k_3[Nu][HOOH][H^+]$. [d] Based on oxidation of Ph_3P by *t*-BuOOH in EtOH (Ref. 19). [e] Estimated value. [f] Constant k_3 not measurable because the nucleophile is protonated in acid solution.

acid catalysed termolecular) with representative nucleophiles. It can be seen that that the range of rate constants is over eleven powers of ten, depending primarily on the nature of the attacking atom of the nucleophile.

It is possible to say that the stronger nucleophiles (i.e., those having high values of E_n, a parameter measuring nucleophilicity) [18] react with HOOH pretty much according to the same trend [19-32] followed in reacting with methyl bromide CH_3Br, a sp^3 carbon substrate. This is also shown by the linear free energy relationship (LFER) in Figure 1.

Figure 1: Linear free-energy relationship between the rates of nucleophilic attack at sp^3 carbon and at hydrogen peroxide.

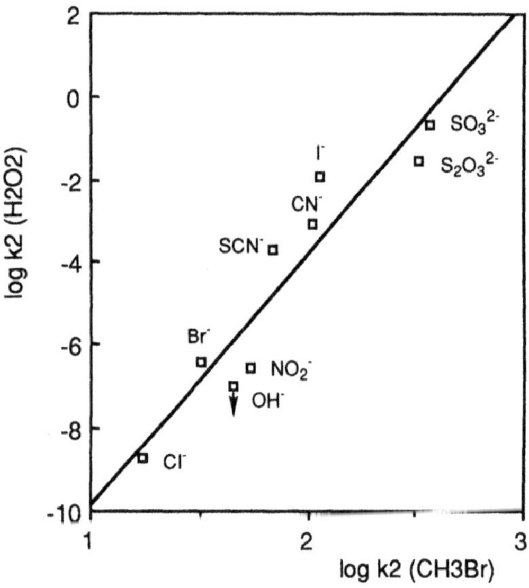

Several kinetic studies have demonstrated [33] that electronic effects of substituents are consistent with the concept that the substrate is donating electrons to electrophilic peroxide oxygen. For example, a Hammett ρ value of ca. -1.0 has been reported [24] for the oxidation of substituted diphenylsulfides with hydrogen peroxide (Eq.6).

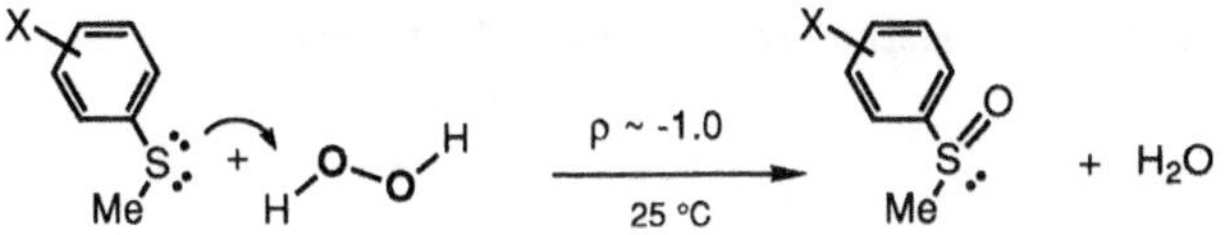

(6)

The reaction in Eq.7, where the three X stands for p-NO_2, H, and p-OMe, yields k_2 values (at 35 °C) of 1.26×10^{-3}, 6.44×10^{-3}, and 12.6×10^{-3}, respectively [34]; this is of course the order of increasing electron release by the substituents.

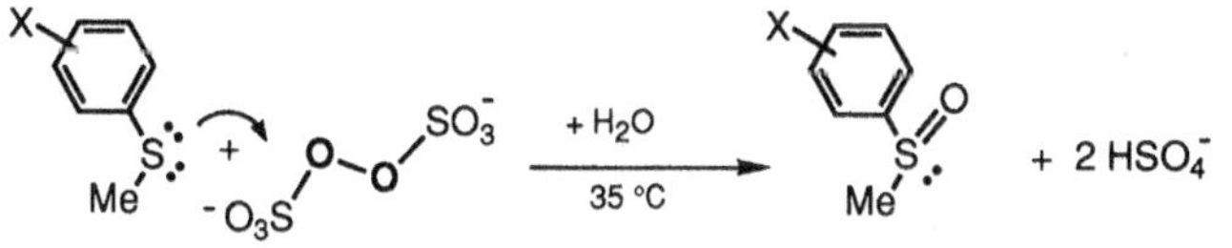

(7)

Also the order of $\Delta H^{\neq}$ measured for the substituted thioanisols is p-OMe < H < p-NO_2, while $\Delta S^{\neq}$ values remain essentially the same [34].
In Table 3, the variation in peroxide YOOH reactivity depending on the nature of Y is presented [25, 33, 35-40]. Again, a large variation of rates is seen, and there is a good correlation of logk with the pK_a of YOH (or YOH_2^+, in the case of H^+ catalysis), i.e. the conjugate acid of leaving group YO^- (or YOH). As these becomes weaker bases, they are also better leaving groups, and rates increase.
What comes out of the selected data in the Tables presented so far are the following conclusions:
i. of all the peroxide examined, HOOH is not a particularly reactive one, thus the need for activation; *ii.* with some nucleophilic substrates (and reductants), "activation" may become necessary, while with others it may not; *iii.* the single most important factor for high reactivity in peroxide YOOH is the low basicity of the leaving group YO^-; *iv.* thermodynamics may be important, but kinetic control is dominant; *v.* learning to govern these factors, selectivity can be high in competitive oxidations.

In heterolytic reactions of the type considered so far (eq. 5), the nuclear event is the transfer of an electrophilic oxygen atom to the substrate. In some cases the actual transfer has been established by mean of ^{18}O isotope labeling of the peroxide [33].

Table 3: Rates of halide oxidation by some peroxide species.[a]

ZOOH	pK_a^{ZOH} [b]	$\log k_2$ [c]: I^-	$\log k_2$ [c]: Br^-	Ref.
ClO-H [d]	~ -7	8.5	3.5	35
H_2O^+-O-H	- 1.7	4.0[e]	1.1[e]	25
$^-$O-S(=O)(=O)-OO-H	2.0	3.2	0.0	36,37
HO-P(=O)(HO)-OO-H	2.0	3.2	0.9	36,37
Me-C(=O)-OO-H	4.7	~ 3.0	- 0.8	33
HO-P(=O)($^-$O)-OO-H	7	1.8	- 1.5	36,37
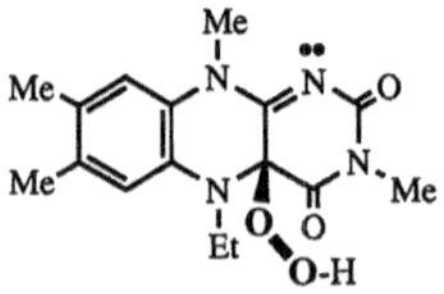	~ 9.4	0.78		38
HOO-H	15.7	-2.0	-6.4	25
MeOO-H	~16	- 2.6		39
EtOO-H	~17	- 2.7		39
Ph-C(Me)(Me)-OO-H	~19	- 3.2		40
Me_3COO-H	~20	- 3.1		39

[a] At 25 °C, aqueous solution. [b] Acidity constant of the corresponding reduced species. [c] Units of k_2 are $M^{-1}s^{-1}$. [d] Not a peroxide. [e] After correction for estimated pK_a value of H_3O^+ , i. e. - 4.7.

Heterolytic cleavage of a peroxide can also be accomplished by a quite different method. Conversion of a peroxide ZOOH to its conjugate anion by a base (Eq. 8) can be followed by addition to a atom centre S of the substrate, having an open orbital to form an intermediate ; the latter then breaks down to give the observed oxidation product SO (Eq. 9).

$$ZOOH + B: \rightleftharpoons ZOO^- + BH^+ \quad (8)$$

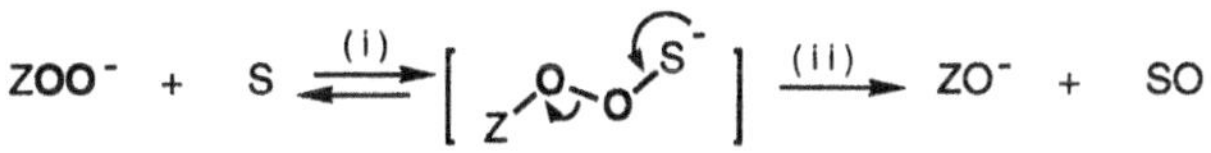

(9)

In Eq. 9 , step (i) has the peroxide anion acting as a nucleophile and the substrate acting as an electrophile, and can be rate determining. Notice however that in step (ii), leading to the oxidized product, the roles are reversed.

Electrophilic activation and nucleophilic activation of hydrogen peroxide are discussed in more details below.

2.2 Electrophilic activation

We have mentioned that, when hydrogen peroxide acts as an elecrophile, it interacts with a reductant S: having a pair of electrons in an open orbital. Some such reductants are Br^-, R_2S:, R_3N:, etc. . In the previous Section, a model has been presented which fits the evidence available (Eq. 5); this has the nucleophile electron pair attacking one peroxide oxygen, with a transition state involving Nu-O bond formation and O-O bond breaking to occur in a concerted fashion.

In terms of frontier molecular orbitals (FMO) [41], the HOMO of the nucleophile interacts with the first LUMO of H_2O_2, i.e. the σ^* of the O-O bond [42], to form a new bond.

As is seen in Figure 2, this process is similar in several ways (e.g., the stereoelectronic requirement for linearity of Nu:--O--O in the t.s.) to nucleophilic attack at carbon in methyl halide. Several pieces of evidence support a close parallelism between nucleophilic attack at peroxide O-O bond and at sp^3 carbon [33, 43]. For instance, substituted anilines $XC_6H_4NH_2$

Figure 2. Frontier orbitals for nucleophilic displacement at the peroxide O-O bond and at sp^3 carbon.

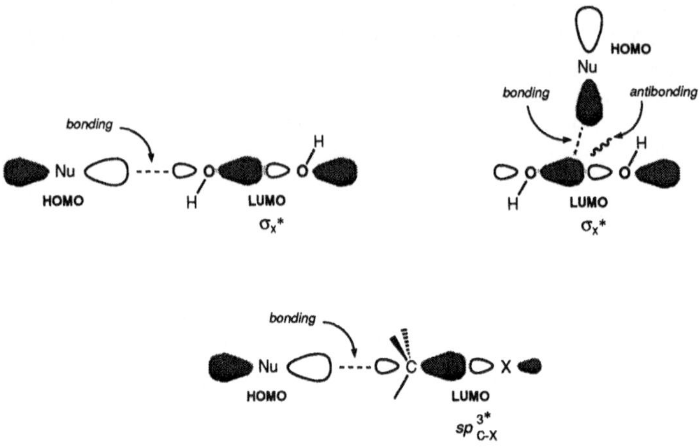

attack peroxide oxygen (of peroxyacetic acid) in the same order of rates (yielding a negative Hammett ρ value) as they attack methyl tosylate MeOTs [44]. The selectivity is quite different for the two electrophiles, i.e. peroxide oxygen and sp^3 carbon; in fact, with halide ions the relative rates $k(I^-) > k(Br^-) > k(Cl^-)$ with peroxides covering a range of nearly 10^6, whereas with CH_3Br the range is only 250 [43].
From these examples and many others, it is possible to describe the characteristics of electrophilic oxidation by peroxides, i.e. nucleophilic displacement on peroxide O-O bond, as follows [33]:

- the most reactive nucleophiles are strongly polarizable (e.g., I^-, Br^-,R_2S, R_3P), whereas the most basic ones (e.g. HO^-, F^-) are unreactive, and alkenes poorly reactive;

- the direction of electron motion in the t.s. is favored when the leaving anion ^-OY is less basic (i.e. when Y is more electron-withdrawing), which makes ^{-}OH a poor leaving group (hence H_2O_2 is less reactive with respect to peroxyacids);

- the characteristic rate law is order one each in peroxide and the substrate (second order overall), i.e. $v = k_2[\text{peroxide}][\text{Nu}]$; when, as in the case of H_2O_2 and alkyl hydroperoxide, the peroxide is scarcely reactive, a third order term is often found with acid as added concentration, i.e., $v = k_2[\text{peroxide}][\text{Nu}] + k_3[\text{peroxide}][\text{Nu}][\text{HA}]$.

On this ground, we can bring out the practical ways by which electrophilic oxidations using H_2O_2 can be activated. In fact, addition of a strong acid to the level of 1 M H^+ often raises the rate by about two powers of ten. It has been shown [33] that this occurs by specific acid catalysis, involving two steps (Eq.s 10 and 11).

$$\text{HOOH} + \text{H}^+ \rightleftharpoons \text{HOOH}_2^+ \qquad (10)$$

(11)

Here, the rate increases because the leaving group is now the weak base water instead than the strong base hydroxide ion.

In strong acids, a sizeable fraction of H_2O_2 is expected to be protonated, giving the hydroperoxonium ion $H_3O_2^+$ (Eq 10); thus hydroxylation of π-donor systems, such as aromatic hydrocarbons [45] and alkenes [46] becomes feasible. The concept of H_2O_2 activation by strong acids is pushed to its limit when concentrated hydrogen peroxide is employed in "superacid", weakly nucleophilic solvents such as FSO_3H ("magic acid"), FSO_3H-SbF_5-SO_3, or FSO_3H-SbF_5-SO_2ClF [47]. This enabled Olah and co-workers to carry out successfully even the hydroxylation of normally resistant substrates, such as branched-chain saturated hydrocarbons [47]. The first step of the electrophilic hydroxylation most likely consists in a O-atom insertion by protonated hydrogen peroxide into the C-H bond of the alkane.

Strong acid catalysis aside, changing the reaction solvent can also be useful. In fact, if the solvent is hydroxylic (e.g., ROH), it can participate as a third particle in the t.s. to transfer in a cycle a proton to the leaving group [27], for example as shown in **3**.

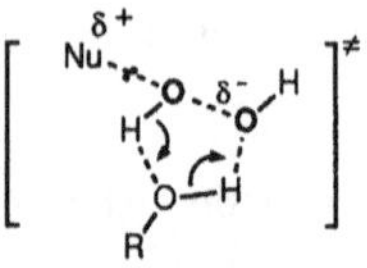

3

The advantage gained by this solvent participation consists in the fact that endothermic charge separation (cf. Eq. 5, Sect. 2.1) is largely avoided [26]. For sulfide oxidation by H_2O_2, the order of rates in hydroxylic solvents has been found to be related to autoprotolysis constants, e.g. AcOH > H_2O > MeOH > *t* -BuOH, with the range covering nearly four powers of ten [26]. The concept of decreased charge separation in the t.s. bears also on the oxidation of nucleophiles by peroxycarboxylic acids, but the mode of solvent expression differs [48]. If the peroxyacid is present in a non-polar, non-basic solvent (e.g., benzene, CCl_4, $CHCl_3$, etc.), it exists in the ground state as an internally H-bonded cyclic form (**4**) [49]; this can react directly with the substrate, moving the proton to the carbonyl oxygen while the nucleophile is attacking the outer oxygen. Thus, charge separation in the t.s. is again minimized.

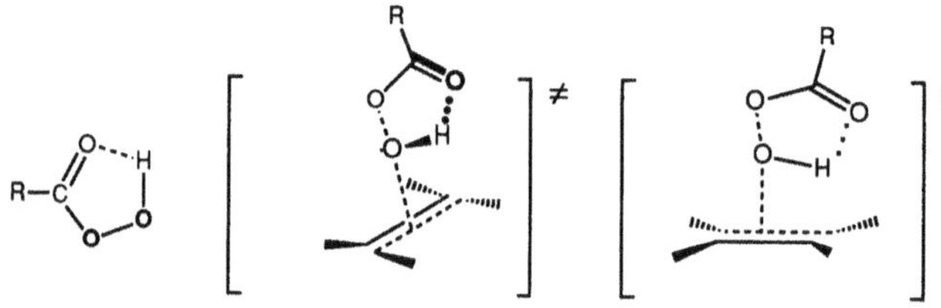

4 4' 4"

For reductants such as olefins the t.s. would have a structure (sometimes called Bartlett's butterfly), which leads concertedly to neutral epoxide and carboxylic acid. Stereoelectronic requirements lead one to envisage two limiting arrangement of the reactants in the t.s., i.e. *spiro* (**4'**) or planar (**4"**). The *spiro* orientation has been favoured by calculations [50a,b], and it has been supported by studies on primary and secondary kinetic H/D isotope effects [51]. However, it is likely that the planar t.s. lies close in energy to its *spiro* counterpart; in fact, in epoxidations involving certain conformationally constrained peracids [50c], a planar arrangement in the t.s. might become favored [50d].

2.3 Nucleophilic activation

At first sight it might seem a contradiction that one compound can be an electrophilic oxidant and a reactive nucleophile (an electron-pair donor) at the same time. However, it can be seen this to be possible for protic peroxide species ZOOH, if the two properties intervene sequentially, rather than competitively, as shown in Eq. 9 (Sect. 2.1). For instance, nucleophilic attack by ZOOH or its peroxyanion at an unsaturated moiety Y=X (e.g., a carbonyl) of the substrate leads to different intermediate adducts (Eq.s 12 and 13). These might have sufficient kinetic stability to

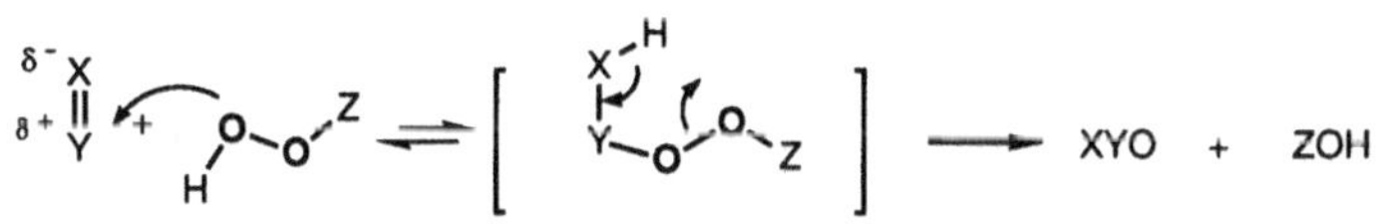

(12)

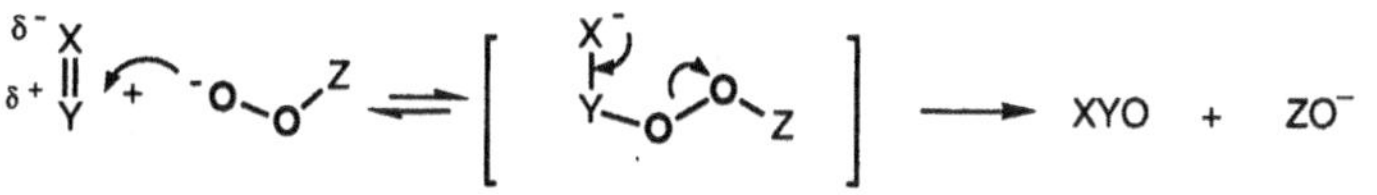

(13)

be detected spectroscopically, or even be isolated. In most cases, however, the adducts would rather break down to products in a rapid subsequent step with O-O bond cleavage, in a process amounting to electrophilic oxygen transfer to an electron-pair of the substrate. An appropriate example is the well-known Baeyer-Villiger reaction [52], allowing the transformation of ketones into esters, as in Eq. 14.

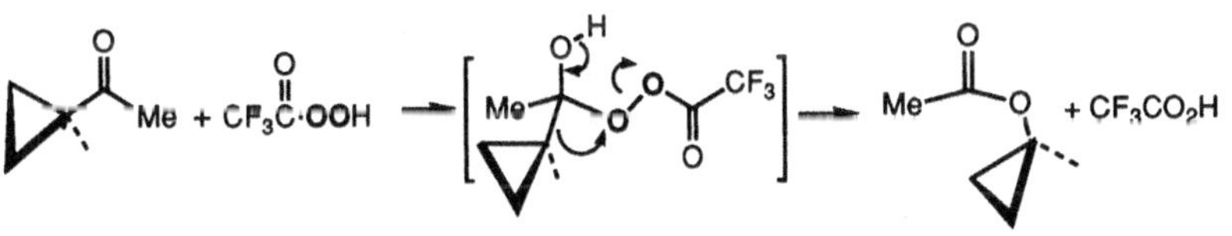

(14)

Notice that here the substrate electron-pair to which electrophilic oxygen is transferred in the second step is tied up in a C-C bond. Yet another pertinent example is the "nucleophilic" oxidation of sulfoxides to sulfones by *t* -BuOO⁻

[53], or by peroxyacids in alkaline media [54] where the peracid is present largely as the peroxyanion [Eq. 15].

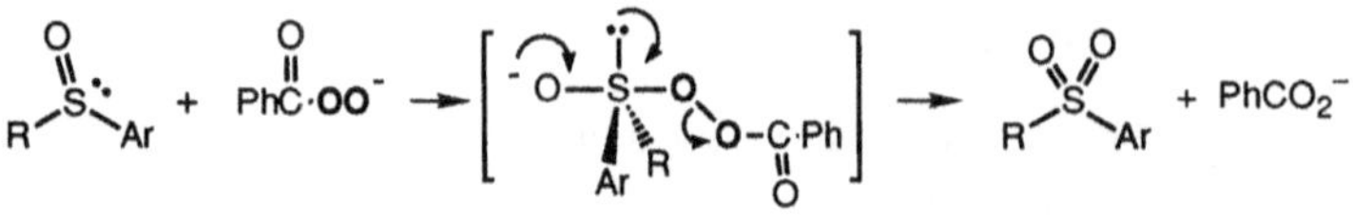

(15)

Of course the peroxyanion ZOO⁻ is a better nucleophile than the undissociated peroxide ZOOH, so that it is considerably more effective than the latter in carrying out the addition step (which is usually the slower step). A prime requisite of this type of activation is the removal of the proton on peroxide oxygen of ZOOH. Thus a quantity very important with respect to "alkaline activation" is the pK_a for ionization of the peroxide (Eq. 16). As the electronegativity of group Z increases,

$$ZOOH + H_2O \rightleftharpoons ZOO^- + H_3O^+ \quad (16)$$

the pK_a, and hence the pH of half-neutralization of ZOOH are lowered. Some pK_a values [55] for peroxyacids and for hydroperoxides (including H_2O_2) are given in Table 4.

It can be seen that for substituted benzoic acids electron-withdrawing groups lower pK_a values and electron-releasing ones raise them; however, the range is rather narrow, going from 7.93 for the *p* -OMe to 7.14 for the *p* -NO_2. Aliphatic peroxyacids have pK_a values of the same magnitude. For hydrogen peroxide and of hydroperoxides, pK_a values are of course lower than those of peroxyacids, but again the dependence on structure is small.

For a given peroxide ZOOH, the degree of "alkaline activation" is a compensating function of the electronic effect exercised by the group Z. At constant pH, the fraction of peroxide molecules in anionic form increases with electronegativity of Z. On the other hand, the nucleophilic reactivity of ZOO⁻ decreases as the electronegativity of Z increases. For instance, this can be seen clearly in a study [56] on the nucleophilic reactivity of peroxybenzoate anions toward p nitrophenyl acetate (Eq. 17).

Table 4: Ionization constants of some protic peroxide species.[a]

A. *Aromatic and Aliphatic Peroxyacids*

	pK_a		pK_a
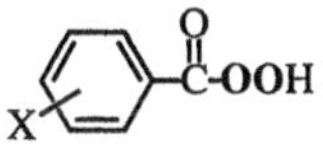		H-C(=O)-OOH	7.1
X= *p* -MeO	7.93	$ClCH_2$-C(=O)-OOH	7.2
H	7.64	Me-C(=O)-OOH	8.2
m -Cl	7.53	$MeCH_2$-C(=O)-OOH	8.1
p -NO_2	7.14	Me_3C-C(=O)-OOH	8.2
m -NO_2	7.23		
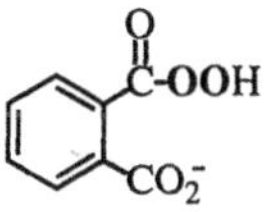	~8.5[b]		

B. *Hydroperoxides*

	pK_a
HOOH	11.75
MeOOH	11.08
$MeCH_2OOH$	11.05
Me_2CHOOH	11.45
Me_3COOH	12.25
Ph-C(Me)(Me)-OOH	12.6

[a] Most but not all of the values are at 25 °C, in water; for a more complete list of ionization constants see Ref. 55. [b] In aqueous dioxane.

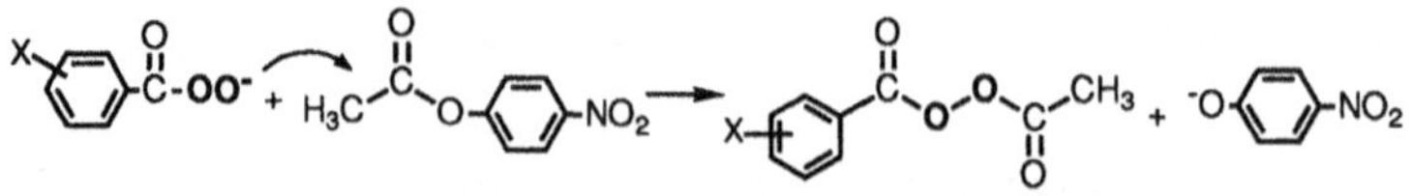
(17)

In fact, the influence of the substituent group X on pK_a is larger in magnitude and opposite in sign to its influence on rates, as expressed by $\log k_2$ values. The nucleophilic reactivity of peroxyanions ZOO^- and HOO^- is dramatically high, as it is shown below. Neutral hydrogen peroxide HOOH is also an excellent nucleophile, having a reactivity considerably higher than H_2O; however, the difference is far less relevant than that existing between HOO^- and HO^-. These high reactivities are examples of the so called "alpha effect" [57], which is the abnormal nucleophilic reactivity of a species when the atom adjacent ("alpha") to the atom reaction center has a non-bonded pair of electrons in the valence shell.

The benchmark on which comparisons are made is the nucleophilicity of the reacting species toward the proton, i.e. its basicity. Based on pK_a of their conjugate acids, i.e. H_2O (pK_a 15.7) and H_2O_2 (pK_a 11.7), hydroxide ion is ten thousand-fold more basic than perhydroxide HOO^-, and yet the latter is remarkably more reactive in nucleophilic displacements. Actually, with respect to basicity, the nucleophilic reactivities of the two anions [57-63] are inverted, as it is seen by inspection of data in Table 5.

The difference in reactivity toward tetrahedral carbon is about a factor of 20, with perhydroxide having the higher value. Towards trigonal carbon the ratios are of the order of two powers of ten, whether the substrate atom is a carbonyl or aromatic carbon (3rd and 4th entry, Table 5), and even larger for digonal (linear) carbon substrates. The phenomenon of large rate enhancements on passing from HO^- to HOO^- persists when substrates have reaction sites other than carbon (last three entries, Table 5). It deserves emphasis the fact that peroxoanions ZOO^-, where Z is not hydrogen (e.g., t-$BuOO^-$, $ArCO_3^-$), also show very high nucleophilic reactivity [57, 58].

Table 5: Relative reactivities of hydroperoxide and hydroxide ion toward some organic substrates.

Substrate[a]	Reaction center[b]	k_2HOO- [c]	k_2HO- [c]	Ratio[d]	Ref.
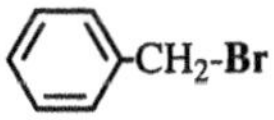	sp^3-C	5.4×10^{-2}	1.6×10^{-3}	34	60
$^-O_2CCH_2$-**Br**	sp^3-C	4.7×10^{-3}	3.6×10^{-4}	13	58
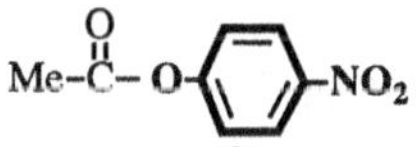	sp^2-C	1.1	15	73	58
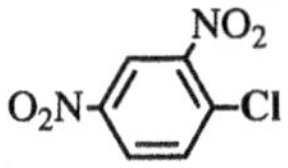	sp^2-C(Ar)	0.33	1.1×10^{-3}	330	58
N≡C-**Cl**	sp -C	8.8×10^3	12	730	61
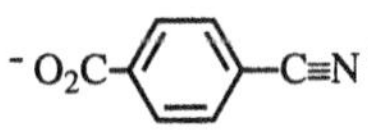 [e]	sp -C	0.11	9.5×10^{-5}	1200	58
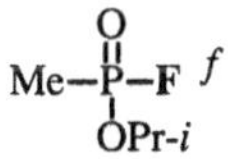 [f]	sp^3-P	1.5×10^3	25	60	62
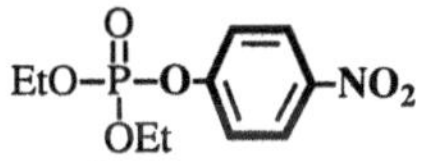 [g]	sp^3-P	1.4	1.0×10^{-2}	140	63
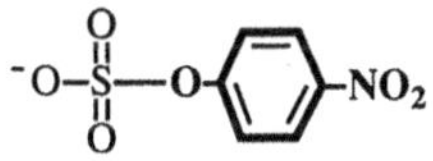	sp^3-S	3.3×10^{-5}	2.7×10^{-3}	120	58

[a] The leaving group is shown in boldface. [b] Hybridization at the atom site of attack on the substrate. [c] In $M^{-1}s^{-1}$. [d] I.e., (k_2HOO-/k_2HO-). [e] Not a nucleophilic displacement; after nucleophilic addition, hydrolysis leads to the amide. [f] Sarin. [g] Paraoxon.

As a rule, nucleophilic displacement by hydrogen peroxide at a carbonyl carbon is not a simple one-step process. In many cases there is an addition/elimination process, which takes the form of a $B_{AC}2$ or $A_{AC}2$ mechanism [59], as outlined in Scheme 1. Here the intermediates (in square brackets) are generally not isolated.

Scheme 1: Addition/Elimination Mechanisms for Nucleophilic Displacements at Carbonyl Carbon by Hydrogen Peroxide Species.

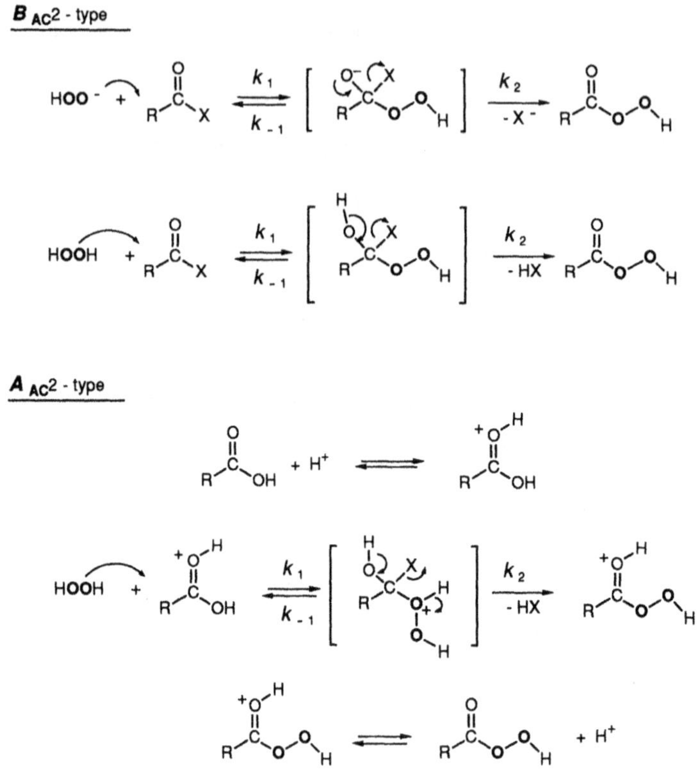

The addition/elimination mechanism has certain consequences. For instance, in $B_{AC}2$ displacements, if the leaving group X^- (or HX) is less basic than the entering nucleophile HOO^- (or HOOH), the first step (k_1) is rate-determining; on the other hand, when the leaving group is more basic than the nucleophile, then the second step (k_2) is rate-determining. In the situation of similar basicities, the overall rate is not determined by one step, because the two steps are comparable in magnitude of rate.

Activation of hydrogen peroxide by conversion to its conjugate anion can be very important. For example, it will be seen further on that peroxyimidic acids RC(NH)OOH generated from HOO^- and nitriles RC≡N are very useful *in situ* oxidants. Also, the reaction of perhydroxide anion with alkenes carrying an electron-withdrawing functionality W (e.g., carbonyl C=O, -S(O)R,-SO_2R, -NO_2, etc.), leading to relatively stable carbanions (Eq. 18), provides a way of obtaining epoxides, both regiospecifically and stereoselectively [64].

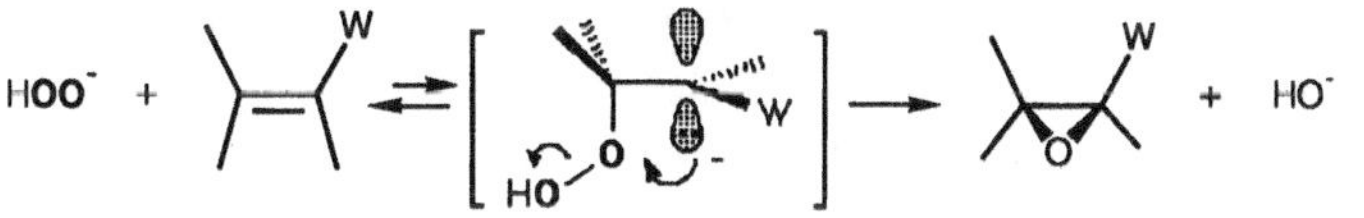

(18)

Normally, this would not be feasible by the usual electrophilic oxidation using peroxyacids.

3. Forming Reactive Derivatives of Hydrogen Peroxide

3.1 Structural types

On ground of the general principles of peroxide reaction mechanisms laid down in the previous Sections, it becomes easy to define the strategies for hydrogen peroxide activation. In fact, by inspection of representative data in Table 2 and Table 3 (Sect. 2.1), one is quickly convinced that the most effective way to activate HOOH is to make a significant change in the basicity of the leaving group upon transforming HOOH into ZOOH, where Z is strongly electron-withdrawing. This way the poor leaving group of hydrogen peroxide, i.e. HO^-, would be changed into a good leaving group ZO^-.

In some industrial processes this is accomplished turning H_2O_2 into peroxyacetic acid $CH_3C(O)OOH$ (PAA), potassium peroxymonosulfate $K^{+-}OSO_2OOH$ (Curox®, Oxone®, Caroat®), magnesium peroxymonophthalate (H48®), and several other peracid species. On laboratory scale, it is done using variable amounts of organic compounds, including formic acid HCO_2H plus strong acids, trifluoroacetic acid CF_3CO_2H, polymer attached arsenic acid, *et cetera*.

Almost invariably, the activators act as a consequence of their having two properties: one property is ease of replacement of a ligand by the OOH functionality (as in Eq. 19), and the other is weak basicity of the leaving anion (e.g., trifluoroacetate).

 (19)

The data in Table 2 are important in showing in which cases H_2O_2 activation is necessary or it is not. Most sulfur compounds, phosphines, and phosphites react rapidly with H_2O_2, so that for these reductants hydrogen peroxide activation is unnecessary. On the other hand, olefins and organonitrogen nucleophiles react only slowly with H_2O_2, hence activation is essential.

In Table 6, we give a short list of typical peroxide species that can be generated by the reaction of H_2O_2 with organic compounds, most often in a single step. In the majority of cases, these can be expected to be significantly more reactive than H_2O_2 in yielding peroxidic O-transfer, due to the mentioned leaving group effect .

In laying down Table 6, a choice of H_2O_2-derived peroxide species [65-67] was made with an eye to those systems more likely to result in applications of practical value. For instance, notice that obvious H_2O_2 derivatives, such as dialkyl peroxides ROOR, peroxy esters $R^1C(O)OOR^2$, and diacyl peroxides RC(O)-OO-C(O)R, were not included; this is because normally their reactivity would not be characterized by heterolysis, but rather by homolysis of the peroxide bond, which must be promoted thermally, photochemically, or by transition metal-ion catalysis. In fact, the combination of steric effects and poor leaving group ability makes simple dialkyl peroxides ROOR almost unreactive in heterolytic O-transfers to most substrates [65].

3.2 Hydroperoxides

With respect to dialkyl peroxides, alkyl hydroperoxides ROOH are distinctly more reactive; they can be generated through nucleophilic displacement on suitable substrates by H_2O_2 in alkaline media (Eq.s 20 and 21).

$$H_2O_2 + HO^- \rightleftharpoons HOO^- + H_2O \qquad (20)$$

$$HOO^- + R_3C{-}X \longrightarrow R_3C{-}OOH + X^- \qquad (21)$$

In this transformation, the leaving group X^- should not be one that is easily oxidizable (e.g., Br^-, I^-), since this would bring about an unwanted reduction of the hydroperoxides. Also, substrate structure and/or reaction conditions should be chosen as to minimize the intrusion of consecutive reactions leading to formation of dialkyl peroxides (Eq.s 22 and 23).

Table 6: Some types of active peroxide species derived from H_2O_2.

Usual precursor species	Peroxide	Nomenclature
R-OH, R-X [a]	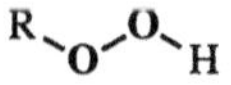	alkyl hydroperoxide, aryl alkyl hydroperoxide
$R_2C{=}O$	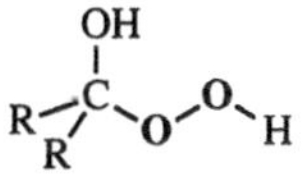	α-hydroxy hydroperoxide
$R_2C(OR^1)_2$	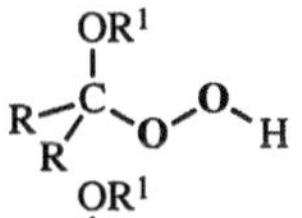	α-alkoxy hydroperoxide
$RC(OR^1)_3$	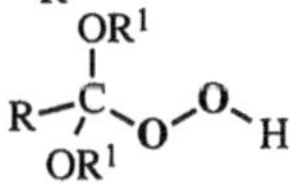	α,α'-alkoxyhydroperoxide
$R_2C{=}NHR^1$	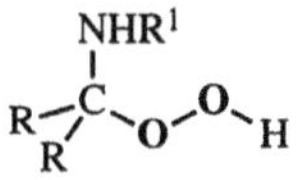	α-amino hydroperoxide
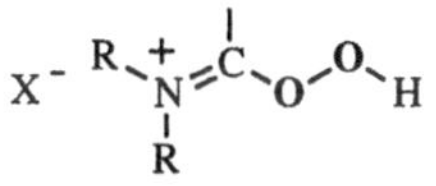	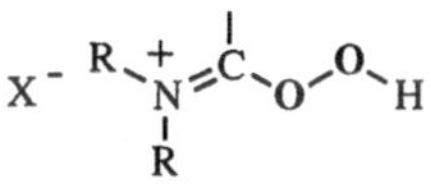	(hydroperoxymethylene) dialkylammonium
R-C(=O)-OH , R-C(=O)-X [b] (R C(=O) O)$_3$B , etc.	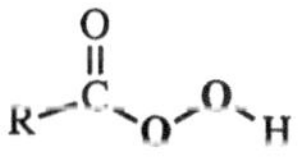	peroxycarboxylic acid
R-C≡N	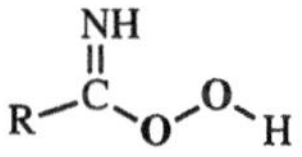	peroxycarboimidic acid
ArO-C≡N	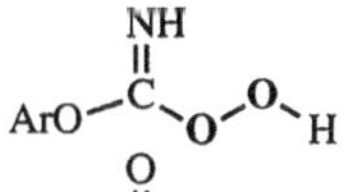	O-aryl peroxycarbamidic acid
ArO-C(=O)-X [c]	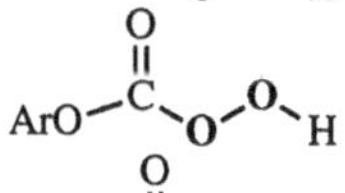	O-aryl monoperoxycarbonic acid
R-N(R(H))-C(=O)-X	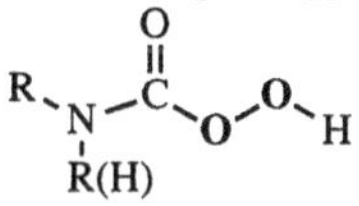	N,N-dialkyl (or N-alkyl) peroxycarbamidic acid

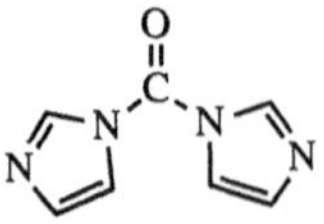	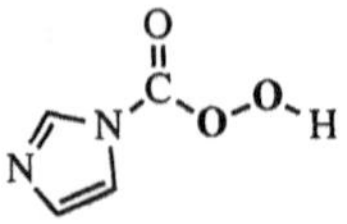	(1H-imidazolid)peroxy-carboxylic acid
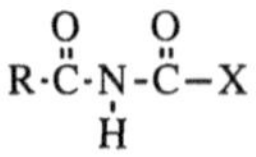	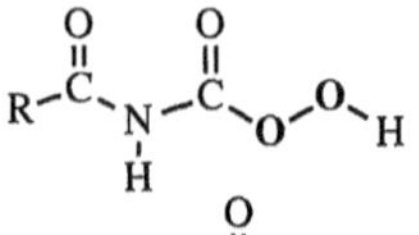	N-acyl peroxycarbamidic acid
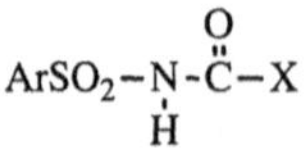	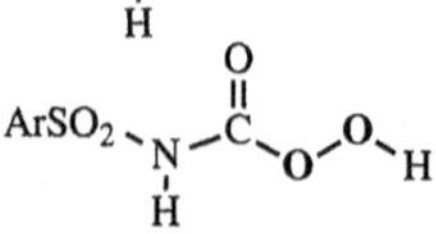	N-arylsulfonyl peroxy-carbamidic acid
R_nM-OH, R_nM-X [d]	$R_nM-O-O-H$	organoelement hydroperoxide

[a] Most commonly, X = Cl, OMs, OTs, with R = primary, secondary, or tertiary aliphatic alkyl, alkenyl, aryl alkyl, etc. [b] With X = O-C(O)R, OR', OAr, NH-C(O)R), Cl, 1H-imidazolid, and others. [c] E.g., X = Cl, 1H-imidazolid, etc.. [d] For instance, R_nM = R3Si, R3Sn, PhSe, etc.

$$R_3C\text{-}OOH + HO^- \rightleftharpoons R_3C\text{-}OO^- + H_2O \quad (22)$$

$$R_3C\text{-}OO^- + R_3C\text{-}X \longrightarrow R_3C\text{-}OO\text{-}CR_3 + X^- \quad (23)$$

Actually these two important limitations are quite general and apply to most systems of generating active H_2O_2 derivatives for practical purposes; namely, the substrate should not contain a leaving group or a functionality that is easily oxidizable, and consecutive reactions leading to less reactive peroxide species should be avoided.

The method above applies satisfactorily to the preparation of primary and secondary hydroperoxides; an example illustrating the S_N2 characteristics of these reactions is the synthesis of optically active 2-heptenyl hydroperoxide [68], which occurs with inversion of configuration upon displacement of methanesulfonate MsO^- (Eq. 24)

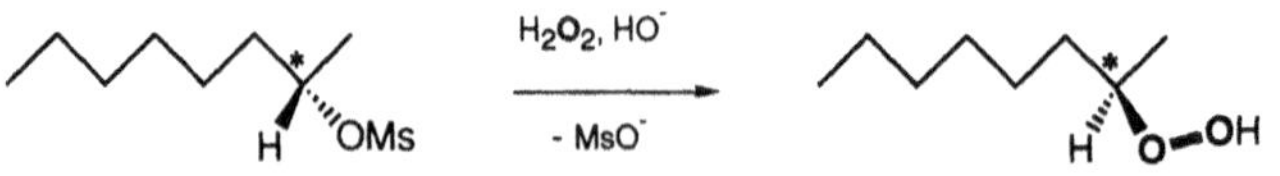

S (+) *R* (-) (24)

Tertiary alkyl hydroperoxides can be generated by perhydrolysis from the alcohols employing concentrated (50-90%) hydrogen peroxide, most commonly assisted by strong-acid catalysis (Eq. 25).

$$R_3C{-}OH \xrightarrow[-\ H_2O]{H_2O_{2(conc)},\ H^+} R_3C{-}OOH \qquad (25)$$

For instance, hydroperoxides **5** [69] and **6** [70] can be obtained from the corresponding alcohols and concentrated H_2O_2 in the presence of H_2SO_4 and H_3PO_4, respectively, while *p* -toluenesulfonic acid (TsOH) was employed as catalyst in the synthesis of S(+)-**7** [71].

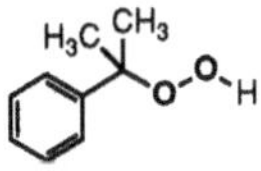

5

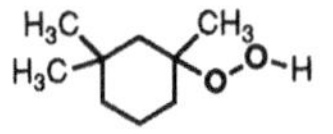

6

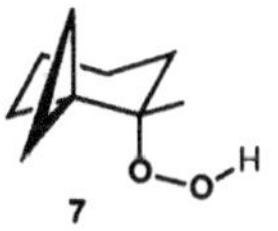

7

The generation of optically active hydroperoxides renders the transformation HOOH → ROOH synthetically useful if one wishes to carry out enantioselective oxidations. Neverthless, from the standpoint of enhancing reactivity this is in general not productive since most alkoxide ions are leaving groups even poorer than hydroxide (cf., Table 3, Sect. 2.1).

There are, however, notable exceptions. In fact, biologically important hydroperoxyflavins, such as 4a-hydroperoxy-5-ethyl-3-methyl-lumiflavin (**8**), are excellent electrophilic O-atom transfer agents toward organosulfur and organonitrogen substrates [38]. As an example, kinetic studies have shown that the 4a-hydroperoxyflavin **8**, although considerably less reactive than a typical peroxyacid (such as *m* -chloroperoxybenzoic acid), it is more reactive than hydrogen peroxide in oxidation of 1,4-thioxane $O(CH_2CH_2)_2S$ and of N,N-dimethylbenzylamine $PhCH_2NMe_2$ by factors of over 7,000 and almost 36,000, respectively [38]. This has recently enabled to carry out successfully the flavin-catalyzed oxidation of a number of substrates using hydrogen peroxide [72] (Eq. 26). No doubt the efficiency of 4a-hydroperoxyflavins (Fl-OOH) as electrophilic oxidizers is due to the good leaving group ability of $Fl\text{-}O^-$; this is suggested by the pK_a value of ca. 9.4 estimated [38a] for 4a-hydroxyflavines Fl-OH (e.g., **8'**), which is markedly lower than the pK_a of water (the reduced form of hydrogen peroxide).

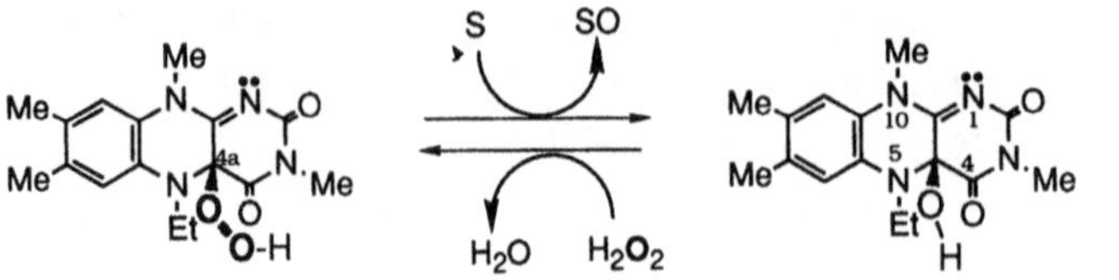

(26)

8 8' (pK_a~ 9.4)

The greater ability of Fl-O⁻ to support a negative charge (hence its good leaving group ability) can be ascribed [38] to the electronegativity of the ring nitrogens at positions 1, 5, and 10, as well as to the carbonyl at position 4. It has also been speculated [67b] that the 4a-hydroperoxyflavins posses structural features as to allow cyclic proton trasfer in the t.s., which would assist O-transfer to nucleophilic substrates (Cf. Sect. 2.2)

The same internal proton transfer mechanism has been invoked [67] for oxidations using α-hydroperoxyesters, e.g. **9** [67a]. This is useful in rationalizing the fact that remarkably

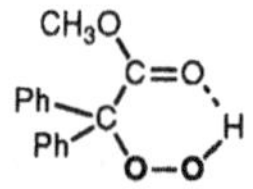

9

effective epoxidations can be carried out using α-hydroperoxy esters ROC(O)-CR_2-OOH or α-hydroperoxy ketones RC(O)-CR_2-OOH [73a-c], as well as α-hydroperoxy nitriles N≡C-CR_2-OOH [73d]. It is appropriate to mention on passing that the omission of these reactive functionalized hydroperoxides from Table 6 is solely justified by the method which is commonly followed for their generation, i.e. base-catalyzed oxygenation of carbonyls or nitriles rather than reaction of H_2O_2 with suitable precursors. This rationale also holds for α-azohydroperoxides R-N=N-CR_2-OOH [73e], since these reactive derivatives are most commonly generated upon autooxidation or photooxidation starting with hydrazones.

One should be warned that in these, as well as in other cases below, the said cooperative internal proton-transfer mechanism may be difficult to support adequately, and especially so when the reactions are carried out in water or hydroxylic solvents. In fact, we have seen (Sect. 2.2) that, by hydrogen bonding, protic solvent molecules can partecipate in the t.s. of peroxidic oxygen transfer [26]. That a number of the α-heterosubstituted hydroperoxide species listed in Table 6 can indeed exist in the intramolecular H-bonded form *in the ground state* is well supported spectroscopically in non-protic solvents [67].

The α-hydroxy or α-alkoxy hydroperoxides show enhanced reactivity with respect to H_2O_2, also due to the electron-withdrawing effect displayed the OH or OR group on the α carbon atom, which increases the propensity for electrophilic O-atom transfer. These species can be formed by making H_2O_2 perform nucleophilic displacements on suitable precursors, such as ortho esters (Eq. 27).

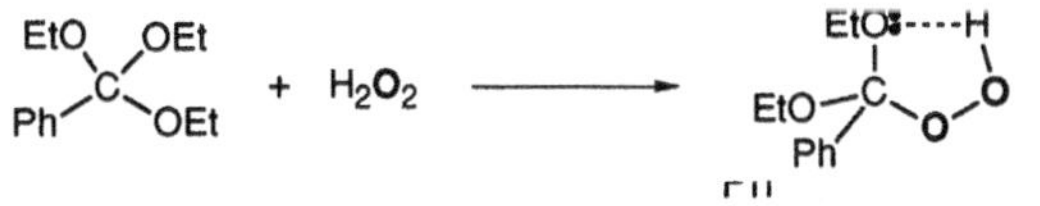

(27)

The reaction of 30% H_2O_2 with Vilsmeier reagent **10** (prepared by the reaction of N-methyl pyrrolidone (**12**) with $POCl_3$) [74] affords a nicely efficient hydroperoxy ammonium salt (**11**) *in situ* ; this yields alkene epoxidations in fair to very good yields (Eq. 28). A limitation, consisting in the concurrent generation of HCl (which would reduce the oxidant), can be circumvented by running the reaction in a CH_2Cl_2/water biphasic system with phase transfer catalysis (PTC), removing the HCl to the aqueous phase. Reaction times are short and the epoxides obtained are only accompanied by small amounts of dichloro byproducts [74].

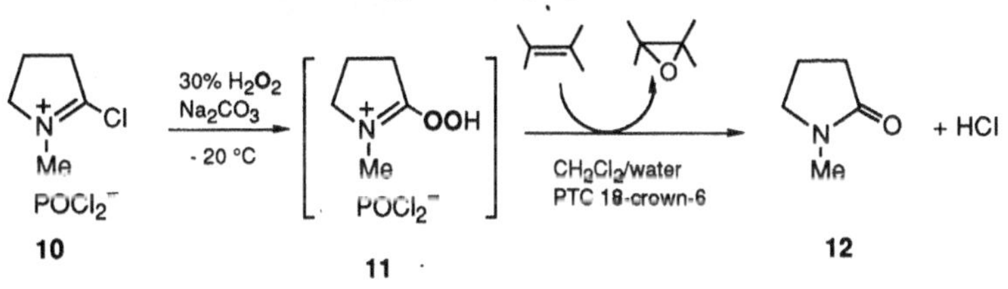

(28)

Other than by nucleophilic displacement, a more frequent way of generating α-heterosubstituted hydroperoxides is by H_2O_2 nucleophilic addition to sp^2-carbon substrates, such as aldehydes or ketones (Eq. 29).

(29)

Table 7: Formation constants of carbonyl adducts $RR'C(OH)_2$ and $RR'C(OH)OOH$

Carbonyl substrate, RR'C=O	$K_f^{H_2O\,a,b}$	$K_f^{HOOH\,a,b}$	Ref.
$H_2C=O$	41	$7. \times 10^4$	75a
$CH_3CH=O$	2.6×10^{-2}	48	75a
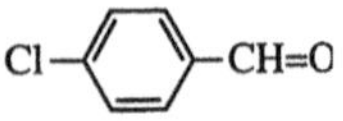	--	0.88	75a
	--	0.18	75a
$(CH_3)_2C=O$	$2.5 \times 10^{-5\,c}$	$5.3 \times 10^{-2\,c}$	75b

[a] In M^{-1}, at 25.0 °C, unless noted otherwise. [b] When the formation constants are equilibrium constants from water solutions, the observed constant have been divided by 55.5 M. [c] At 35.0 °C.

Inspection of data in Table 7 reveals that the formation constants of such H_2O_2 adducts [75] are several orders higher than those measured for nucleophilic addition of H_2O, yielding gem-diols $R_2C(OH)_2$. The new peroxide species thus formed can be quite reactive. For instance, the hydroperoxide **13** derived from hexafluoroacetone (Eq. 30) is capable of alkene epoxidations under mild conditions [76].

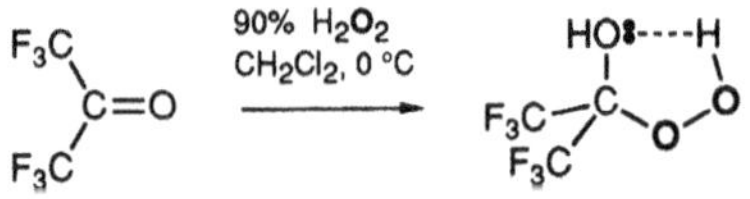

13 (30)

In generating α-hydroxy hydroperoxides care must be excercised to adopt the appropriate reaction conditions in order to avoid further addition of the hydroperoxide to its parent carbonyl compound. In fact this would lead to α,α' dihydroxy peroxides **14**; from these, in turn, the highly sensitive dimeric and trimeric peroxides, **15** and **16** respectively, might form by way of dehydrative cyclization [77].

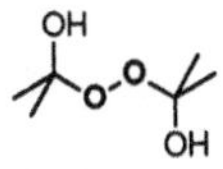 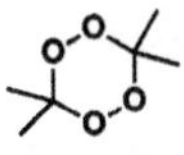 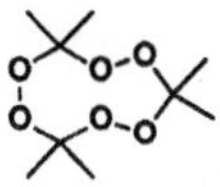

14 **15** **16**

Under suitable conditions, even reactive α,α'-dihydroperoxide peroxides such as **17** can be generated upon reaction of concentrated H_2O_2 with ketones [66b] (Eq. 31).

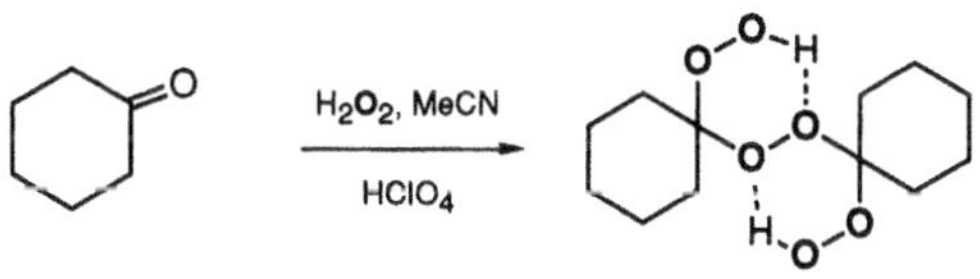

17

(31)

The addition of H_2O_2 to the C=N bond of imines, giving rise to α-amino hydroperoxides (e.g., Eq. 32) [78], is conceptually analogous to H_2O_2 addition to carbonyls. Along the lines discussed above, here too it can be expected that the newly formed species present enhanced reactivity with respect to hydrogen peroxide.

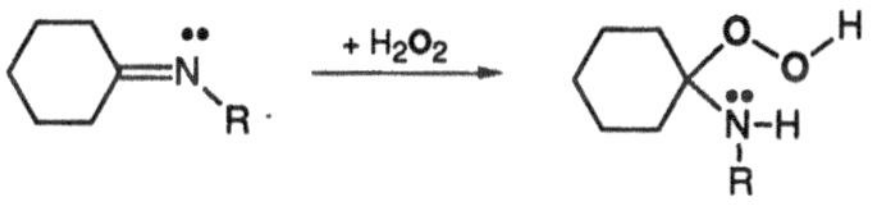

(32)

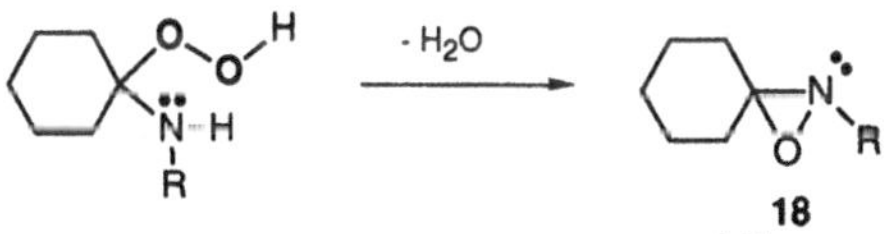

18

(33)

The possibility exists that such species evolve yielding oxaziridines, as e.g. **18** in Eq. 33. Oxaziridines, in turn, are known to be excellent O-atom transfer agents [79].

The transformation of hydrogen peroxide into a organoelement hydroperoxide [66c] (Table 6), as in eq.s 34 and 35, is another instance where increased reactivity may derive primarily from the leaving group effect, i.e. R_nM-O^- (e.g., Me_3SnO^-, Ph_3SiO^-) vs. HO^-.

$$Me_3Sn\text{-}OH + H_2O_2 \xrightarrow{-H_2O} Me_3Sn\text{-}\mathbf{OOH} \quad (34)$$

$$Ph_3Si\text{-}NH_2 + H_2O_2 \longrightarrow Ph_3Si\text{-}\mathbf{OOH} \quad (35)$$

The bis(trimethylsilyl) derivative of hydrogen peroxide (**19**) can be synthetized [80] as in Eq. 36, employing hydrogen peroxide in the form of its 1,4-diazabicyclo[2.2.2]octane (DABCO) adduct (**20**) [81].

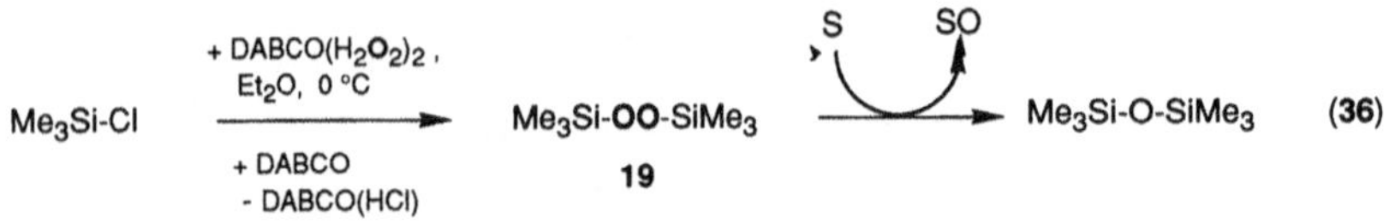

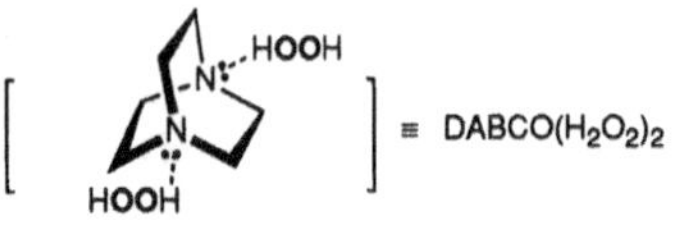

20

Bis(trimethylsilyl)peroxide (**19**) is a good electrophilic oxidant; for example, in the oxidation of organic sulfides it has been estimated to be more effective than H_2O_2, *t* -BuOOH, and *t* -BuOO-*t* -Bu by factors of 40, 120, and over 2×10^4, respectively [82]. This fact can be in a good measure ascribed to the leaving group effect (cf., Me_3SiOH: $pK_a \sim 10$; H_2O: $pK_a = 15.7$; *t* -BuOH: $pK_a \sim 19$). However, in the silyl peroxide **19**, reduced steric effects (with respect to t-BuOOBu-t), and – more important – a migration akin to proton transfer may also contribute significantly [82]. In fact, because $Me_3SiOSiMe_3$ is the reduction product (Eq. 36) and the rate of transfer from **19** is pretty fast, the hypothesis was advanced [82] that one Me_3Si group (the "organic proton") in the silyl peroxide might assist the heterolytic O-O bond breaking by undergoing concomitant 1,2-migration from one peroxide oxygen to the other. This would give rise to bridged transition states like **21** [82].

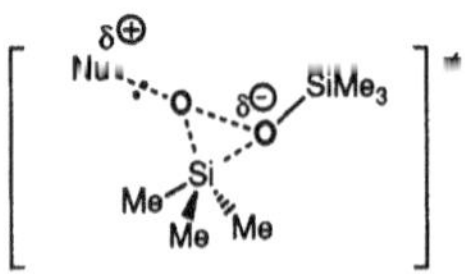

21

It can be seen that this would assist the reaction in one way that protons do (cf. **3**, Sect. 2.2), and endothermic charge separation is largely avoided. Of course, bis(trimethylsilyl)peroxide (**19**) is a very convenient derivative of H_2O_2 also because it allows one to carry out oxidations in organic solvents. Furthermore, its reduction product bis(trimethyl)siloxane $(Me_3Si)_2O$ is quite volatile and easily removed from reaction mixtures, thus simplifying product isolation.

The generation of organoboron hydroperoxides, e. g. **22**, from suitable precursors and H_2O_2 is impaired by the propensity of these compounds to undergo rearrangement with O-atom insertion into the B-C bond [83a,b] (Eq. 37).

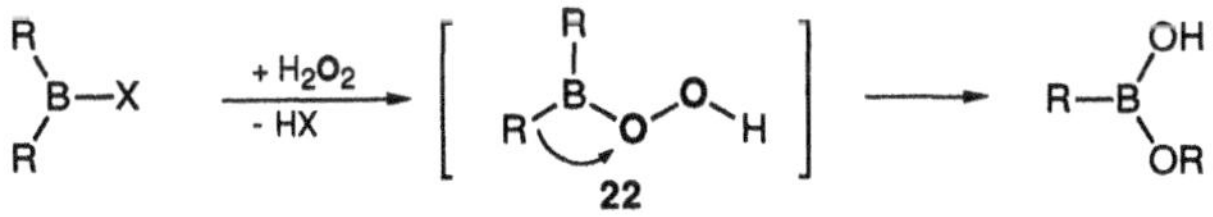

22

(37)

This is most pronounced when the boron bears aryl substituents, e.g. R = Ph. However, one might conceivably generate organoboron hydroperoxides such as **23** or **24** that are protected against rearrangement, starting with borate esters (Eq. 38).

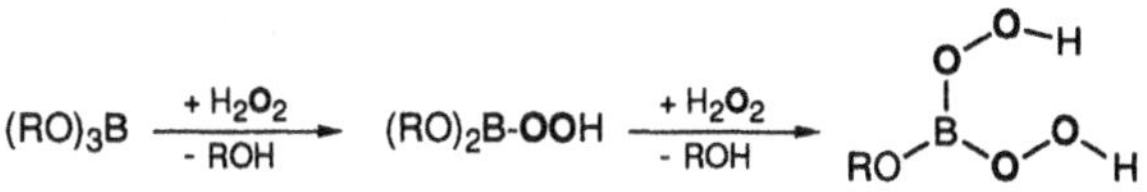

23 **24**

(38)

Indeed, bis(alkylperoxy)alkoxy- and tri(alkylperoxy)boranes, $R'OB(OOR)_2$ and $B(OOR)_3$ respectively, are known [83b,c], and can be obtained as isolated species.

It is likely that organoboron hydroperoxides which are effective O-transfer agents might be *in situ* generated upon addition of H_2O_2 to metaborate esters (Eq. 39).

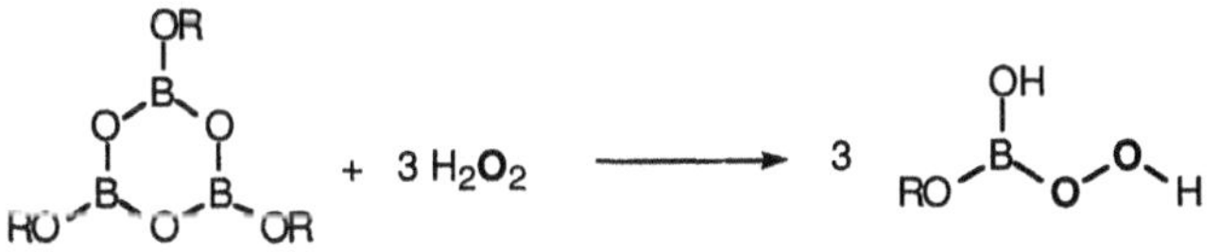

(39)

This is analogous to the inorganic boron hydroperoxide claimed to arise from the reaction between H_2O_2 and metaboric acid [84] (Eq. 40), and thought to be the active species in an important industrial epoxidation process.

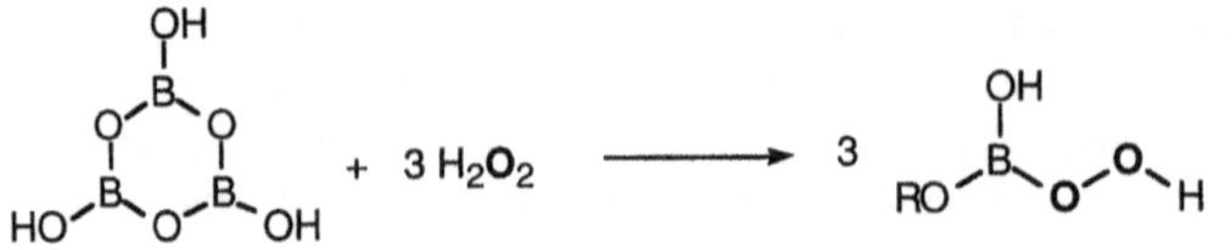

(40)

In Sect. 2.3 we have seen that addition of alkaline H_2O_2 to "activated" alkenes, such as α,β-unsaturated ketones leads to α'-hydroperoxo α-carbonylcarbanion intermediates having short lifetime, since their collapse into epoxides is fast. On the other hand, the direct addition of H_2O_2 to "unactivated" olefins is uneffective, so that the partnership of an efficient electrophilic species , e. g. $Hg(OAc)_2$, is required [85]. However, from the standpoint of subsequent O-atom transfer to other substrates, the hydroperoxides thus derived are of little practical use.

3.3 Peroxyacids

By and large, the most effective and frequently used means of H_2O_2 activation not involving metals is that of generating a peroxyacid RC(O)OOH, by way of $B_{AC}2$ or $A_{AC}2$ displacement at a sp^2 carbon compound RC(O)X (cf., Sect. 2.3, Scheme 1). That the leaving group X^- (or HX) not be an easily oxidizable species counts among the few restrictions affecting this important method. Another is that the temperature and pH dependent decomposition [14] of the peroxyacid thus generated does not compete with O-atom transfer to organic or inorganic substrates. Thus, in H_2O_2 activation, perhydrolysis of suitable carbonyl precursors is a method of broad scope, having diversified and important applications.

These include several commercial bleaching processes. For instance, both tetraacetylethylene diamine (TAED, **25**) and pentaacetylglucose (PAG, **26**) are employed in bleach formulations [12,13] as precursors of peracetic acid (Eq.s 41 and 42), an oxidizer five powers of ten more reactive than H_2O_2; this allows the bleach to occur at a temperature significantly lower than with H_2O_2. It should be noted that, in the process shown by Eq. 42, significantly less peroxyacetic acid than stoichiometric is released in practice, since perhydrolysis of PAG is less than complete [12].

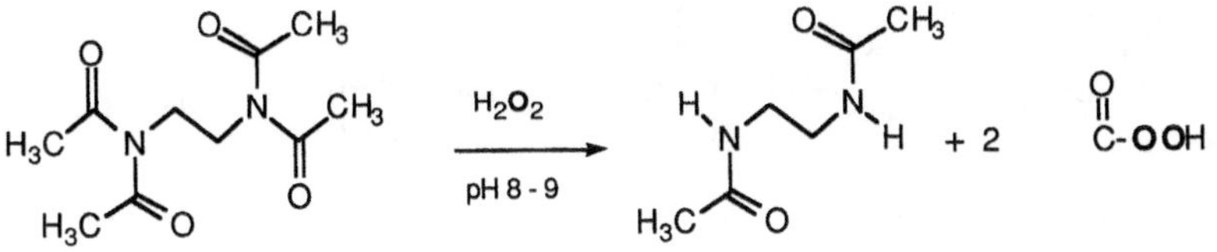

25 (TAED) (41)

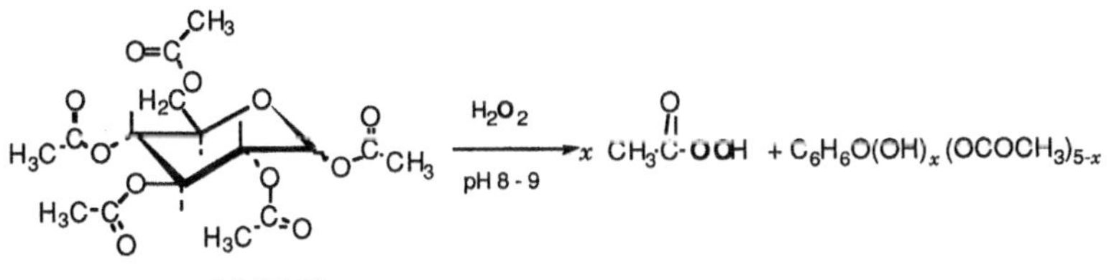

26 (PAG) (42)

Perhydrolysis of ester **27** generates peroxynonanoic acid (Eq.43); this gives excellent bleach performance on textiles, most likely due to the surface active properties of the peracid and of its precursor.

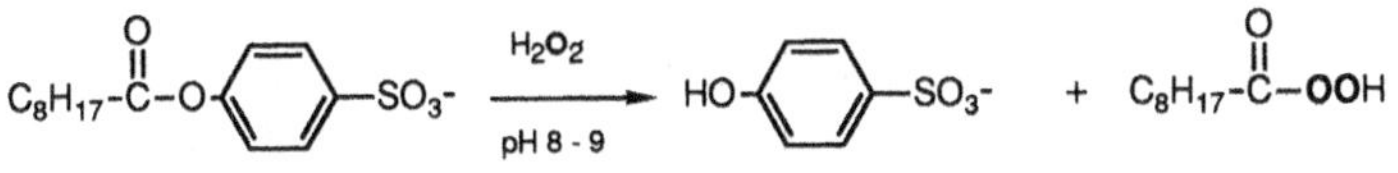

27 (43)

Other effective precursors of peroxyacids upon perhydrolysis under mild conditions are the acyl dialkylphosphates (Eq. 44) and the N-acylimidazoles (Eq. 45), while N,N'-carbonylimidazole is an obvious candidate to form the elusive [67c] diperoxycarbonic acid (Eq. 46).

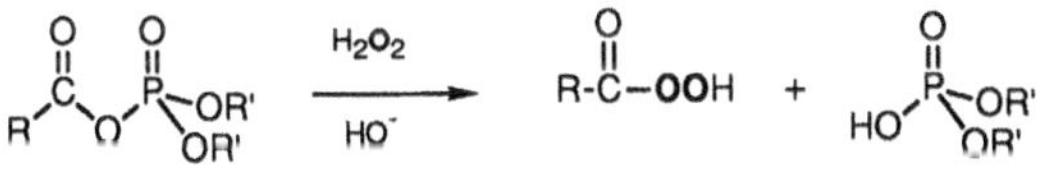

(44)

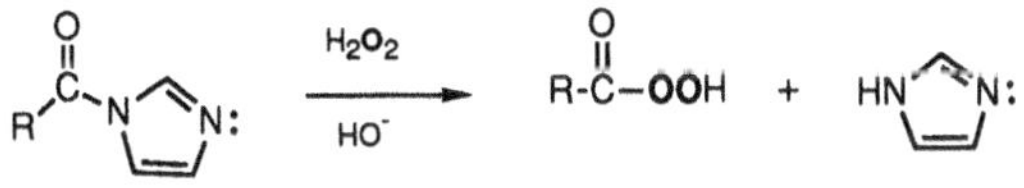

(45)

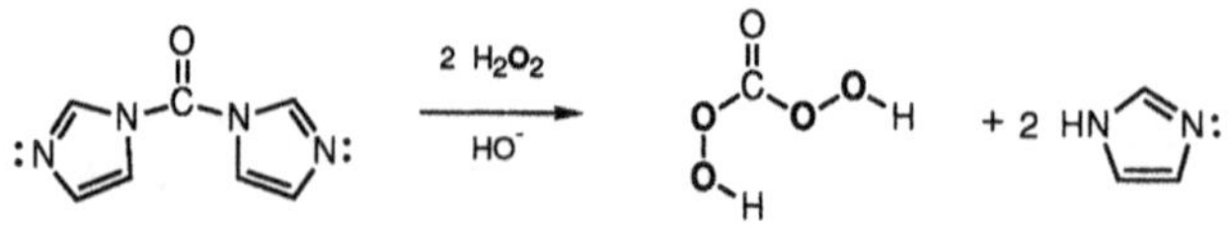

(46)

Perhydrolysis of anhydrides [66c] under carefully controlled conditions offers yet another method of generating peroxyacids, as in the example reported in Eq. 47.

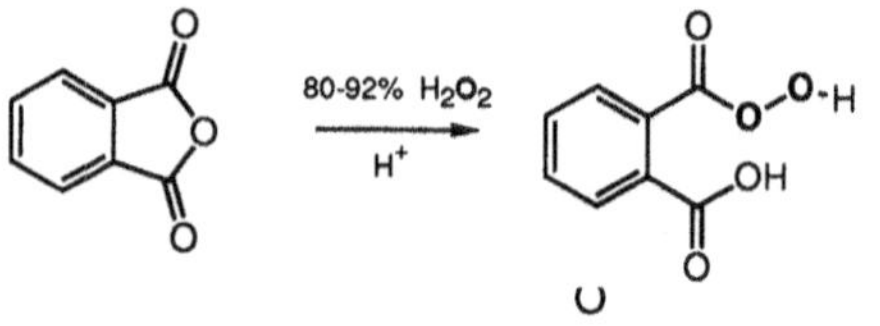

(47)

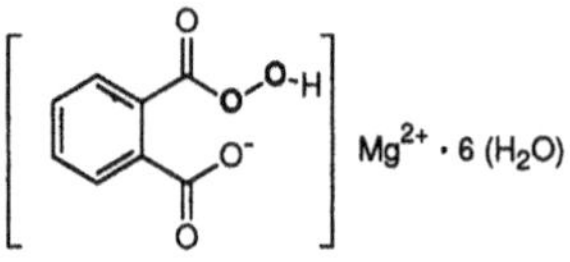

28

The hydrated magnesium salt of monoperoxyphthalic acid (**28**), named H48®, has been commercialized as a bleaching agent and disinfectant [12].

As illustrated by the example in Eq. 47, the perhydrolysis of anhydrides is usually carried out with concentrated H_2O_2, catalyzed by strong acids, e.g. H_2SO_4 or $HClO_4$.

Prepared by this method (Eq. 1, Sect. 1), peroxyacetic acid is commercially available as a ca. 40% solution in acetic acid. Starting with the parent anhydrides and 70-80% H_2O_2, concentrated solutions of peroxyacetic or peroxytrifluoroacetic acid in CH_2Cl_2 (or other chlorinated solvents) can be obtained [66c].

Peroxyacids can also be formed upon reaction of carboxylic acid with 30 to 92% H_2O_2 in the presence of a strong acid catalyst (Eq. 48).

$$\text{R-}\overset{\overset{\text{O}}{\|}}{\text{C}}\text{-OH} \quad + \quad H_2O_2 \quad \overset{H^+}{\rightleftharpoons} \quad \text{R-}\overset{\overset{\text{O}}{\|}}{\text{C}}\text{-OOH} \quad + \quad H_2O \qquad (48)$$

With some carboxylic acids, however, this reaction can be quite slow. Aliphatic or aromatic peroxyacids which are water-insoluble can be synthesized using an excess of concentrated H_2O_2 in methanesulphonic acid; the latter serves as a catalyst and effective dehydration agent, as well as solvent [86].

Perhydrolysis is also well suited to the generation of organoelement peroxyacids, as in the examples given in Eq. 49 and Eq. 50.

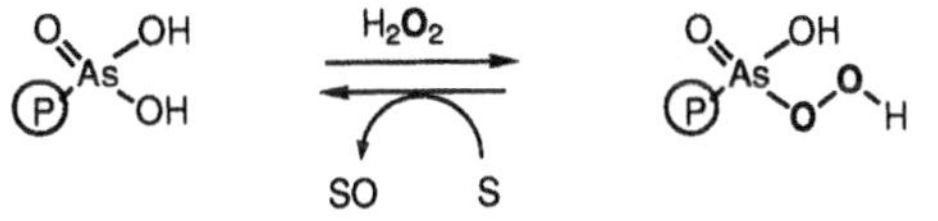

(49)

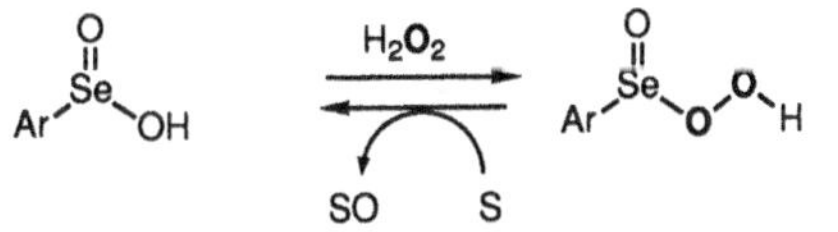

(50)

In both of the cases, an acid catalyst is not required to speed peracid formation since the precursor species involved are themselves strong acids. Either the peroxyacid derived from arsonated polystyrene (Eq. 49) and peroxyseleninic acid (Eq. 50) are effective O-transfer agents; in fact, both systems have been employed in epoxidations [87,88].

Polymer-supported peroxycarboxylic acids can also be prepared by perhydrolysis, as in Eq. 51.

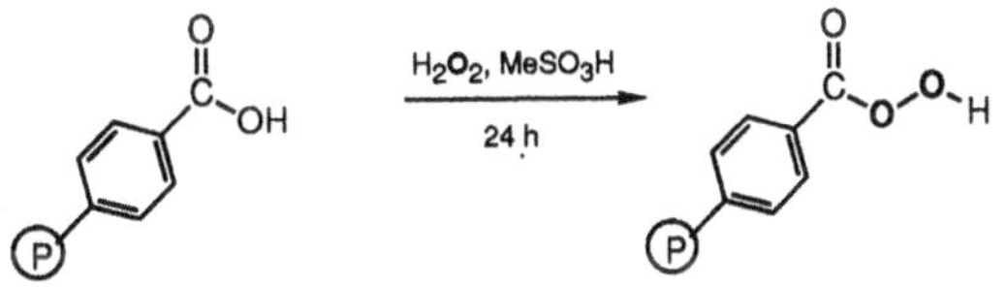

(51)

In this way, the transformation of up to 85% of the carboxylic groups of the polymer can be achieved; the reagent is useful in the synthesis of acid sensitive epoxides [89].

Other reagents eminently suited to serve this purpose are the peroxyimidic acids (**29**) arising from the addition of H_2O_2 to nitriles [90,91], as it is shown in Scheme 2.

Scheme 2: Mechanism of formation of peroxyimidic acid from the addition of hydrogen peroxide to nitriles.

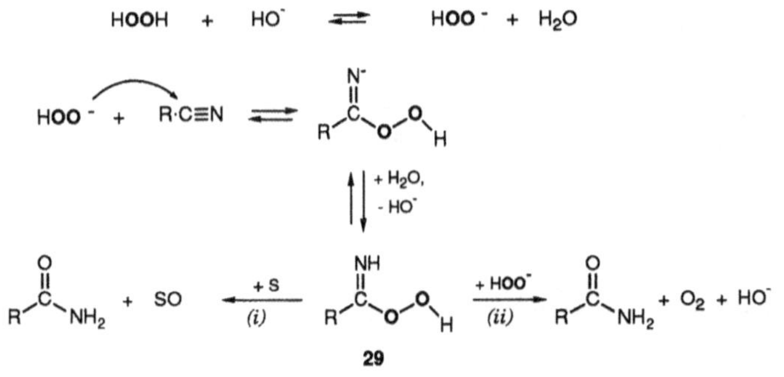

Under suitable conditions, the peroxyimidic acid (**29**) can survive capture by hydroperoxide anion (Scheme 2, path *ii*) and becomes available for O-atom transfer to even weakly nucleophilic substrates. Indeed, the nitrile-hydrogen peroxide system, first described by Payne [90], can be adapted to achieve olefin epoxidation under mild conditions on a molar scale [92a]. It is worthy of note that this relatively safe and inexpensive epoxidation procedure avoids the handling of organic peroxyacids. Trichloroacetonitrile $Cl_3CC{\equiv}N$ (an electron-deficient nitrile) and H_2O_2 affords a highly efficient O-transfer agent for epoxidations; in this case, a biphasic system consisting of CH_2Cl_2 and buffered (pH 6.8-7.0) 30% aqueous H_2O_2 can be used to advantage [92b].

Isocyanates, e.g. Ph-N=C=O, as well as diakylcarbodiimides R-N=C=N-R, can also serve in activating H_2O_2 to perform epoxidations under mild conditions [67,93,94]. Several other examples [67] of H_2O_2 activation by organic compounds that are dehydrating agents closely resemble, or fall into categories analogous to the cases so far presented.

In some instance, the organic is believed to serve merely as a carrier of anhydrous H_2O_2, much like the DABCO-hydrogen peroxide complex (**20**) mentioned in the previous Section. This seems to be the case with the stable urea-H_2O_2 adduct (UHP), which has been employed to generate peroxyacids *in situ* upon reaction with a carboxylic anhydride in a solvent convenient for epoxidations such as CH_2Cl_2 [95].

In view of the key role of epoxides as intermediates in synthesis [96], it is hardly surprising that olefin epoxidation be the benchmark where the efficiency of a H_2O_2 derivative is most often tested. Of course, to achieve substantial stereoselectivities in this important transformation is of great significance.

In general, since the alkenes act as nucleophiles attacking the peroxide O-O bond (cf. Sect. 2.2), for structurally simple olefins the rate of epoxidation is enhanced by increasing alkyl substitution; this leads to the familiar sequence of reactivity: $R_2C{=}CR_2 > R_2C{=}CHR > R_2C{=}CH_2 \sim$ *cis* -RCH=CHR ~ *trans* -RCH=CHR > $RCH{=}CH_2$. Thus, the scene seems to be entirely dominated by the electron-donating properties of the substituents, while the steric effects of the alkyl groups are not important. Indeed, inspection of t.s. structures **4'** and **4"** for alkene epoxidation by peroxyacids reported in Sect. 2.2, reveal that there is little opportunity for steric interaction between the R group of the peracid and alkyl substutuents attached to the alkene π bond.

An unfortunate corollary of this situation is that, in epoxidations using conventional peroxyacids, the *cis/trans* stereoselectivity and enantioselectivity is poor [67, 96]. This is not surprising given the absence of steric interactions, which would be needed in building up discriminating t.s. geometries. Only with allylic alcohols substantial stereoselectivity can be seen, most likely because the OH functionality serves to establish cooperative H-bonds with the epoxidation agent in the t.s. [50d, 96].

Asymmetric epoxidation of simple alkenes using optically active peroxyacids is low, often in the range from 3 to 7% enantiomeric excess (e. e.) [67]; the highest e.e. recorded for a simple alkene is 9.2%, when carefully purified purified (+)-monoperoxycamphoric acid (**30**) was employed [97].

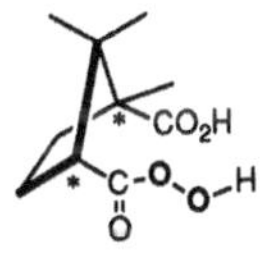

30

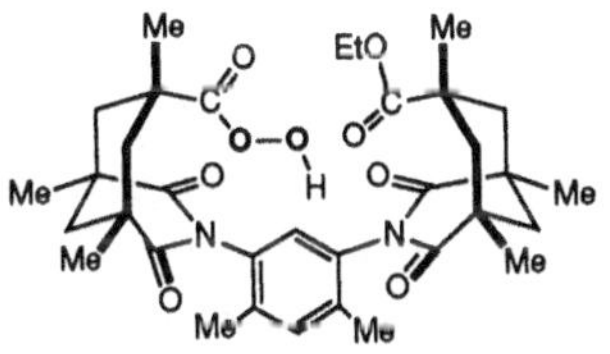

31

As for *cis/trans* stereoselectivity using conventional peroxyacids (e.g., MCPBA), rate ratios with a number of olefin pairs reflect just a slight preference for the *cis* -stereomer (usually from 1.2 to 2.2 fold faster than the *trans*) [67]. However, Rebek and co-workers have been able to employ conformationally constrained peroxyacids (e.g., **31**) that yield significantly higher *cis* selectivities (k_{cis}/k_{trans} up to 7.8) in epoxidations [50c]. It is noteworthy that the C-shaped peroxyacid **31** can be generated *in situ* by treating its precursor acid chloride-ester with excess 90% H_2O_2 in CH_2Cl_2, in the presence of pyridine [50c].

The case above illustrates the fact that in epoxidations steric effects can be gradually brought into play by modeling of the peracid structure. It seems likely that cases will be found wherein the electronic effects – which normally dictate selectivity in epoxidations – will be ultimately overwhelmed by the steric ones. Obviously, the attainment of a t.s. having restricted degrees of freedom seems essential in order to achieve high stereo- and enantioselectivities.

4. Some Special Systems

4.1 Activation by single electron transfer

In Sect. 2.2 we have seen that the great majority of bimolecular electrophilic oxidations by H_2O_2 could be recounted as a two-electron nucleophilic attack by the electron donor substrate (hereby noted as S:) at the O-O bond. However, the oxidation might also be envisaged as involving single-electron transfers (Scheme 3). The difference between the two mechanisms (i.e, direct S_N2 and ET) is less significant than one might appreciate at first glance. In fact, according to the Configuration Mixing (CM) model developed by Pross and Shaik [98] the polar and the ET mechanism merge as two sides of the same coin, and should be regarded at the extremes of a continuum, with borderline cases having characteristics of both types of mechanism. In simple terms, one might describe a direct displacement at the O-O bond as an inner-sphere process accompanied by ligand transfer, with one HO group of hydrogen peroxide having the role of the bridging ligand; on the other hand, the ET process is substantially an outer-sphere mechanism (Scheme 3).

The idea that single-electron transfers may be involved in a variety of bimolecular processes (including several peroxide reactions) has become a popular one; to deal with this important topic is beyond the scope of this review, so that the unacquainted reader is urged to consult the available literature [99] in order become familiar with the subject.

For the ET path to be easily accessible in the reaction of hydrogen peroxide with nucleophiles (and electron-donors), the solution ionization potential of the donor – in conjunction with the electron affinity of H_2O_2 – should result in a thermodynamically favored process. This can be estimated from the simplified Rehm-Weller relationship in Eq. 52.

Scheme 3: Leakage of ET mechanism from S_N2 Displacements at O-O bond of Hydrogen Peroxide.

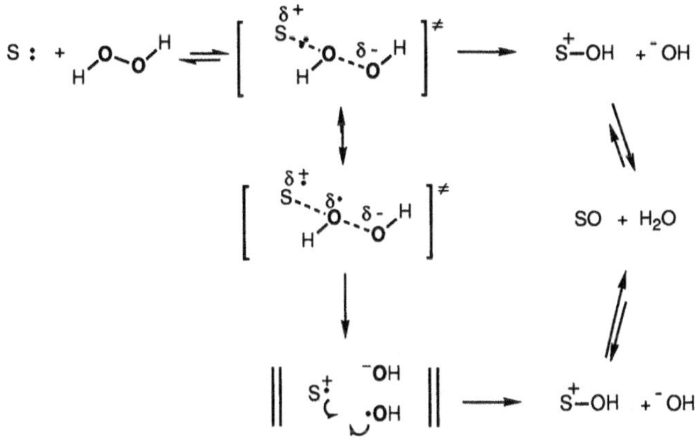

$$\Delta G^{o'}\ (\text{kcal mol}^{-1}) = 23.06\,[\,E^{o}_{ox(S/S^{+\bullet})} - E^{o}_{red(H_2O_2)}\,] + C \qquad (52)$$

In Eq. 52, $\Delta G_{ET}^{o'}$ is the corrected [99] standard free energy for the ET process, and C is a constant accounting for a number of other interactions including electrostatic stabilization [99]. Since the potential of the (irreversible) reduction in the couple $H_2O_2/(HO^{\bullet}+HO^{-})$ in water is ca. +0.7 V, it is seen that $\Delta G_{ET}^{o'} < 0$ with substrates having standard oxidation potentials lower than +0.7 V. For these estimates to apply to reactions in organic solvents, one is faced with the uneasy task of converting the reduction potential of H_2O_2 for aqueous solutions to values for the given solvent, often resulting in the introduction of somewhat arbitrary correction terms. Based on the values known for other peroxide species [100], a reasonable estimate is that the reduction potential of H_2O_2 in aprotic solvents decreases by at least 0.4-0.6 V, thus ranking hydrogen peroxide as a only moderate ET oxidant .

In hydrogen peroxide reactions, therefore, the condition that a ET outer-sphere process be thermodynamically favoured (i.e., $\Delta G_{ET}° < 0$) is easily met when metal ions are used that have two near and easily accessible oxidation states (e.g., Fe^{2+}/Fe^{3+}, Co^{2+}/Co^{3+}). With organic compounds this is by far less common. Even with substrates which are good electron-donors, in a continuum proceeding from a classic S_N2 displacement, a purely ET process would lie at one extreme that is most often out of reach.

However, in the reaction of certain organic peroxides with good electron-donors, product studies and fast spectroscopic monitoring of transient intermediates have brought about convincing evidence for a predominant ET process [101,102]. Here, a classic example is the strongly chemiluminescent reaction between cyclic peroxides $(ArCOO)_2$ (e.g., phthaloyl) and polycyclic aromatic hydrocarbons ArH with low $E_{ox}°$ (< 1.4 V) [102], wherein the first step is a dissociative ET process yielding $ArH^{\cdot+}$ and $[ArCOO^{-} \cdot OOCAr]$. On the other hand, in several debated cases involving organic peroxides there might exist enough evidence to envisage the development of radical character just in the transition state [101, 103]. Of course probing such transition state structures is difficult, since "partial" and "complete" charge-transfer formulations lead to qualitatively similar predictions. Even the most obvious parameter that might imply an ET mechanism in a peroxide-nucleophile reaction, that is the yield of scavangeable radicals, might not be a reliable probe; in fact, systems involving an initial straightforward S_N2 displacement may lead to an intermediate undergoing subsequent fast homolysis [101,103]. These drawbacks notwithstanding, there are a few reactions of H_2O_2 with organic substrates which clearly show the earmarks of a distinct ET process. One is the oxidation of the cation radical $PQ^{\cdot+}$(**32**) derived from herbicide Paraquat (PQ^{2+}, i.e. 1,1'-dimethyl-4,4'-bipyridilium dichloride).

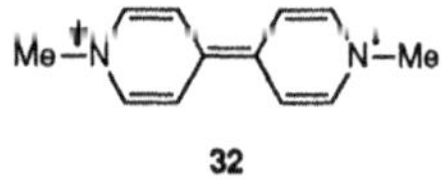

32

33

$$PQ^{\dagger} + H_2O_2 \longrightarrow PQ^{2+} + HO^{\bullet} + HO^{-} \qquad (53)$$

The reaction in Eq. 53 has been the subject of careful kinetic and spectroscopic studies [104]; it follows second-order kinetics, yielding $k_2 = 2.0\ M^{-1}s^{-1}$ at 25 °C, and $\Delta H^{\neq} = 22$ kcal mol^{-1}; worth of notice is the fact that a positive $\Delta S^{\neq}$ value of +17.5 cal K^{-1}mol^{-1} was recorded, clearly pointing to a dissociative process consistent with the one in Eq. 53, generating HO• radicals. Similar results were obtained for the reaction of H_2O_2 with the cation radical **33** derived from Diquat, another powerful herbicide [104]. Of course, hydroxyl radical HO• is a very strong oxidant, capable to oxidize neutral molecules (e.g. ArH) to their radical cations ($ArH^{\bullet +}$ + H_2O) by multistep one-electron processes [5, 6, 105].

The example above might suffice to suggest that transition metal ions are not necessarily the only choice for generating hydroxyl radicals from H_2O_2 under mild conditions. It is entirely possible that – through a judicious selection of suitable organic donors – one would be able to carry out much of the chemistry characteristic of H_2O_2 oxidations via hydroxyl radicals HO• without employing metal catalysts.

4.2 Generation of singlet oxygen

Careful kinetics and ^{18}O labeling studies have demonstrated that the polar (non-radical) decomposition of hydrogen peroxide, similar to the analogous decomposition of peroxyacids, is pH dependent. In analogy to peroxyacid decomposition, it has been envisaged to be a bimolecular reaction involving the nucleophilic attack of the hydroperoxide anion HO_2^- at the undissociated hydrogen peroxide molecule (Eq. 54); however, it has been pointed out there is no firm evidence to substantiate the postulation in the H_2O_2 case [106a].

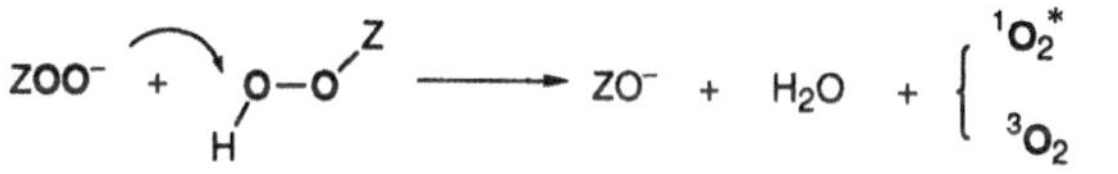

(54)

If radical processes can be avoided, the spin conservation rule requires that the molecular oxygen produced be in a singlet state, e.g. $^1O_2^*$ ($^1\Delta_g$). Indeed, evidence has been presented suggesting that this reaction yields singlet oxygen in a ratio $^1O_2^*/^3O_2$ in excess of 3% [106b].

As for the reaction mechanism, the reaction above may find a parallel in the well studied hypochlorite/hydrogen peroxide method of singlet oxygen generation [107], as shown by eq.s 55-58.

$$HOOH + {}^{-}OCl \rightleftharpoons HOO^{-} + HOCl \tag{55}$$

$$HOO^{-} + Cl\text{-}OH \rightleftharpoons [HOO\text{-}Cl] + {}^{-}OH \tag{56}$$

$$HO^{-} + [HOO\text{-}Cl] \rightleftharpoons [{}^{-}OO\text{-}Cl] + H_2O \tag{57}$$

$$[{}^{-}O\text{-}O\text{-}Cl] \longrightarrow {}^{1}O_2^{*} + Cl^{-} \tag{58}$$

Foote and co-workers were able to confirm that the reactivity of ${}^{1}O_2^{*}$ produced chemically by this way is identical with that of singlet oxygen formed in the photosensitization of ground-state dioxygen ${}^{3}O_2$. Several books and detailed reviews have been devoted to the chemistry of singlet oxygen [108], so that it is appropriate that our reader refers to the abundant literature on this topic in order to gain insight into the many useful applications of oxidations via ${}^{1}O_2^{*}$.

Consistent with the boundaries defined by the purpose of this review, here we shall limit ourselves to consider briefly a few systems which have the potential of generating ${}^{1}O_2^{*}$ *in situ* from the reaction of H_2O_2 with a suitable partner (HO_2^-, in the case of hydrogen peroxide self-decomposition in Eq. 54). Along these lines, suffice it to say here that the H_2O_2/ClO^- method is of limited value for practical preparative purposes, because of the low solubility of the reagents in organic solvents, and also because of the high alkalinity of the medium that renders the method unsuitable for alkali-sensitive products. Although the solubility problem can be circumvented by the adoption of a biphasic system, the ${}^{1}O_2^{*}$ yields are usually low, and quite often the onset of competing radical reaction cannot be avoided.

Since the maximum rate for H_2O_2 self-decompostion (sometimes termed disproportionation) in Eq. 54 is attained at a quite high pH value (i.e., ca. pH 11.3, its pK_a value), the limitations affecting the H_2O_2/ClO^- method would be even more of a problem for chemical production of ${}^{1}O_2^{*}$ by the H_2O_2 decomposition path. Furthermore, the uncatalysed decomposition of H_2O_2 is unsuitably slow at room temperature (cf., $k_2 < 5\times10^{-7}\ M^{-1}s^{-1}$ at pH 11, assuming second-order in peroxide) [106]. However, one might conceive of employing peroxyanions other than HO_2^- in order to trigger a more rapid O_2 release. For instance, a second-order rate constant of about $2\times10^{-2}\ M^{-1}s^{-1}$ at 0 °C has been measured for the reaction in Eq. 59.

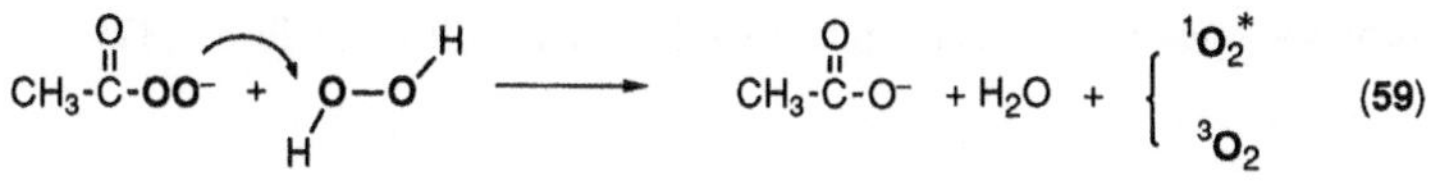

The mechanism for peroxyacid decomposition implies formation of singlet oxygen; this was demonstrated by Evans and Mason, who found that the yield of $^1O_2^*$ was essentially quantitative

More recently, it has been shown by Rebek and co-workers [67c] that singlet oxygen is efficiently produced in the reaction of 98% H_2O_2 with carbonylditriazole **34** in THF, most likely via the *in situ* generation of the elusive (1H-1,2,4-triazolid)peroxycarboxylic acid **35**, undergoing fast reaction with further hydrogen peroxide (Eq. 60).

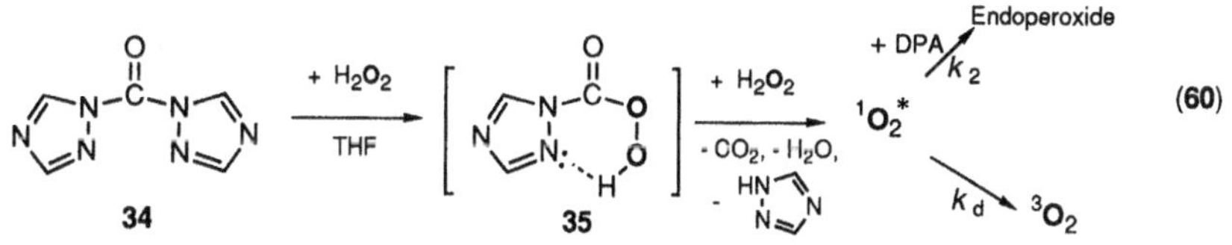

The $^1O_2^*$ may then decay to ground-state oxygen or be trapped by diphenylanthracene (DPA).

In Eq. 61 it is shown yet another system wherein $^1O_2^*$ might have been produced, i.e. the reaction of H_2O_2 with iodosylbenzene (in **36**: X = H), or with 4-iodosylbenzene-1-sulphonate (in **36**: X = SO_3^-) in aqueous media [110].

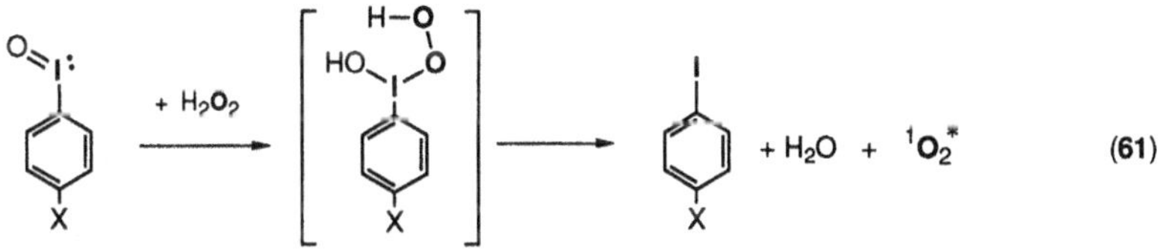

In this case singlet oxygen (yield up to 46%) was trapped using a water-soluble anthracene derivative, i. e. anthracene-9,10-bis-ethanesulphonate.

4.3 Formation of hyperenergetic peroxide species

An interesting aspect of hydrogen peroxide chemistry is its potential to serve as a reagent in the synthesis of small-ring peroxides, such as 1,2-dioxetanes

(**37**), 1,2-dioxetanones (α-peroxylactones) (**38**), and 1,2-dioxetandione (**39**) [111].

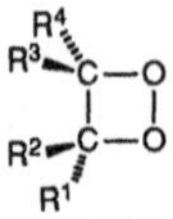

37

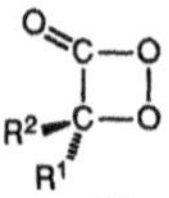

38

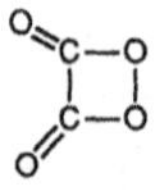

39

Actually, the latter - i.e. the carbon dioxide dimer **39** - has not yet been obtained as an isolated entity to the date of writing this review. On the other hand, authentic samples of representative dioxetanes (**37**) and of α-peroxylactones could be isolated by applying skillful synthetic techniques [111]. For instance, a number of substituted dioxetanes were obtained by the route outlined in Eq. 62, using concentrated H_2O_2 in the presence of an elecrophilic bromine (Br^+) carrier, e.g. 1,3-dibromo-5,5-dimethylhydantoin (DDH) [112].

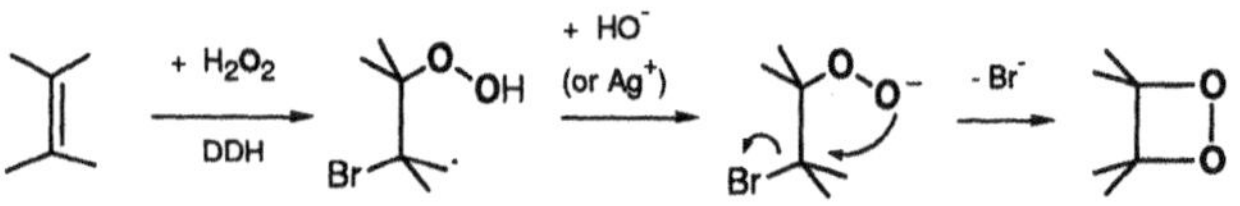

(62)

A most remarkable feature of the chemistry of four-membered ring peroxides **37-39** is their ability to yield excited-state carbonyl fragments upon thermal decay [111], as illustrated by the example in Eq. 63.

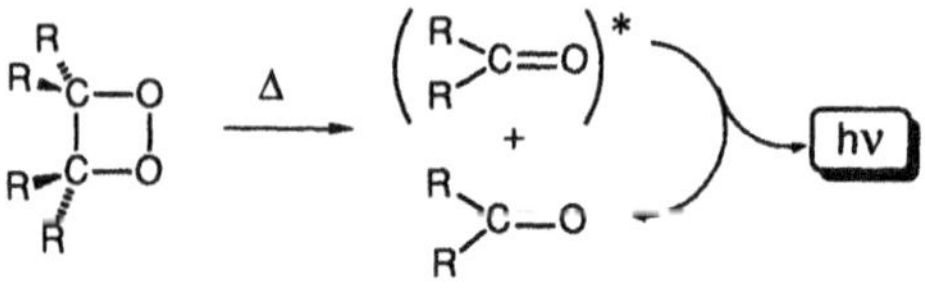

(63)

In such cases chemiluminescence is often observed either directly, or by energy-transfer to suitable fluorophors.

Excellent reviews are available which illustrate in detail the theoretical implications and the practical applications of the hyperenergetic peroxide species above in chemistry and in biochemistry [111]. Therefore, the coverage of this fascinating area of peroxide chemistry is not intended, as it is well beyond the limited scope of this chapter. However, by examining the few cases presented below, the reader might have a lead into this enlightened (often in its

Scheme 4: Energy-transfer Luminescence of Dioxetanedione Formed During Alkaline Perhydrolysis of Oxalyl Derivatives

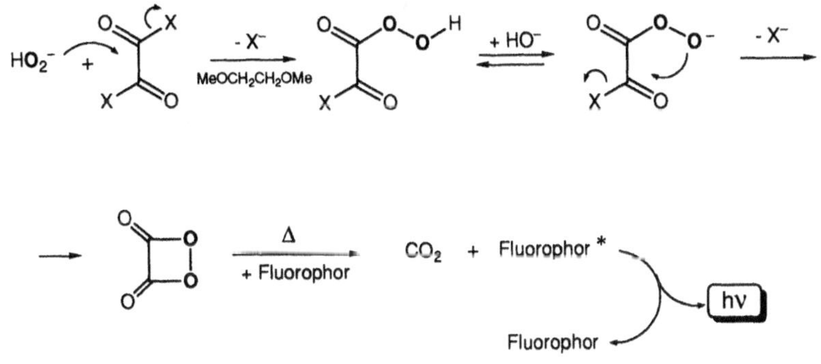

literal meaning!) avenue to exploiting H_2O_2 potential. One example is the reaction of hydrogen peroxide with oxalyl derivatives in Scheme 4. Using the 2,4-dinitrophenyl diester of oxalic acid (in Scheme 4, X = $OC_6H_3(NO_2)_2$), concentrated H_2O_2 in alkali, and a fluorophor (e.g., 9,10-bis(phenylethynyl)anthracene), the system has some commercial applications, such as in the manufacture of "chemical" torches [113].

Another example is the well known Lucigenin chemiluminescent reaction in Eq. 64 [114], involving H_2O_2 and diacridinium salts (**40**).

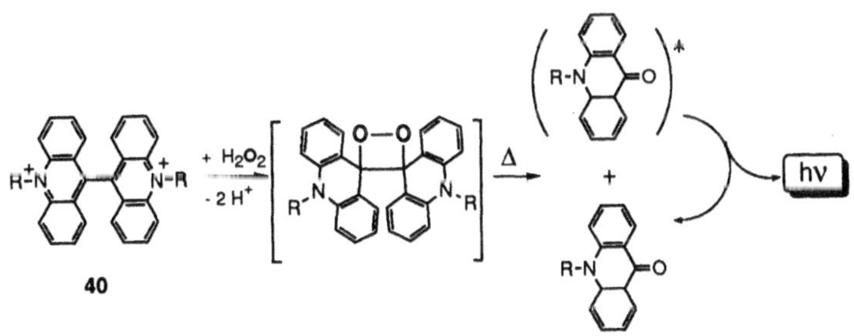

40

(64)

Turning to three-membered ring peroxides, Martin and co-workers [115] reported evidence concerning the formation of a reactive intermediate, most likely having structure **41**, in the reaction of anhydrous H_2O_2 with certain alkoxysulfuranes (Eq. 65).

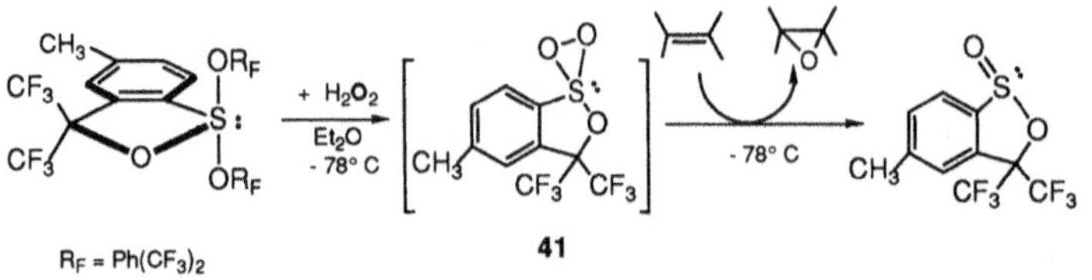

(65)

One could not leave the subject of cyclic peroxides without mentioning the dioxiranes (**42**), i.e. the smallest ring peroxides containing carbon, which are the most powerful electrophilic O-atom transfer agents to-date [116]. These can be generated [117] by the reaction of potassium peroxymonosulfate (caroate, an inexpensive inorganic peroxyacid) and simple ketones, as sketched in Scheme 5.

The system generating dioxiranes *in situ* has been employed to carry out a variety of selective oxidations and oxyfunctionalizations of organic substrates [116,117]. A few authentic representatives of the dioxirane family, possessing sufficient volatility and kinetic stability, could be isolated and fully characterized spectroscopically [118]. Then, the availability of dioxiranes in the isolated form further broadened their scope to perform a variety of synthetic transformations. Dioxirane chemistry has been the subject of recent reviews [116,119] which have illustrated the potential of these new reagents.

Because they are powerful oxidants, yet highly selective under mild conditions, and easily prepared from cheap starting materials, the dioxiranes are nearly ideal electrophilic O-atom transfer agents. From the point of view of large scale operations, however, one problem is the formation of bulk amounts of sulfate salts, the reduction product of caroate. However, a good leaving group such as sulfate ion is needed in order to make the initial ketone-caroate adduct undergo ring closure to dioxirane (cf. Scheme V). To this end, hydrogen peroxide would be unsuitable because it provides a very poor leaving group, i.e. HO^-.

No doubt the search for a practical method to generate dioxiranes *in situ* using some reactive modification of H_2O_2 in lieu of caroate poses an interesting challenge to future research in the field of hydrogen peroxide activation.

Scheme 5: Dioxirane Generation and *In Situ* Oxidation of Organic Substrates by the System Caroate-Ketone

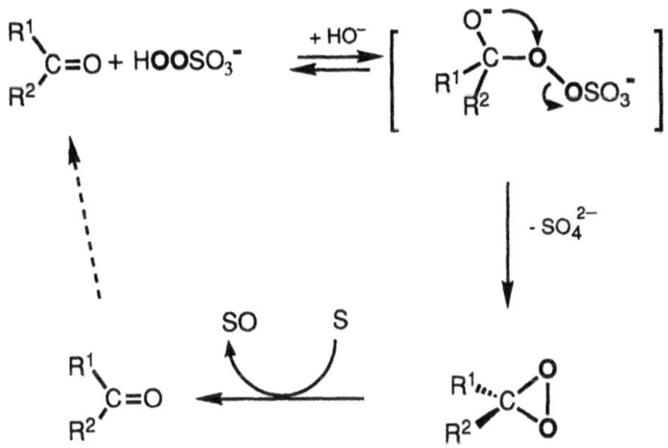

42

5. Conclusions

In common laboratory and industrial practice, activation of hydrogen peroxide by transition-metal ions is rightfully popular. However, it is well recognized that in some cases the method may lead to disposal and environmental problems; partial loss of active oxygen is another of its shortcomings.

We hope this chapter has served to point out that reaction of H_2O_2 with organic compounds of choice can often provide a viable alternative to metal activation. In this respect, we have emphasized that methods for H_2O_2 activation are based on two important principles; namely, *i.* enhancement of leaving group capability upon transformation of H_2O_2 into a new species ZOOH, and *ii.* conversion of hydrogen peroxide (or of the newly formed species ZOOH) by loss of proton, thus making a very reactive nucleophile. Along these lines, we have seen that there are a number of organic activators, which can produce a wide range of peroxide reactivities.

Given the relevance of the subject, the search for ever more effective, safe, and economical activation methods is never ending. In this connection, it seems that a new opportunity has recently arisen from the discovery that energy rich small-ring peroxide intermediates can be formed *in situ* by simple procedures. Thus, we believe that deeper studies on hydrogen peroxide activation by organic compounds are warranted.

Acknowledgements

Financial support by the National Research Council of Italy (CNR, Progetto Finalizzato-CFS II) and by the Italian Ministry for University, Scientific and Technological Research (MURST 40) is gratefully acknowledged. The group of R. C. in Bari is also grateful to Montefluos S.p.A. (Milan, Italy) for generous funding; thanks are also due to Dr. Fulvio Burzio and Dr. Massimo Malavasi of Montefluos for many helpful discussions and ideas concerning peroxide activation.

J. O. E. is pleased to thank Interox R.&D. and Brown University for continuing support of peroxide research.

References

1. R. A. Sheldon and J. K. Kochi, '*Metal-Catalyzed Oxidations of Organic Compounds* ', Academic, New York 1981 and references therein.
2. (a) J. A. Connor and E. A. V. Ebsworth, *Adv. Inorg. Chem. Radiochem.*, **6**, 279 (1964). (b) L. Vaska, *Acc. Chem. Res.*, **9**, 175 (1976).
3. J. O. Edwards, Ed. '*Peroxide Reaction Mechanisms* ', Interscience, New York, 1962.
4. Y. Ogata, K. Tomizawa and K. Furuta in '*The Chemistry of Functional Groups, Peroxides* ', Ed. S. Patai, Wiley, New York 1983, Chapter 21, p. 711.
5. M. Rodgers and E. L. Powers '*Oxygen and Oxy-radicals in Chemistry and Biology* ', Academic, New York 1981.
6. (a) J. O. Edwards and R. Curci, Chapter 4 in this book. (b) C. Walling, *Acc. Chem. Res.*, **8**, 125 (1975). (c) Ref. 1, Chapter 3.
7. J. P. Schirmann and S. Delavarenne '*Hydrogen Peroxide in Organic Chemistry* ', Informations Chimie, Paris 1979.
8. W. M. Weigert, Ed. '*Wasserstoffperoxid und seine Derivate* ', Hüthig, Heidelberg 1978.
9. D. Swern, Ed. '*Organic Peroxides* ', Wiley-Interscience, New York 1970-1973, Vol. 1, 2, and 3. See also references therein.
10. D. Cremer in '*The Chemistry of Functional Groups, Peroxides* ', Ed. S. Patai, Wiley, New York 1983, Chapter 1 and references therein.
11. (a) P. R. Ortiz de Montellano '*Cytochrome P-450, Structure, Mechanism, and Biochemistry* ', Plenum, New York 1986 (b) Ref. 1, Chapter 8.
12. F. Burzio, *Chimica Oggi* , **8**, 23 (1990).
13. A. H. Gilbert, *Detergent Age* , **4**, 32 (1967).
14. R. Curci and J. O. Edwards, in '*Organic Peroxides* ', Ed. D. Swern, Wiley-Interscience, New York 1970, Vol. 1, Chapter IV.
15. W. Edl and H. Kropf in '*Methoden der Organischen Chemie (Houben-Weyl)* ', Ed. H. Kropf, Thieme, Stuttgart 1988, Vol. E13/1, pp 917-930 and references therein..
16. P.D. Bartlett and J.E. Leffler, *J. Amer. Chem. Soc.*, **72**, 3030 (1950).

17. A.C. Baldwin in '*The Chemistry of Functional Groups, Peroxides* ', Ed. S. Patai, Wiley, New York 1983, Chapter 3, p 97 and references therein.
18. J. O. Edwards, *J. Amer. Chem. Soc.*, **76**, 1540 (1954).
19. (a) R. Hiatt, R. J. Smythe, and C. McColeman, *Can. J. Chem.*, **49**, 1707 (1971). (b) R. Hiatt and C. McColeman, *ibid.*, **49**, 1712 (1971).
20. M. R. Hoffmann, *Envir. Sci. Tech.*, **11**, 61 (1977).
21. P. M. Mader, *J. Amer. Chem. Soc.*, **80**, 2634 (1958).
22. (a) E. Abel, *Monatsh. Chem.*, **28**, 1239 (1907). (b) K. Sandved and J. B. Holte, *Kgl. Norske Videnskab Selskabs Forh.*, **11**, 189 (1938); through *Chem. Abstr.*, **33**, 4556 (1939).
23. M. Hoffmann and J. O. Edwards, *Inorg. Chem.*, **16**, 3333 (1977).
24. G. Modena, *Gazz. Chim. Ital.*, **88**, 834 (1959).
25. (a) H. A. Liebhafski and A. Mohammad, *J. Amer. Chem. Soc.*, **55**, 3977 (1933). (b) A. Mohammad and H. A. Liebhafski, *J. Amer. Chem. Soc.*, **56**, 1680 (1934) (c) M. C. R. Symons, *J. Chem. Soc.*, 273 (1955). (d) F. Bell, R. Gill, D. Holden, and W. F. K. Wynne-Jones, *J. Phys. Colloid Chem.*, **55**, 874 (1951). (e) B. Surfleet and P. A. H. Wyatt, *J. Chem. Soc. A* , 1564 (1967).
26. M. A. P. Dankleff, R. Curci, J. O. Edwards, and H.-Y. Pyun, *J. Amer. Chem. Soc.*, **90**, 3209 (1968).
27. O. Masson, *J. Chem. Soc.*, **91**, 1449 (1907).
28. (a) I. R Wilson and G. M. Harris, *J. Amer. Chem. Soc.*, **82**, 4515 (1960). (b) I. R Wilson and G. M. Harris, *J. Amer. Chem. Soc.*, **83**, 286 (1961).
29. S. D. Ross, *J. Amer. Chem. Soc.*, **68**, 1484 (1946).
30. A. S. Gordon in '*Third Symposium on Combustion, Flame, and Explosion Phenomena*', Williams and Wilkins, Baltimore 1949, p. 493.
31. J. O. Edwards and J. J. Mueller, *Inorg. Chem.*, **1**, 696 (1962).
32. (a) E. R. S. Winter and H. V. A. Briscoe, *J. Amer. Chem. Soc.*, **73**, 496 (1951). (b) P. Baertschi, *Experientia*, **7**, 215 (1951). (c) M. Dole, P. R. De Forest, G. R. Muchow, and C. Comte, *J. Chem. Phys.*, **20**, 961 (1952). (d) C. A. Bunton and D. R. Llewellyyn, *Research* , **5**, 142 (1952). (e) M. Anbar, *J. Amer. Chem. Soc.*, **83**, 2031 (1961). (f) M. Anbar and S. Guttman, *J. Amer. Chem. Soc.*, **83**, 2035 (1961). (g) J. Halperin and H. Taube, *J. Amer. Chem. Soc.*, **74**, 380 (1952). (h) M. C. R. Symons, *Chem. Ind. (London)*, 1480 (1960).
33. J. O. Edwards in '*Peroxide Reaction Mechanisms* ', Ed. J. O. Edwards, Interscience, New York, 1962, pp 67-106; see data and references collected in this review article.
34. C. Srinivasan, P. Kuthalingam, and N. Arumugam, *Can. J. Chem.*, **56**, 3043 (1978).
35. M. W. Lister and P. Rosenblum, *Can. J. Chem.*, **41**, 3013 (1963).
36. D. H. Fortnum, C. J. Battaglia, S. R. Cohen, and J. O. Edwards, *J. Amer. Chem. Soc.*, **82**, 778 (1960).
37. F. Secco and M. Venturini, *J. Chem. Soc. Dalton Trans.*, 1410 (1976).
38. (a) T. C. Bruice, *J. Chem. Soc. Chem. Commun.*, 14 (1983). (b) T. C. Bruice in '*Flavins and Flavoproteins* ', Eds. R. C. Bray, P. C Engel and S. E. Mayhew, de Gruyter, Berlin 1984, p 46 and references therein. (c) C. Kemal, T. W. Chan, and T. C. Bruice, *Proc. Natl. Acad. Sci. U.S.A.*, **74**, 405 (1977). (d) S. Ball and T. C. Bruice, *J. Amer. Chem. Soc.*, **101**, 4017 (1979); *J. Amer. Chem. Soc.*, **102**, 6498 (1980).
39. R. D. Cadle and H. Huff, *J. Phys. Chem.*, **54**, 1191 (1950).
40. H. Boardman and G. E. Hulse, *J. Amer. Chem. Soc.*, **75**, 4272 (1953).
41. For example, see: I. Fleming '*Frontier Orbitals and Organic Chemical Reactions* ', Wiley, New York 1976.

42. (a) R. D. Bach, C. L. Willis, and J. M. Domagala, in *Applications of MO Theory in Organic Chemistry* , Ed. I. G. Csizmadia, Elsevier, New York 1977; Vol. 2, p 221 and references therein. (b) B. M. Gimarc '*Molecular Structure and Bonding, The Qualitative Molecular Orbital Approach* ', Academic, New York 1979, Chapters 6-8.
43. E. J. Behrman and J. O. Edwards, *Prog. Phys. Org. Chem.*, **4**., 117 (1967) and references therein.
44. (a) K. B. Ibne-Rasa and J. O. Edwards, *J. Amer. Chem. Soc.*, **84**, 763 (1962). (b) K. B. Ibne-Rasa, *J. Chem. Ed.*, **44**, 89 (1967).
45. M. E. Kurtz and G. J. Johnson, *J. Org. Chem.*, **36**, 3184 (1971).
46. T. N. Baker, III, G. J. Mains, M. N. Sheng, and J. G. Zajacek, *J. Org. Chem.*, **38**, 1145 (1973).
47. G. Olah, D. G. Parker, and N. Yoneda, *Angew. Chem., Int. Ed. Engl.*, **17**, 909 (1978) and references therein.
48. (a) R. Curci, R. A. Di Prete, J. O. Edwards, and G. Modena, *J. Org. Chem.*, **35**, 740 (1970); (b) P. Renolen and J. Ugelstad, *J. Chim. Phys.*, **57**, 634 (1969).
49. (a) R. Kavcic and B. Plesnicar, *J. Org. Chem.*, **35**, 2033 (1970). (b) P. D. Bartlett, *Rec. Chem. Prog.*, **18**, 111 (1957).
50. (a) R. D. Bach, A. L. Owensby, C. Gonzales, and H. B. Schlegel, *J. Amer. Chem. Soc.*, **113**, 2338 (1991). (b) B. Plesnicar, M. Tasevski, and A. Azman, *J. Amer. Chem. Soc.*, **100**, 743 (1978). (c) J. Rebek, L. Marshall, R. Wolak, and J. McManis, *J. Amer. Chem. Soc.*, **106**, 1170 (1984). (d) For an excellent discussion of the transition state geometries in epoxidations, see: M. G. Finn and K. B. Sharpless in '*Asymmetric Synthesis* ', Ed. J. D. Morrison, Academic, New York 1985, Vol 5, Chapter 8.
51. R. P. Hanzlik and G. O. Shearer, *J. Am. Chem. Soc.*, **97**, 5231 (1975).
52. For reviews, see e.g.: (a) H. O. House '*Modern Synthetic Reactions* ', 2nd ed., Benjamin, Menlo Park CA 1972, pp 321-329. (b) G. R. Krow, *Tetrahedron*, **37**, 2697 (1981).
53. R. Curci, F. Di Furia, and G. Modena, *J. Chem. Soc. Perkin Trans. 2* , 603 (1978).
54. (a) R. Curci and G. Modena, *Tetrahedron Lett.*, 1749 (1963). (b) R. Curci and G. Modena, *Tetrahedron*, **22**, 1227 (1966).
55. W. M. Richardson, in *The Chemistry of Functional Groups, Peroxides* , Ed. S. Patai, Wiley, New York 1983, Chapter 5, p 130 and references therein.
56. D. M. Davies and P. Jones, *J. Org. Chem.*, **43**, 769 (1978).
57. (a) J. O. Edwards and R. G. Pearson, *J. Am. Chem. Soc.*, **84**, 16 (1962). (b) N. J. Fina and J. O. Edwards, *Int. J. Chem. Kinet.*, **5**, 1 (1973). (c) R. Curci and F. Di Furia, *Int. J. Chem. Kinet.*, **7**, 341 (1975). (d) S. Hoz, *J. Org. Chem.*, **47**, 3545 (1982). (e) E. Buncel and S. Hoz, *Tetrahedron Lett.*, **24**, 4777 (1983).
58. J. E. McIsaac, L. R. Subbaraman, J. Subbaraman, H. A. Mulhausen, and E. J. Behrman, *J. Org. Chem.*, **37**, 1037 (1972).
59. (a) C. K. Ingold '*Structure and Mechanism in Organic Chemistry* ', 2nd ed., Cornell Univ., Ithaca NY 1969, p 1129. (b) For an overview, see: J. March '*Advanced Organic Chemistry, Reactions, Mechanisms, and Structure* ', 3rd ed., Wiley, New York 1985, p 334 and references therein.
60. R. G. Pearson and D. N. Edgington, *J. Am. Chem. Soc.*, **84**, 4607 (1962).
61. J. O. Edwards, T. E. Erstfeld, K. M. Ibne-Rasa, G. Levey, and M. Moyer, *Int. J. Chem. Kinet.*, **18**, 165 (1986).
62. L. Larsson, *Svensk. Kem. Tidskrift* , **70**, 405 (1958).
63. J. Epstein, M. M. Demek, and D. H. Rosenblatt, *J. Org. Chem.*, **21**, 796 (1956).

64. (a) R. Curci and F. Di Furia, *Tetrahedron Lett.*, 4085 (1974). (b) R. Curci , F. Di Furia, and M. Meneghin, *Gazz. Chim. Ital.*, **108**, 123 (1978).
65. R. A. Sheldon, in *The Chemistry of Functional Groups, Peroxides* , Ed. S. Patai, Wiley, New York 1983, Chapter 6, p 161.
66. For recent reviews giving extensive literature coverage, see: (a) E. Hägel, H. Kropf, and S. Munke in '*Methoden der Organischen Chemie (Houben-Weyl)* ', Ed. H. Kropf, Thieme, Stuttgart 1988, Vol. E13/1, pp. 140-167. (b) H. Kropf and W. Nüremberg, *ibid.*, pp. 521-576, 592-596.. (c) M. Dankowski and G. Prescher, *ibid.*, pp. 763-786. (d) W. Edl and H. Kropf, *ibid.*, pp. 905-908. (e) E. Bolick and J. Dahlmann, *ibid.*, pp. 949-956.
67. (a) J. Rebek, *Heterocycles* , **15**, 517 (1981). (b) J. Rebek and R. McCready, *J. Am. Chem. Soc.*, **102**, 5602 (1980). (c) J. Rebek, R. McCready, S. Wolf.; A. Mossman, *J. Org. Chem.*, **44**, 1485 (1979).
68. H. Williams and S. Mosher, *J. Am. Chem. Soc.*, **76**, 3495 (1954).
69. H. Ross and R. Hüttel, *Chem. Ber.*, **89**, 2641 (1956).
70. H. Kropf and C.-R. Bernert, *Justus Liebigs Ann. Chem.*, **751**, 109 (1971).
71. R. C. Ronald, S. M. Ruder, and T. S. Lillie, *Tetrahedron Lett.*, **28**, 131 (1987).
72. S.-I. Murahashi, T. Oda, and Y. Masui, *J. Am. Chem. Soc.*, **111**, 5002 (1989).
73. (a) M. Avramoff and Y. Sprinzak, *J. Am. Chem. Soc.*, **85**, 1655 (1963). (b) D. A. Konen, L. S. Silbert and P. E. Pfeffer, *J. Org. Chem.*,**40**, 3253 (1975). (c) Y. Sawaki and Y. Ogata, *J. Am. Chem. Soc.*, **97**, 6983 (1975). (d) S. J. Selikson and D. S. Watt, *J. Org. Chem.*, **40**, 267 (1975). (e) D. W. Dixon, in *Advances in Oxygenated Processes* , Ed. A. L. Baumstark, JAI, Greenwich CT 1988, Vol 1, Chapter V pp.179-205 and references therein.
74. J. Rodriguez and J.-P. Dulcere, *J. Org. Chem.*, **56**, 469 (1991).
75. (a) E. G. Sander and W. P. Jencks, *J. Am. Chem. Soc.*, **90**, 6154 (1968). (b) M. C. V. Sauer and J. O. Edwards, *J. Phys. Chem.*, **75**, 3377 (1971); **76**, 1283 (1972).
76. (a) R. P. Heggs and B. Ganem, *J. Am. Chem. Soc.*, **101**, 2484 (1979). (b) B. Ganem, R. P. Heggs, A. J. Biloski, and D. R. Schwartz, *Tetrahedron Lett.*, **21**, 685 (1980).
77. For instance, see: (a) J. R. Sanderson, R. J. Wilterdink, and A. G. Zeiler, *Synthesis* , 479 (1976). (b) H. Kropf and M. Maher-Detweiler in '*Methoden der Organischen Chemie (Houben-Weyl)* ', Ed. H. Kropf, Thieme, Stuttgart 1988, Vol. E13/1, pp. 736-755 and references therein.
78. (a) E. Höft and A. Rieche, *Angew. Chem., Int. Ed. Engl.*, **4**, 524 (1965). (b) E. G. E. Hawkins, *J. Chem. Soc., C* , 2686 (1969).
79. F. A. Davis and M. S. Haque, in *Advances in Oxygenated Processes* , (A. L. Baumstark, Ed.), JAI, Greenwich, CT, 1990; Vol 2, Chapter II; pp. 61-116 and references therein.
80. (a) P. G. Cookson, A. G. Davies, and N. Fazal, *J. Organomet. Chem.*, **99**, C 31 (1975). (b) M. Taddei and A. Ricci, *Synthesis* , 633 (1986). (c) B. Bressel and A. Blaschette, *Z. Anorg. Allg. Chem.*, **377**, 182 (1970).
81. A. A. Oswald and D. L. Guertin, *J. Org. Chem.*, **28**, 651 (1963).
82. R. Curci, R. Mello, L. Troisi, *Tetrahedron*, **42**, 877 (1986).
83. (a) H. G. Kuivila and A. G. Armour, *J. Am. Chem. Soc.*, **79**, 5659 (1957). (b) A. G. Davies and R. B. Moodie, *J. Chem. Soc,* 2372 (1958). (c) Y. A. Aleksandrov and.V. P. Maslennikov, *J. Organomet. Chem. Libr.*, **3**, 103 (1977).
84. M. Pralus, J. C. Lecoq, and J. P. Schirmann, in *Fundamental Research in Homogeneous Catalysis* , (M. Tsutsui, Ed.), Plenum, New York, 1979; Vol. 3, p. 327.

85. A. J. Bloodworth and M. E. Loveitt, *J. Chem. Soc., Perkin Trans. 1* , 1031 (1977).
86. L. S. Silbert, E. Siegel, and D. Swern, *J. Org. Chem.*, **27**, 1336 (1962); *Org. Synth.*, **43**, 93 (1963).
87. S. E. Jacobson, F. Mares, and P. M. Zambri, *J. Am. Chem. Soc.*, **101**, 6946 (1979).
88. P. A. Grieco, Y. Yokoyama, S. Gilman, and M. Nishizawa, *J. Org. Chem.*, **42**, 2034 (1977).
89. C. R. Harrison and P. Hodge, *J. Chem. Soc., Chem. Comm.* , 1009 (1974).
90. (a) G. B. Payne, P. H. Deming, and P. H. Williams, *J. Org. Chem.*, **26**, 659 (1961). (b) G. B. Payne, *Tetrahedron* , **18**, 763 (1962).
91. J. E. McIsaac, R. E. Ball, and E. J. Behrman, *J. Org. Chem.*, **36**, 3048 (1971).
92. (a) R. D. Bach and J. W. Knight, *Org. Synth.*, **60**, 63 (1981). (b) L. A. Arias, S. Adkins, C. J. Nagel, and R. D. Bach, *J. Org. Chem.*, **48**, 888 (1983).
93. N. Matsumura, N. Sonoda, S. Tsutsumi, *Tetrahedron Lett.*, 2029 (1970).
94. S. Krishnan, D. G. Kuhn, and G. A. Hamilton, *Tetrahedron Lett.*, 1369 (1977).
95. M. S. Cooper, H. Heaney, A. J. Newbold, and W. R. Sanderson, *Synlett.*, 533 (1990) and references therein.
96. For example, see: A. S. Rao, S. K. Paknikar, and J. G. Kirtane, *Tetrahedron* , **39**, 2323 (1983).
97. W. H. Pirkle and P. L. Rinaldi, *J. Org. Chem.*, **42**, 2080 (1977).
98. (a) A. Pross and S. S. Shaik, *Acc. Chem. Res.*, **16**, 363 (1983). (b) A. Pross, *Acc. Chem. Res.*, **18**, 212 (1985) and references.
99. For a monographic book giving an excellent coverage of this topic, see: L. Eberson, '*Electron Transfer Reactions in Organic Chemistry* ', Springer-Verlag, Berlin 1987, and references therein.
100. (a) L. Eberson, *Chem. Scr.*, **20**, 39(1982). (b) V. L. Antonovskii, Z. S. Frolova, T. T. Shleina, M. M. Buglanova, *Zh. Obshch. Khim.*, **39**, 368 (1969). (c) L. S. Silbert, in '*Organic Peroxides* ', Ed. D. Swern, Wiley-Interscience, New York 1971, Vol. 2, Chapter VII, pp 755-778 and references therein.
101. (a) C. Walling, *J. Am. Chem. Soc.*, **102**, 6854 (1980). (b) G. B. Schuster, *J. Am. Chem. Soc.*, **101**, 5851 (1979) and references therein.
102. G. B. Schuster, *Acc. Chem. Res.*, **12**, 366 (1979).
103. (a) W. A. Pryor and W. H. Hendrickson, Jr., *J. Am. Chem. Soc.*, **105**, 7114 (1983). (b) A. E. Miller, J. J. Bischoff, C. Bizub, P. Luminoso, and S. Smiley, *J. Am. Chem. Soc.*, **108**, 7773 (1986) and references therein.
104. G. Levey, A. L. Rieger, and J. O. Edwards, *J. Org. Chem.*, **46**, 1255 (1981).
105. K. Sehested and J. Holcman, *J. Phys. Chem.*, **82**, 651 (1978).
106. (a) E. Koubek, M. L. Hagget, C. J. Battaglia, K. M. Ibne-Rasa, H.-Y. Pyun, and J. O. Edwards, *J. Am. Chem. Soc.*, **85**, 2263 (1963). (b) D. F. Evans and M. W. Upton, *J. Chem. Soc., Dalton Trans.*, 2525 (1985).
107. C. S. Foote, S. Wexler, W. Ando, and R. Higgins, *J. Am. Chem. Soc.*, **90**, 975 (1968) and references.
108. For instance, see: A. A. Frimer and L. M. Stephenson in '*Singlet O_2* ', Ed. A. A. Frimer, CRC, Boca Raton FL 1985, Vol. 1 and 2 and references therein.
109. D. F. Evans and M. W. Upton, *J. Chem. Soc., Dalton Trans.*, 1151 (1985).
110. D. F. Evans and M. W. Upton, *J. Chem. Soc., Dalton Trans.*, 1141 (1985).
111. For instance, see: W. Adam in '*Small Ring Heterocycles.- Part 3* ', Ed. A. Hassner, Wiley, New York 1985, Chapter IV, pp. 351-429 and references therein.

112. K. R. Kopecky, P.A. Lockwood, J. E. Filby, and R. W. Reid, *Can. J. Chem.*, **51**, 468 (1973). (b) K. R. Kopecky, J. E. Filby, C. Mumford, P.A. Lockwood, and J. Y. Ding, *Can. J. Chem.*, **53**, 1103 (1975).
113. M. M. Rauhut, *Acc. Chem. Res.*, **2**, 80 (1969).
114. J. R. Totter and G. E. Philbrook, *Photochem. Photobiol.*, **5**, 177 (1965).
115. L. D. Martin and J. C. Martin, *J. Am. Chem. Soc.*, **99**, 3511 (1977).
116. For reviews, see: (a) R. Curci in '*Advances in Oxygenated Processes* ', Ed. A. L. Baumstark, JAI, Greenwich CT 1990, Vol. 2, Chapter I, pp 1-59. (b) W. Adam, R. Curci, and J. O. Edwards, *Acc. Chem. Res.*, **22**, 205 (1989). (c) R. W. Murray, *Chem Rev.*, **89**, 1187 (1989).
117. (a) J. O. Edwards, R. H. Pater, R. Curci, and F. Di Furia, *Photochem. Photobiol.*, **30**, 63 (1979). (b) R. Curci, M. Fiorentino, L. Troisi, J. O. Edwards, and R. H. Pater, *J. Org. Chem.*, **45**, 4758 (1980). (c) G. Cicala, R. Curci, M. Fiorentino, and O.Laricchiuta, *J. Org. Chem.*, **47**, 2670 (1982). (d) R. Curci, M. Fiorentino, and M. R. Serio, *J. Chem. Soc., Chem. Comm.*, 155 (1984).
118. (a) R. W. Murray and R. Jeyaraman, *J. Org. Chem.*, **50**, 2847 (1985). (b) L. Cassidei, M. Fiorentino, R. Mello, O. Sciacovelli, and R. Curci, *J. Org. Chem.*, **52**, 699 (1987). (c) R. Mello, M. Fiorentino, O. Sciacovelli, and R. Curci, *J. Org. Chem.*, **53**, 3890 (1988).
119. For a review of recent developments, see: W. Adam, L. Hadjiarapoglu, R. Curci, and R. Mello in '*Organic Peroxides* ', Ed. W. Ando, Wiley, New York in press, Chapter IV.

CHAPTER 4

Fenton Type Activation and Chemistry of Hydroxyl Radical

John O. EDWARDS # and Ruggero CURCI *

Department of Chemistry, Brown University, Providence, RI 02912, U.S.A.

* *Dipartimento di Chimica, Università di Bari, Via Amendola, 173; 70126 Bari, Italy*

1. Introduction

1.1. Brief background

Nowhere among the multitude of organic compounds is the need for peroxide activation more apparent than among the simple compounds of C, H and O: hydrocarbons, carboxylic acids, alcohols and ethers. In these molecules there are no nucleophilic sites or other types of centers that can react under mild conditions with hydrogen peroxide HOOH. This is well illustrated by the case of polycarboxylic acids which Fenton [1] in 1894 was trying to oxidize with HOOH when he found the strong promotion by ferrous ion. Forty years later Haber and Weiss [2] suggested that hydroxyl radical HO• is formed by the step

$$Fe^{2+} + HOOH \xrightarrow{H^+} Fe^{3+} + H_2O + HO\cdot \quad (1)$$

and this oxidizing intermediate attacked the reluctant acids. Further development of the radical intermediate concept was made in three series of articles by Barb et al. [3], by Merz and Waters [4], and by Walling et al [5].

The possibility that the oxidizing intermediate is an iron(IV) particle such as the ferryl ion FeO^{2+} was introduced as early as 1932 [6]. It has been given strong support by several groups [7], particularly when the iron atom is attached to a porphyrin ring. This dichotomy in mechanism type is the source of lively discussion which continues right up to the present time.

During the 1960's and 1970's, great emphasis was placed on obtaining, by radiolytic techniques that have been developed during and after World War II,

G. Strukul (ed.), Catalytic Oxidations with Hydrogen Peroxide as Oxidant, 97–151.

the very high rate constants for HO• oxidations. The past two decades has seen the research interest shift to the toxicity of oxygenated radicals in living systems.

In principle one can compare reactivities of the oxidizing intermediate formed by the Fenton reagent with those for HO• formed by a process (e.g., radiolysis) wherein iron is excluded; this is not as simple a test as might appear. The iron in its states Fe^{2+} and Fe^{3+} is known to reduce and oxidize various organic radicals thereby complicating both rate and stoichiometry for the HO• reaction. These secondary effects of the iron, shown clearly by Walling and coworkers [5], seriously complicate comparisons of systems said to generate HO•.

The question of oxidizing intermediate nature is not limited to the reaction of Fe^{2+} with HOOH, of course. It is brought up again in the reaction of iron(II) complexes (such as $Fe(EDTA)^{2-}$) and of other metal ions with HOOH. For this reason it seems appropriate to discuss the iron situation in some detail, to give some comparisons of relative rates for the radical produced by Fenton's reagent with HO• generated by energetic radiation, and then to look at the other reducing metal ions.

The authors will use the hydroxyl radical HO• as their unifying core for discussions of the oxidizing intermediate in these two systems even though the matter remains open to debate.

To a significant extent the disagreements mentioned above stem from a simple rate law for the initiation step of the Fenton reaction and for the overall reaction under some conditions. This law is

$$\text{Rate} = k\,[Fe^{2+}][HOOH]$$

To fit such a simple rate law, a number of postulations can be put forth. The two types of mechanisms to be discussed both fit the law and each has its proponents. This should not be taken as criticism of either group because both recognize the difficulty in resolving the problem and both sides must be heard.

Given that the rate controlling step provides no strong clues, chemists who seriously investigate the Fenton type systems must rely on rates of secondary steps (all of which are very rapid) and on selectivity measurements. The selectivity data turn out to involve often unsuspected side reactions, and this is discussed with specific illustration below.

It had been hoped that spin-trap reagents such as DMPO (dimethylpyrrolidine-N-oxide) would provide conclusive results on radical nature. In the event, the results are useful but rarely definitive.

1.2. The oxidizing species

Although the weight of evidence now favors HO• as the particle which attacks the organic substrate in the Fenton system itself, other oxidizing species may be formed in other Fenton type systems. The studies of Johnson and coworkers [8] on copper(I) reactions with HOOH provide a fine example of an alternative possibility.

It is important, therefore, to consider a number of variables that are involved in the step giving rise to the oxidizing species (which may not be a radical in the way this word is used by many chemists).

One of the variables that bear on the nature of the oxidant is the reduced metal species, so these metals will be mentioned individually. Also, the ligands attached to the metal ion must be considered; it would be a poor assumption to say that because $Fe^{2+}(aq)$ produces HO• so must also $Fe(EDTA)^{2-}$. The oxidizing species formed must also depend on the peroxide employed. A reduced metal ion plus peroxomonosulfate ion $HOOSO_3^-$ can form either HO• or SO_4^{-}• or both as has been shown by Gilbert et al [9].

Important to the discussion of Fenton type systems and to the chemistry of HO• are several quite different methods for radical generation. Some of these are chemical in nature but only incompletely understood; for example, this is the situation for peroxonitrous acid. Others are important as sources of HO• in cell tissue where the radical exerts its high toxicity; a case in point is the chromate and glutathione combination. Two others are physical in nature; e.g., radiation with energetic particles and peroxide photolysis. These optional methods, needed for the establishment of rate constants of HO• with substrates and for our understanding of the role of radicals in biology, are also delineated in the sections below. Again, as in the case of Fe^{2+} and HOOH, the results must be analyzed with caution.

1.3. Thermodynamic considerations

All of the reactions discussed herein are governed by the free energy changes and from the standard free energy values electrode potentials can be calculated; also some potentials are measurable. The present level of knowledge of E^0 values is useful but incomplete.

Reduction of dioxygen O_2 to H_2O by successive single electron transfers provides

$$O_2 \underset{H^+}{\overset{e^-}{\rightarrow}} HOO\cdot \underset{H^+}{\overset{e^-}{\rightarrow}} HOOH \underset{H^+}{\overset{e^-}{\rightarrow}} H_2O + HO\cdot \underset{H^+}{\overset{e^-}{\rightarrow}} 2\,H_2O \qquad (2)$$

three oxidizing intermediates, namely hydroperoxyl radical $HOO\cdot$, hydrogen peroxide and hydroxyl radical. The latter two are of greatest interest here, but it is necessary to remember that O_2 and $HOO\cdot$ (or its conjugate base) are normally present (albeit in small quantities) in a Fenton type system and in living tissue.

It is particularly unfortunate for the present context that the free energy for $HO\cdot$ is poorly known because electrode potentials for half-reactions involving $HO\cdot$ are calculated from this free energy. Potentials for the three steps of Scheme 1 are related; that for step I which is complete reduction of HOOH to H_2O (+ 1.763 volts) is well established.

Scheme 1:

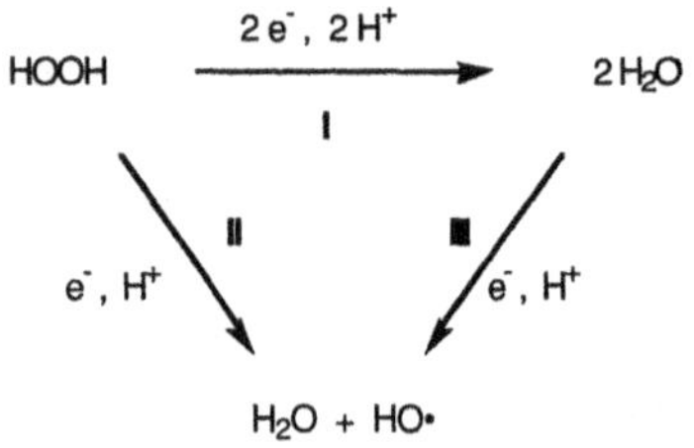

The sum of the potentials for the two single electron transfers (steps II and III) must be 3.526 volts (twice that for step I). The literature data are consistent with the proper sum, but the individual E^0 values for I and II vary. An often quoted value for step I is 0.71 volts. Assuming this is correct, it can be said that any metal reductant whose oxidation potential is more positive than -0.7 volts can reduce HOOH to $HO\cdot$ at 1M concentrations. The always small concentration of $HO\cdot$ adds a further driving force, and this may be what brings ferrous ion and vanadyl ion VO^{2+} into play as one electron reductants in Fenton type systems. If the reduction potential for HOOH is more positive than 0.7 volts as may well be the case, then the concentration effect may be less important.

In Table 1, electrode potentials for oxygen-containing molecules and for metal ions relevant to this review are given. These are obtained from three sources [10, 11, 12] and include some calculated values made here.

Table 1: Electrode Potentials[a].

Half-cell reactions	Potentials (volts)
A: Oxygenated Particles	
$HO\bullet + H^+ + e^- \rightarrow H_2O$	2.81[b]
$HOOH + 2\,H^+ + 2\,e^- \rightarrow 2\,H_2O$	1.763
$HOO\bullet + H^+ + e^- \rightarrow HOOH$	1.44
$O_2(g) + 4\,H^+ + 4\,e^- \rightarrow 2\,H_2O$	1.229
$HOOH + H^+ + e^- \rightarrow HO\bullet + H_2O$	0.71[c]
$O_2(g) + 2\,H^+ + 2\,e^- \rightarrow HOOH$	0.695
$O_2(g) + H^+ + e^- \rightarrow HOO\bullet$	-0.046
B: Metal Ions	
$Ag^{2+} + e^- \rightarrow Ag^+$	1.980
$Co^{3+} + e^- \rightarrow Co^{2+}$	1.808
$Ce^{4+} + e^- \rightarrow Ce^{3+}$	1.61[d]
$FeO^{2+} + 2\,H^+ + e^- \rightarrow Fe^{3+} + H_2O$	> 1.5
$Mn^{3+} + e^- \rightarrow Mn^{2+}$	1.51
$Cu^{3+} + e^- \rightarrow Cu^{2+}$	~ 1.2
$VO_2^+ + 2\,H^+ + e^- \rightarrow VO^{2+} + H_2O$	1.00
$Fe^{3+} + e^- \rightarrow Fe^{2+}$	0.771
$MnO^{2+} + 2\,H^+ + e^- \rightarrow Mn^{3+} + H_2O$	~ 0.7
$Cu^{2+} + e^- \rightarrow Cu^+$	0.153
$TiO^{2+} + 2\,H^+ + e^- \rightarrow Ti^{3+} + H_2O$	0.099
$Cr^{3+} + e^- \rightarrow Cr^{2+}$	-0.408
$Eu^{3+} + e^- \rightarrow Eu^{2+}$	-0.429

[a] For aqueous acid at 25°C. [b] Value 2.38 volts also mentioned [10]. [c] Value 1.14 volts also mentioned [10]. [d] Value depends strongly on anion nature and concentration.

2. Iron Ions and Hydrogen Peroxide

2.1 Ferrous ion and peroxide

When HOOH is added to excess Fe^{2+} in aqueous acid, a clean stoichiometry obtains.

$$HOOH + 2\,Fe^{2+} + 2\,H^+ \rightarrow 2\,Fe^{3+} + 2\,H_2O \qquad (3)$$

The rate law

$$-d\,[HOOH]\,/\,dt = k\,[HOOH][Fe^{2+}]$$

suggests the mechanism with eq. 4 being rate determining.

$$HOOH + Fe^{2+} \rightarrow Fe(OH)^{2+} + HO\cdot \qquad (4)$$

$$HO\cdot + Fe^{2+} \rightarrow Fe(OH)^{2+} \qquad (5)$$

$$Fe(OH)^{2+} + H^+ \rightarrow Fe^{3+} + H_2O \qquad (6)$$

An alternative mechanism

$$HOOH + Fe^{2+} \rightarrow FeO^{2+} + H_2O \qquad (7)$$

$$FeO^{2+} + Fe^{2+} + H^+ \rightarrow Fe(OH)^{2+} + Fe^{3+} \qquad (8)$$

$$Fe(OH)^{2+} + H^+ \rightarrow Fe^{3+} + H_2O \qquad (9)$$

invokes the ferryl ion FeO^{2+} as the oxidizing intermediate, yet fits the rate law equally well. The rate constant k is 41.4 $M^{-1}s^{-1}$ at 20.2°C with small variations for acidity and ionic strength [13]. The temperature dependence in 0.1 M $HClO_4$ is given by

$$k = 5.3 \times 10^8 \exp(-9450\,/\,RT)$$

A detailed study of this redox reaction was carried out by Barb et al [3a]. The mixture was found to initiate vinyl polymerization; when the monomer was present, the stoichiometric ratio for Fe^{2+} and HOOH dropped to 1:1. If the HOOH was in excess, then peroxide decomposition with dioxygen release accompanied the redox reaction. Addition of oxidants Fe^{3+} and Cu^{2+} caused the rate of O_2 evolution to increase up to a saturation rate and the oxidants appeared to have rates which increased with pH. Fluoride ion inhibited the

Fe^{3+} enhancement. A radical mechanism was proposed [3a] following the earlier suggestions of Haber and Weiss [2]. The steps postulated by Barb et al [3] (using numeration designated for the present review) were as follows:

$$Fe^{2+} + HOOH \xrightarrow{k_1} Fe^{3+} + OH^- + HO\cdot \quad (10)$$

$$Fe^{2+} + HO\cdot \xrightarrow{k_2} Fe^{3+} + OH^- \quad (11)$$

$$HOOH + HO\cdot \xrightarrow{k_3} H_2O + HOO\cdot \quad (12)$$

$$Fe^{2+} + HOO\cdot \xrightarrow{k_4} Fe^{3+} + HOO^- \quad (13)$$

$$Fe^{3+} + HOO\cdot \xrightarrow{k_5} Fe^{2+} + H^+ + O_2 \quad (14)$$

When Cu^{2+} was present, the following two steps were included;

$$Cu^{2+} + HOO\cdot \xrightarrow{k_6} Cu^+ + H^+ + O_2 \quad (15)$$

$$Cu^+ + Fe^{3+} \xrightarrow{k_7} Cu^{2+} + Fe^{3+} \quad (16)$$

(The ferryl mechanism would be similar; the main difference would be that FeO^{2+} formed in the rate step would react rapidly with Fe^{2+} in the second and with HOOH in the third).

2.2. Fenton's reagent revisited

Walling et al [5] updated the experiments of Merz and Waters [4] in which stoichiometric measurements played a large role. It was now possible for competitive rates from Walling's experiments with Fenton's reaction to be compared with the tidal wave of radiolysis and photolysis constants just emerging in 1970. In order to minimize side reactions such as O_2 evolution, a small known amount of HOOH was added to a well-stirred solution containing excess of Fe^{2+} and organic substrate. The amount of Fe^{2+} remaining after complete addition was then analyzed, and from the ratio $\Delta[Fe^{2+}] / \Delta[HOOH]$ it was possible to obtain rate constants for the various steps.

The new steps attributable to organic substrates (following the numeration pattern given above) are as follows:

$$HO\cdot + R_iH \xrightarrow{k_{8i}} H_2O + R_i\cdot \quad (17)$$

$$HO\cdot + R_jH \xrightarrow{k_{8j}} H_2O + R_j\cdot \quad (18)$$

$$HO\cdot + R_kH \xrightarrow{k_{8k}} H_2O + R_k\cdot \quad (19)$$

$$R_i\cdot + Fe^{3+} \xrightarrow{k_9} Fe^{2+} + \text{product} \quad (20)$$

$$2\,R_j\cdot \xrightarrow{k_{10}} \text{product (dimer)} \quad (21)$$

$$R_k\cdot + Fe^{2+} \xrightarrow[H^+]{k_{11}} Fe^{3+} + R_kH \quad (22)$$

The various organic particles undergoing hydrogen abstraction (or HO• addition in the case of olefins) gave radicals which subsequently could be oxidized by Fe^{3+}, could undergo dimerization, or could be reduced by Fe^{2+}.

Walling et al [5] in their studies oxidized a variety of organic molecules using the Fenton reagent. The results were successfully interpreted on the basis of HO• being the oxidizing species formed in the first step. This was and remains amply justified by the fact that the rate constants (k_{8i}, k_{8j} and k_{8k}) for the step involving organic molecule and HO• are in satisfactory agreement with constants for the same step as obtained from radiolysis experiments. Comparison constants are given in Table 2 for six alcohols and four solvents. In no case is the difference between the constant obtained from the Fenton reagent and that obtained from radiolysis experiments as much as a factor of two. It is worth noting that individual values for the reaction

$$HO\cdot + (CH_3)_2CHOH \rightarrow H_2O + R\cdot \quad (23)$$

where R• contains the sum of isomeric radicals, range from 1.3×10^9 to 3.2×10^9 $M^{-1}s^{-1}$ (except for an outlyer at 6.5×10^9) covering seventeen measurements given in the compilation of Farhataziz and Ross [14].

For k_1 and k_2, Walling et al [5] used the values 76×10^8 $M^{-1}s^{-1}$ and 3×10^8 $M^{-1}s^{-1}$. Their k_8 type constants fell in the range from 10^7 to 10^{10} $M^{-1}s^{-1}$ for a variety of organic molecules.

A point which must be emphasized is that all of the fast constants (k_2, k_8 etc.) are dependent on either relative rates or difficult absolute measurements. Therefore the level of agreement between different investigators in some instances is no better than a factor of two and can be as poor as a power of ten.

The data of Walling et al. appear to be the most comprehensive set of rates and stoichiometries generated by a single group of investigators. Therefore they are presumed by these writers to provide a "matched set" of numbers in an area of chemistry beset with complications. They will be used in the fourth section to describe much of the organic chemistry of HO• and to show some of the complexities. The account of Walling [5e] should be required reading for any student initiating a study of free radical oxidations. The title of this subsection is that of the Walling account [5e].

Table 2: Comparison of Hydroxyl Rate Constants for the Reaction

$$HO\cdot + RH \rightarrow H_2O + R\cdot$$

at Room Temperature.

Compound	$k \times 10^{-9\,a}$	$k \times 10^{-9\,b}$
CH_3OH	1.2	0.9
CH_3CH_2OH	2.1	1.85
$(CH_3)_2CHOH$	2.0	2.2
$CH_3CH_2CH_2OH$	3.4	2.6
$(CH_3)_2CHCH_2OH$	4.2	3.4
$(CH_3)_3COH$	0.57	0.53
CH_3COCH_3	0.097	0.088
CH_3COOH	0.018	0.027
THF	2.0	2.7
DMF	1.8	1.7

a Based on the relative rate k/k_{Fe} with $k_{Fe^{2+}} = 3 \times 10^8\ M^{-1}s^{-1}$ [5]. *b* First three entries are values used [14] for normalizing constants; balance are averages for radiolysis experiments only.

2.3. Ferric ion and HOOH

Even if one starts with a single oxidation state of iron in the presence of HOOH, one soon has both states. Ferric ion can oxidize HOO• (and/or its conjugate ion $O_2^{-}\cdot$) as seen in the k_4 step above; also it can oxidize HOOH itself more slowly. (It even may be oxidized by HO•, but that does not concern us here.) Thus the mechanism of HOOH oxidation by Fe^{3+} must be discussed in this section; only high points of the ferric-centered catalysis of HOOH decomposition will be touched upon because an excellent review is available [15]. It is interesting to note that the decomposition of HOOH by Fe^{3+} and its complexes has raised strikingly similar mechanism questions to the Fenton system.

For the case of [HOOH] >> [Fe^{3+}], the rate law in dilute aqueous acid is

$$\frac{-d[HOOH]}{dt} = k\left\{\frac{[Fe^{3+}][HOOH]}{[H^+]}\right\}$$

As Brown et al [15] have said "A rate law of such simplicity can be derived from a variety of conceptually different hypothetical mechanisms". In order to sort out these possible mechanisms, one must rely on rates at widely variable reactant ratios.

A distinct adduct which can be written as $Fe(OOH)^{2+}$ is formed in aqueous acidic solution by an equilibrium

$$Fe^{3+}(aq) + HOOH \overset{K_{12}}{\rightleftharpoons} Fe(OOH)^{2+} + H^+ \qquad (24)$$

rapidly established [16]. It seems reasonable to the writers as it did to Barb et al. [3b] that this complex leads to the rate determining step

$$Fe(OOH)^{2+} \overset{k_{13}}{\rightarrow} Fe^{2+} + HOO\cdot \qquad (25)$$

in the slow decomposition.

At low concentrations of reactants, the rate dependence at constant acid concentration would be cleanly second order and there would be only small amounts of the adduct at equilibrium. However, as higher concentrations of reactants are introduced, there would be gradual switch to overall first order. Even if the actual rate step is bimolecular

$$Fe^{3+} + HOO^- \rightleftharpoons Fe^{2+} + HOO\cdot \qquad (26)$$

the same saturation type of kinetics would be seen. The HOO· could then be oxidized by a second Fe^{3+} as in step 14 of the Barb type mechanism.

Kremer and Stein [17] put forth a mechanism without radicals, specifically

$$Fe^{3+} + HOO^- \rightleftharpoons CI \qquad (27)$$

$$CI \rightarrow CII \qquad (28)$$

$$CII + HOOH \rightarrow products + Fe^{3+} \quad (29)$$

The present authors interpret CI to be the complex $Fe(OOH)^{2+}$, CII to be a perferryl ion FeO^{3+} and the products to include O_2. This mechanism would, of course, fit the observed rate law. The advantages of this mechanism have been carefully delineated in Brown et al [15].

It now appears that the mechanism for ferric ion catalyzed decomposition of HOOH has been resolved in a fashion akin to that applied to the mechanism of the Fenton reaction [18]. Organic compounds retard the decomposition by competing with HOOH for HO•; the measured competition ratio was then compared with values from other experiments including those of radiation chemistry and from the Fenton reaction itself.

The agreement is satisfactory. Walling and Goosen [18] were also able to obtain an initial rate and a chain length for this decomposition.

2.4. Anionic ligand effects

Qualitative predictions of rate effects due to complexation of the iron ion by simple anions are possible. The rate constant for the presumed first step in the Fenton process (eq. 4) must reflect the change in oxidation state of the iron center. Addition of an anion (e.g., F^-) to the ferrous ion

$$Fe^{2+} + F^- \rightleftharpoons FeF^+ \quad (30)$$

should enhance its oxidizability by the neutral peroxide. The rate constant is expected to increase, as has been found by Wells and Salam [19].

The initial step in the ferric centered catalysis of decomposition involves reduction of iron(III) to iron(II) and this should be more difficult when the iron has an attached anionic ligand. Indeed, it was found by Jones et al [20] that $FeSO_4^+$ and FeF^{2+} are catalytically inactive in the decomposition.

Large ligands including EDTA, NTA and ATP occupy several metal coordination positions and have oxidizable constituents. Therefore the elementary electrostatic arguments used for F^-, SO_4^{2-} etc., are not easy to apply to large ligands. Nevertheless, they are important and they will be discussed below along with factors such as coordination positions occupied and ligand field stabilization.

3. Physical Generation of Hydroxyl Radical

3.1. Basis for discussion

In order to establish the intermediacy of HO• in the Fenton reaction, it is necessary to have an unquestioned method for generating this radical with a minimum of complications. A combination of a number of techniques, largely developed since World War II, has given us the necessary information. Although the use of physical processes avoids the problem of oxidation and/or reduction by a metal ion, there have been some technical problems that had to be sorted out prior to obtaining reliable rate constants for HO• reacting with organic substrates.

Among the techniques that now are commonly employed in radical chemistry are: electron spin resonance (ESR), nuclear magnetic resonance (NMR), chromatography, flow systems (both continuous and stopped), photolysis, and radiolysis. ESR is particularly important in radical chemistry due to its high sensitivity and to the signal splittings from hyperfine coupling. NMR along with the several types of chromatography has made possible identification and quantitation of small amounts of product compounds. Flow systems make feasible fast mixing and inspection. These four techniques do not take part or contribute to the chemical reactions, so they will not be discussed further in this review.

On the other hand, both photolysis and radiolysis contribute directly to the initiation of reactions, including the controlled release in solution of HO•. For this reason, they are discussed here but only in respect to their chemical actions.

3.2. Photolysis

Peroxides have a broad absorption band beginning above 300 nm, it increases monotonically in extinction coefficient as wavelength decreases at least as far down as 200 nm. At the mercury vapor line (254 nm), the extinction coefficient is about 20 $M^{-1}cm^{-1}$ for all peroxides not having an additional chromophore. Peroxoanions XOO^- absorb in a similar way but the extinction coefficient is a factor of ten larger.

A number of possible photolysis primary steps have been suggested [21], but the consensus now favors the simple homolytic scission of the oxygen-oxygen bond

$$XOOY \xrightarrow{h\nu} XO\cdot + YO\cdot \qquad (31)$$

to form two oxidizing radicals. The photolysis of HOOH has been studied a number of times [21, 22, 23]. In the case of HOOH, both radicals are of course HO•; in the case of $HOOSO_3^-$, the radicals formed are HO• and $SO_4^{-}\cdot$; etc. For organic peroxides, there are a number of side-reactions which occur after the homolysis step but are not relevant to this review. In dilute solution and at high light intensity, HOOH decomposes by the mechanism

$$HOOH \xrightarrow{h\nu} 2\,HO\cdot \qquad (32)$$

$$HO\cdot + HOOH \rightarrow H_2O + HOO\cdot \qquad (33)$$

$$2\,HOO\cdot \rightarrow O_2 + HOOH \qquad (34)$$

which is not a chain process. (A chain decomposition can be induced by light but is not pertinent to the present discussion.) The concentration of HO• can be controlled by variation of wavelength and of intensity. The generation of this radical has a quantum yield of about one.

The photolysis of HOOH in the presence of alcohols (ethanol, 2-propanol, etc.) gives ESR spectra indicating HO• attack at positions α and β to the hydroxyl group. The ratio of attack at the two sites depends somewhat on the medium (relative amounts of alcohol to water, presence of acid, etc.). Much of the more recent work has been carried out in dilute aqueous solution.

3.3. Radiolysis

When aqueous solutions are treated with very energetic particles (X-rays, γ-rays, electron beams), the water molecules are cleaved both homolytically and heterolytically. Three radical species (HO•, $e^-(aq)$, and H•) are formed in the particle track; from there they escape to the bulk solution and react with solutes. The relative yields (G values) for the above three radicals are 2.65, 2.65 and 0.60, respectively.

The hydrated electron can be converted to HO• by reaction 35 which has a rate

$$N_2O + e^-(aq) \xrightarrow{H_2O} N_2 + OH^- + HO\cdot \qquad (35)$$

constant of 5.6 x 10^9 $M^{-1}s^{-1}$ at pH 7 [25]. The amount of H• is less important (< 10% of radicals) so HO• dominates the reaction with organic solutes to form radicals R• which can then be identified and tracked by ESR [26].

The majority of rate constants of HO• with reductants has been measured using radiolysis combined with a fast analytical technique such as ESR or electronic spectra. Many compounds, particularly the alcohols, have been quantitatively studied many times with ever increasing precision and accuracy. For example, the oxidations of ethanol and 2-propanol have rate constants of 2.0 x 10^9 and 2.3 x 10^9 $M^{-1}s^{-1}$, with the ratio of α to β attack being close to 6:1 [27]. It is believed that two significant figures are appropriate for these two alcohol rate constants. Such assurance cannot be given for many of the rate constants in the literature.

4. Chemistry of Hydroxyl Radical

4.1. Premise

In a previous section, the similarity in rate constants with organic compounds for the intermediate formed from the Fenton reagent and that from radiolysis experiments was shown. Comparison of reactivities for HOOH and the different reducing metals with 2-propanol has been made successfully by Czapski et al [28]; both organic radicals $(CH_3)_2C{\bullet}(OH)$ and $CH_3CH(OH)CH_2{\bullet}$ are formed in similar patterns for the four metals tried. It seems safe to conclude then that many of the Fenton type systems generate HO• as the oxidizing intermediate. That being the case, this section will summarize the rates of oxidations by this radical, irrespective of source. Where helpful the product nature will also be given. It is necessary to remember that most of the data refer to water soluble molecules in aqueous solution.

4.2. Rate constant magnitudes

A 1977 compilation [14] of the rate constants for the reaction of HO• with over 600 organic compounds indicates that the vast majority of reactions have second order constants between 10^7 and 10^{10} $M^{-1}s^{-1}$. The half-dozen exceptions with low rate constants either have a strong electron attracting group (e.g., CH_3CN and $HCCl_3$) near the reacting site or are lacking in CH bonds (e.g., urea and oxalic acid).

It is appropriate to ask if the method used for obtaining the rate constant has an influence on the number obtained. Some data for the reaction of HO•

with 1-propanol bearing on this question are given in Table 3. The early work of Merz and Waters seems to have given low values by a factor of about three, but this is not seen in the later values. The rest of the results when suitably corrected do not show deviations attributable to constant determination method.

The reactivity of HO• is, as said, very high. Both hydrogen abstractions from C-H links and additions to unsaturated bonds have constants near 10^9 $M^{-1}s^{-1}$. The faster reactions approach collisional velocities. As a consequence there is little selectivity and this severely limits application of Fenton's reaction to fine synthesis. On the other hand, lack of selectivity along with high reactivity may be a useful attribute for an oxidant used in waste water treatment and for solving pollution problems.

4.3. Solvent trapping

The need to oxidize a large organic molecule can effectively preclude the use of water as a solvent. The choice of an alternative solvent can, therefore, be important in order to minimize solvent trapping of HO•.

Table 3: Constants for hydroxyl and 1-propanol reaction[a].

10^9 x constant ($M^{-1}s^{-1}$)	method	ref.
0.60	Fenton[b]	4
3.4	"	5
2.7	Pulse[c]	14
2.5	"	14
2.7	"	14
2.8	Gamma[d]	14
3.2	"	14
3.0	"	14

[a] All data in water at room temperature. [b] Fenton reagent. [c] Pulse radiolysis. [d] Gamma radiolysis.

In Table 4, twelve potential solvents are listed along with their second order constants *in water* . The most reactive one - DMSO - is known to be an effective trap for HO•. Benzene derivatives, alcohols, DMF and ethers are also effective. There is then a sharp drop in trapping efficiency to acetone, another drop to acetic acid, and still another to acetonitrile. This order of solvent character is borne out by the fact that radical oxidations by the peroxovanadate compound (pic)$VO(O_2)$, where pic is the picolinate anion, are quenched in CH_3OH and DMF but carried out successfully in CH_3CN.

Table 4: Solvent molecule rate constants[a].

molecule	10^9 x k ($M^{-1}s^{-1}$)
DMSO	~ 6; ~ 9
C_6H_6	4.4
$O(CH_2CH_2)_2O$	~ 4
n -C_3H_7OH	3.4
C_2H_5OH	2.1
THF	2.0
DMF	1.8
CH_3OH	1.2
t -C_4H_9OH	0.57
CH_3COCH_3	0.096
CH_3COOH	0.027
CH_3CN	~ 0.006

a All data in water at room temperature. Data compiled by present authors.

4.4. Alcohols and ethers

In Tables 5 and 6 and in two compilations [5, 14], a large number of alcohols are listed. It was found that the rate constant increases with chain length for the homologous *n* -alkanols, rising from 1.2 x 10^9 for methanol to 6.5 x 10^9 $M^{-1}s^{-1}$ for 1-octanol. Electronic effects can be extracted from the rate constants. Walling et al [5] found that each hydrogen on the methylene position in ethanol is more than a factor of ten more reactive than each hydrogen on the methyl group. In other alcohols, the ratio of α to β abstraction is smaller, although the order is maintained.

Methanol is three times as reactive as methane and methylamine is still more reactive, so potentially electron donating groups increase reactivity. As expected, electron attracting groups decrease reactivity; the reactivities of 2,2,2-trifluoroethanol and 2,2,2-trichloroethanol are far less reactive than methanol and ethanol. Acetonitrile and sulfolane, both with electronegative groups, are two of the few compounds with second order constants less than 10^7 $M^{-1}s^{-1}$.

Cupric ion by itself does not interact with either Fe^{2+} or HOOH at a convenient rate but it does react rapidly with some of the organic radicals generated by Fenton's reagent. The Cu^{2+} oxidizes the radicals, in some cases forming ROH from R•, and in others forming an olefin from a β-hydroxyradical. As may be seen in Table 6, it has no observed effect on the methanol oxidation, presumably because the Fe^{3+} present also oxidizes $•CH_2OH$. The Cu^{2+} has only a small effect on the yields from ethanol oxidation, but a definite shift is seen on the 2-propanol oxidation. A dramatic alteration of yields in the $(CH_3)_3COH$ oxidation occurs; the primary radical $(CH_3)_2C(OH)CH_2•$ dimerizes (84.4%) when iron is the only metal present but the glycol is formed predominantly (86.2%) when Cu^{2+} is also in the solution.

Table 5: Relative reactivities of alcohol hydrogens[a].

alcohol	per molecule k_α/k_2	per molecule k_β/k_2	per H k_α/k_2	per H k_β/k_2
CH_3OH	4.30	=	1.43	=
C_2H_5OH	7.32	0.69	3.66	0.23
$CH_3CH(OH)CH_3$	5.73	0.92	5.73	1.15
$CH_3CD(OH)CH_3$	3.42	0.91	3.42	0.15
$CD_3CH(OH)CD_3$	5.43	0.22	5.43	0.036
$(CH_3)_3COH$	=	1.90	=	0.21

a Data from [5a]; the numbers are reactivities relative to the reaction $Fe^{2+} + HO• \rightarrow Fe^{3+} + HO^-$ with constant k_2 taken as equal to $3 \times 10^8\ M^{-1}s^{-1}$.

Insertion of deuterium in place of hydrogen in the alcohol molecules, either in α or β positions, brings about changes in the relative reactivities (see Table 5) thereby indicating kinetic isotope effects related to breaking the C-H bond. For the alpha hydrogen in 2-propanol, k_H/k_D is 1.68; for the beta hydrogen, k_H/k_D is 4.22.

As would be expected from the alcohol results, the α position on aliphatic ethers reacts rapidly with HO• and the radical so formed is readily oxidized to hemiacetal. The rate constants for the limited amount of ether data are about $2 \times 10^9\ M^{-1}s^{-1}$ and show significant variation [14]. Approximately 13% β CH reaction was found for THF [5].

Table 6: Stoichiometries of alcohol oxidations[a].

alcohol	product	% yield Fe	Fe-Cu
CH_3OH	HCHO	81	80
	HCOOH	< 0.5	< 0.5
CH_3CH_2OH[b]	CH_3CHO	72	63
	CH_3COOH	20	22
	$HOCH_2CH_2OH$	=	1.8
	$(HOCH_2CH_2)_2$	1.1	0.13
$(CH_3)_2CHOH$[c]	CH_3COCH_3	93	84
	1,2-propanediol	=	11.2
	2.5-hexanediol	4.4	0.7
$(CH_3)_3COH$[d]	CH_3COCH_3	< 0.1	< 0.1
	$(CH_3)_2C(OH)CH_2OH$	=	86.2
	$[(CH_3)_2C(OH)CH_2]_2$	84.4	1.6

[a] Conditions: 1.25 M alcohol; 0.25 M HOOH; 0.1 M Fe^{2+}; 0 or 0.4 M Cu^{2+}; 30°C; data of reference [5]. [b] Absent: 2,3-dimethyl-2,3-butanediol; 2-methyl-2,4-pentanediol.
[c] Absent: $CH_3CH(OCH_2CH_3)_2$. [d] Absent: $(CH_3)_2CHCHO$.

4.5. Carbonyl compounds

Merz and Waters [4] found that acetone was not oxidized by the Fenton reagent even though it inhibited the oxidation of methanol. The explanation given [4, 5] is that acetone is attacked by HO•, but the resultant radical CH_3COCH_2• is reduced back to acetone by Fe^{2+}. From the competition ratio, the rate of HO• attack on acetone was calculated [5] to be lower than that of methanol by more than one power of ten.

Other carbonyl compounds that are reduced back to the starting compound by Fe^{2+} but compete with CH_3OH for HO• are acetic acid, malonic acid, and succinic acid.

Coupling of two CH_3COCH_2• radicals from acetone was found when order of addition of reagents was inverted (slow addition of Fe^{2+} to HOOH and acetone solution). By the same type of experiment, coupling of $NCCH_2$• radicals from acetonitrile was observed.

The amides DMF and dimethylacetamide DMA react at their N-methyl groups at rates comparable to alcohols [5].

4.6. Unsaturated compounds

Double bonds readily add HO• to form organic radicals. In the case of maleic acid, the step is (eq. 36):

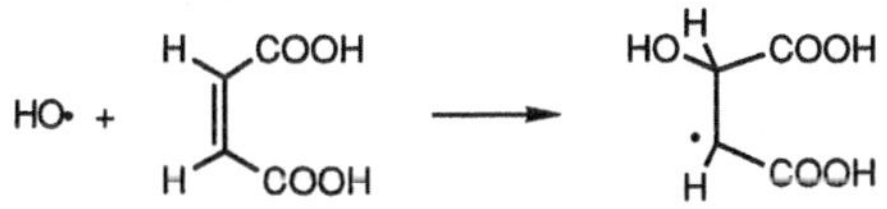

(36)

and the resultant radical is reduced by Fe^{2+} in the acid solution (eq. 37).

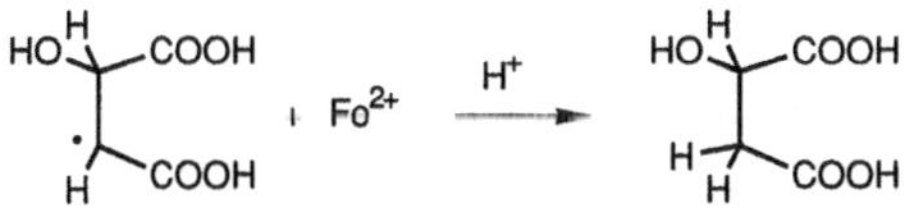

(37)

As is the case with carbonyl compounds, maleic acid inhibits the methanol oxidation thereby allowing estimation of the rate constant for HO• addition to the double bond. Fumaric, acrylic and crotonic acids act in similar fashion to maleic acid.

The nitromethane anion adds hydroxyl with the high rate constant of (8.5 ±1.5) x 10^9 $M^{-1}s^{-1}$ [14]. Additions to 1-butene, butadiene and allyl alcohol also occur rapidly with constants near 10^9 $M^{-1}s^{-1}$

$$HO\cdot + CH_2{=}NO_2^- \rightarrow HOCH_2NO_2^{-}\cdot \qquad (38)$$

4.7. Acetylenes

Kinetic and product studies on the reactions of acetylenes with HO• were carried out by Walling and El-Taliawi [5c]. The rapid addition of HO• to acetylene gives a vinyl radical which is reduced by Fe^{2+} and oxidized by Cu^{2+} (eq. 39, 40).

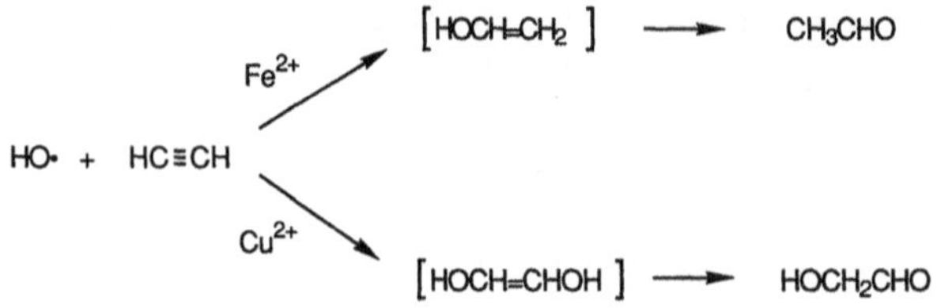

(39)

(40)

In all, six acetylenes were reacted with Fenton's reagent. In every case the vinyl radical was reduced. On the other hand, oxidation by Cu^{2+} was observed only when the vinyl radical had a small group (e.g., hydrogen) at the non-hydroxylated end. Evidence for two radicals was obtained with 3-hydroxy-3-methyl-1-butyne, an unsymmetrical acetylene.

4.8. Aromatic compounds

Delocalization of the double bonds in benzenes does not noticeably decrease the rate of addition of HO•; with benzene, k is 5×10^9 $M^{-1}s^{-1}$. Three intermediates, two with an unpaired electron, are formed (Scheme 2); A is a hydroxycyclohexadienyl radical and B is a radical cation. Electronic effects of substituents on benzene are consistent with HO• being a strong electrophilic partner (see Table 7). The correlation is marked by significant deviations, the most noticeable being that of toluene which is unexpectedly low.

An industrial process which allows homolytic hydroxylation to catechol and hydroquinone using HOOH in combination with iron and cobalt salts has been developed [30a]. Another relevant industrial application consists in the hydroxylation of *o* -hydroxy-azo and *o* -aminoazo compounds selectively in the ortho position by using HOOH and copper(II) acetate [30b,c].

The peroxovanadate complex (pic)$VO(O_2)$, where pic is the picolinato ligand, hydroxylates aromatic compounds. The kinetics are consistent with the intermediacy of a strongly oxidizing radical, possibly HO•. From a series of experiments based on competitions between pairs of substituted benzenes, relative reactivities (from yields) of the aromatics were calculated and a Hammett rho value of -1.6 obtained [30].

In accordance with the results of Walling et al. [5], in the reaction of HO• with phenylacetic acid, the -CH_2COOH substituted cyclohexadienyl radical formed initially may evolve to yield benzyl radical $PhCH_2$• upon loss of CO_2 and H_2O. The yields of products obtained depend on H^+, Fe^{2+}, Fe^{3+}, Cu^{2+} and O_2 (see Table 8). In the absence of metal oxidant, dibenzyl $PhCH_2$-CH_2Ph is the main product. Benzyl alcohol predominates when Fe^{3+} is added; with Cu^{2+}

the main products are phenols except when acetate ion is added and then it is benzyl alcohol.

Scheme 2:

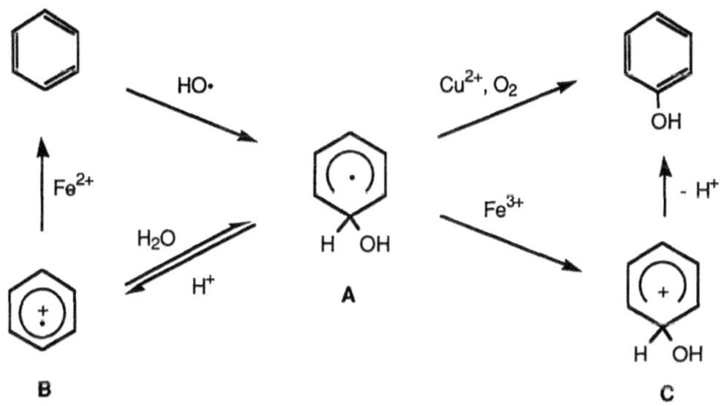

Table 7: Rate constants for reactions of HO• with substituted benzenes[a].

Substituent	σ_p	10^{-9} x k ($M^{-1}s^{-1}$)
$-NH_2$	-0.66	8.9
-OH	-0.46	~ 10
$-OCH_3$	-0.27	6.0
$-CH_3$	-0.17	3
-H	0.00	7.8
-Cl	+0.23	~ 5
-CN	+0.63	3.9[b]
$-NO_2$	+0.78	3.4

[a] Data compiled by authors. No corrections made. [b] The 1,4-dicyanobenzene with two electron-withdrawing groups has k = 7.2 x 10^8 $M^{-1}s^{-1}$ [14].

Table 8: Yields from phenylacetic acid oxidation[a].

		% Yield			% Phenol isomers		
$[Fe^{3+}]$	$[Cu^{2+}]$	BzOH	$(Bz)_2$	phenols	*o*	*m*	*p*
=	=	12	54	1	=	=	=
b	=	2	3	7	66	34	0
0.16	=	60	6	7	91	6	3
=	0.08	27	0	47	60	29	11
=	0.08[c]	4	0	69	45	31	24
=	0.16	17	0	50	58	31	11
=	0.16[d]	58	0	15	77	21	2

[a] 0.05 M $HClO_4$, 0.02 M Fe^{2+}, 0.06 M substrate under N_2 unless indicated. Yields based on HOOH added. Bz = benzyl. Ref. [5]. [b] O_2 saturated. [c] 0.25 M $HClO_4$ added. [d] 0.02 M NaOAc added.

4.9. Oxidative dimerization

The organic radicals formed either by hydrogen abstraction or by addition to multiple bonds can and do react further. As elegantly laid out by Walling et al [5], the main three steps followed by the radical are *i.* oxidation by Fe^{3+} and/or Cu^{2+} *ii.* reduction by Fe^{2+}, and *iii.* dimerization. The last, perhaps better termed oxidative dimerization, occurs most often when the radical has a hydrocarbon functionality. Methyl radical gives ethane. Another example is found in the t-butanol oxidation. The first step is β-abstraction (eq. 41)

$$HO\cdot + (CH_3)_3COH \rightarrow (CH_3)_2C(OH)CH_2\cdot + H_2O \qquad (41)$$

and the second (dimerization) step forms 2,5-dimethyl-2,5-dihydroxyhexane.

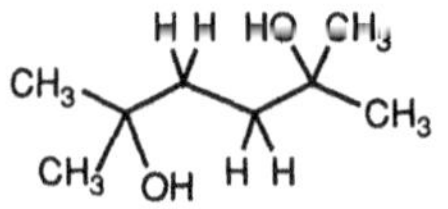

Oxidative dimerization reactions due to the Fenton's reagent have been observed for aliphatic ethers, esters, nitriles and carboxylic acids [31].

Coffmann et al [32] have worked out syntheses of some unusual compounds using radical steps initiated by Fenton's reagent. In many cases, the products obtained were derived from coupling of two organic radicals. Among the conversions accomplished were *i.* oxidative dimerizations of aliphatic carboxylic acids, nitriles, amines, amides, alcohols and ketones, *ii.* additive couplings of 1,3-dienes and monofunctional compounds, and *iii.* a coupling synthesis of carboxylic acids from monofunctional organic compounds with carbon monoxide. An example of the last is the preparation of β-hydroxyisovaleric acid from $(CH_3)_3COH$ and CO.

4.10. Dimethyl sulfoxide

The rate constant for reaction of HO• with DMSO is approximately 8×10^9 $M^{-1}s^{-1}$. Because of its high rate constant and its compatibility with biological tissue, DMSO is used as an HO• trapping agent in biological chemistry. An interesting aspect of its complicated reactivity is the unusual splitting out of an alkyl radical [5, 33] (eq.42).

$$HO\cdot + CH_3SOCH_3 \rightarrow CH_3SO_2H + \cdot CH_3 \qquad (42)$$

4.11. Organic radicals and metal ions

To a degree, the oft-considered possibility that Fenton's reagent did not form HO• arose from lack of agreement between rate constants based on yields and those based on relative rates. Product yields from the Fenton reaction can be misleading because a radical R• which came from the step

$$RH + HO\cdot \rightarrow R\cdot + H_2O$$

can be reduced to RH or to an olefin by Fe^{2+}. Also it can be oxidized by either Fe^{3+} and Cu^{2+} and bring a chain mechanism in play. On the other hand, relative rates from radiolysis or photolysis are less subject to major deviations but have errors in the range of factors of two or three. Only when proper account is taken of these complexities can an appropriate judgement as to the extent of agreement in rates be made.

The agreement between k values from the different experiments for acetone is now well within a factor of two whereas it was initially considered that acetone was not oxidized by Fenton's reagent; the radical $CH_3COCH_2\cdot$ is reduced back to acetone by Fe^{2+}.

Maleic acid presents another type of reaction that can mislead: the acid has no effect on the stoichiometry of the Fe^{2+} and HOOH reaction even though it is hydrated to form malic acid (see above).

Another question which came up involved the various reducing metals that reacted with HOOH. It was found by ESR that the apparent yields varied with nature of the metal. Czapski et al [28] investigated the relative rates for the two radicals $(CH_3)_2(OH)C\cdot$ and $CH_3CH(OH)CH_2\cdot$ formed from 2-propanol when the initiating metals were Fe^{2+}, Ti^{3+}, Cr^{2+} and V^{4+}. (We write the last as VO^{2+} as may be seen below.) The amounts of the two radicals as measured in the ESR cell were dependent on conditions. For example, organic radical formation was not complete when the sample was passed from mixing cell to ESR cell. Also the reaction rates for reduced metal ions with HOOH were not identical. Nevertheless, the pairs of ESR signals in the flow/ESR apparatus gave similar patterns of intensity against HOOH concentration for each metal. On the other hand, when plotted against added oxidizing metal (Fe^{3+}, Ti^{4+}, etc.) concentration, the lines for the two radicals were different. Disappearance of $(CH_3)_2(OH)C\cdot$ was dependent on metal nature in a manner consistent with this radical being oxidized at different rates. It was concluded that radical oxidation rate followed the order $Ti^{4+} < Cr^{3+} < Fe^{3+}$. The other radical showed an intensity that was relatively insensitive of the metal.

It is known that Cu^{2+} oxidizes organic radicals more generally (but not necessarily more rapidly) than does Fe^{3+} [5, 27]. Both rates and stoichiometry can be changed by the addition of Cu^{2+} [34].

Cations other than Fe^{2+} can also reduce organic radicals. In a study of ethanol oxidation by peroxodisulfate, it was found necessary to postulate a termination interaction between Cu(I) and $(CH_3)C\bullet(H)(OH)$ [35]. It now seems probable that the radical is reduced back to ethanol (eq 43)

$$Cu^{+} + CH_3\dot{C}HOH \longrightarrow CH_3CH_2OH \qquad (43)$$

whereas radicals of the type $\cdot CH_2CHR(OH)$ are reduced to olefins [27].

4.12. Inorganic reductants

Far fewer reaction rates of $HO\cdot$ with inorganic compounds and ions than with organic molecules have been reported [14]. Because of this and because of the absence of homologous series, generalizations about rates for the inorganic reductants are difficult to formulate. Also the variations among reported constants for some important reductants are so large that calculation of average rate values gives a misleading impression that the number is useful. For example the reported constants for the step

$$HO\cdot + Cl^- \rightarrow OH^- + Cl \qquad (44)$$

range from < 10^6 to 2×10^{10} $M^{-1}s^{-1}$.

Anions, both monatomic (Br^-, I^-) and polyatomic (SCN^-, N_3^-, CO_3^{2-}, etc.). tend to react rapidly; some have rate constants that suggest diffusion control of rate. An example is I^- for which the constant of $(1.21 \pm 0.08) \times 10^{10}$ $M^{-1}s^{-1}$ has recently been reported [36]. Placement of hydrogen ion on the anion to form the conjugate acid decreases the constant, suggesting that electron transfer from the anion to HO• is an important factor in determining rates.

Reducing cations including those that react with HOOH in Fenton type systems react rapidly with HO•; constants for these are given in the next section. On the other hand, cations of the alkali metals, alkaline earth metals, and many others do not react with HO•. Even Zn^{2+} and Cd^{2+} are reported to react very slowly if at all ($k < 5 \times 10^5$ $M^{-1}s^{-1}$).

4.13. Reaction with radicals

With other radicals, HO• reacts rapidly. Constants for reactions with $O_2^{-}\cdot$, HOO•, H•, and $e^-(aq)$ are about 1×10^{10}, 7×10^9, 2×10^{10} and 3×10^{10} $M^{-1}s^{-1}$, respectively. The rate constant for formation of HOOH from two HO• is 5×10^9 $M^{-1}s^{-1}$.

4.14. Radical interconversions

Many peroxides react with HO•, often with apparent hydrogen atom transfer as in the step

$$HO\cdot + HOOX \rightarrow H_2O + XOO\cdot \qquad (45)$$

Peroxomonosulfate ion $HOOSO_3^-$, HOOH and HOO^- are converted to $\cdot OOSO_3^-$, HOO• and $O_2^{-}\cdot$ with rate constants 3.5×10^8, 4×10^7 and 4×10^9 $M^{-1}s^{-1}$.

Peroxodisulfate ion $S_2O_8^{2-}$ does not seem to react with HO•.

One radical interconversion, not involving HO• but the similar oxidant $SO_4^{-}\bullet$, provokes an interesting question. The presumed hydrogen atom transfer

$$SO_4^{-}\bullet + HOOSO_3^{-} \rightarrow HSO_4^{-} + \bullet OOSO_3^{-} \qquad (46)$$

would be expected to be exothermic and rapid. In fact, to date this step has not been found; the reported rate constant for the step is $< 10^5\ M^{-1}s^{-1}$. Work at Brown University on peroxomonosulfate reactions produced no evidence that this step occurred in any of reactions studied.

4.15. Vinyl polymerizations

The HO• formed in the reaction of Fe^{2+} and HOOH initiates vinyl polymerization (eq. 47, 48).

(47)

(48)

etc.

This initiation [37] can also be carried out with other peroxides including $S_2O_8^{2-}$ and organic hydroperoxides [38].

5. Chemical generation of oxidizing intermediates

5.1. Metal ion activities

A number of metal ions M^{n+} are one electron reductants. These react with HOOH and other peroxides to generate active oxidants capable of converting organic compounds to radicals. As seen in the previous sections, the oxidizing intermediate generated by Fenton's reagent is HO•, but this is not invariably true for other reductants.

In this section the formation of oxidizing intermediates by other metal ions is described. The intent here is to give the reader an awareness of these methods of generation, of their differences, and to provide leading references.

The problems that beset identification of HO• in Fenton's reagent appear for some other reductants and are unresolved.

In Table 9 some data on properties of metal reductants are given (electrode potentials were given in Table I). The values of $k(H_2O)$ refer to the first order exchange constant between aqua ligands in the coordination sphere and bulk solvent; this gives a relative

Table 9: Rate properties of reducing cations[a].

Cation	n[b]	$k(H_2O)$[c]	k(HOOH)[d]	k(HO•)[e]
Ti^{3+}	1	1 x 10^5	~ 10^3	1.2 x 10^9
VO^{2+}	1	6 x 10^2	~ 80	2.5 x 10^8
Cr^{2+}	4	large	~ 2 x 10^4	4.8 x 10^9
Mn^{2+}	5	3 x 10^7	slow	=
Fe^{2+}	6	3 x 10^6	60	2.3 x 10^8
Fe^{3+}	5	8 x 10^3	slow	8 x 10^7
Co^{2+}	7	1 x 10^6	slow	~ 2 x 10^6
Cu^{+}	10	large	4 x 10^3	=
Cu^{2+}	9	large	slow	3 x 10^8
Ag^{+}	10	large	slow	1.5 x 10^{10}
Eu^{2+}	7[f]	=	=	9 x 10^8
Ce^{3+}	1[f]	=	slow	3 x 10^8

[a] Data at room temperature. Potentials in Table 1. [b] Number of electrons in d orbitals except as noted. [c] Data primarily from ref. 39.Units are s^{-1}. [d] Data primarily from ref. 38. Units are $M^{-1}s^{-1}$ [e] Data from ref. 14 plus compilation by authors. Units are $M^{-1}s^{-1}$ [f] In f orbitals.

measure of the capability of the metal ion to undergo inner sphere replacement. Oxidation of metal ions by HOOH, as indicated by k(HOOH), shows how fast the aqueous metal ion is oxidized. This is related to the initiation step and it is believed that the process involves an inner sphere oxidation. The values of k(HOOH) are so much slower than the exchange constants that they reflect the need of energy input required for breakage of a covalent bond (i.e., HO-OH). As expected the reducing metal ions also react with HO•; the constants k(HO•) are large and it is probable that entry into the coordination sphere is not required. For all of these metal ions, one electron oxidation is facile; for some (Cr^{2+}, Fe^{2+}, Mn^{2+}, Cu^{+}, Co^{2+}, Ag^{+}), two electron oxidation also is possible.

It is important to reiterate that the metal ions may also act as reductants toward the organic radicals and that the higher oxidation state(s) may act as organic radical oxidants, and this complicated identification of the oxidizing intermediate.

Also in this section two unusual chemical methods for generating HO• are mentioned; both reappear in biological systems (Section 6).

5.2. Aqueous iron(II)

The ferrous ion in water is present as an octahedral aqua complex with six d electrons in high spin configuration. It is oxidized conveniently by HOOH in the HO• generation step. Its large rate constant with HO• is important because many substrates have their HO• rates compared with those of Fe^{2+} in order to evaluate the substrate and HO• constant. Farhataziz and Ross [14] used the value 2.3×10^8 and Walling et al [5] used 3×10^8 $M^{-1}s^{-1}$ for Fe^{2+} in their conversions of relative rates to constants for other substrates.

5.3. Ligated iron(II)

Ferrous ion forms a variety of complexes, ranging from those of simple anions (mentioned above) to large ligands with high dentation. The ease of oxidation of iron(II) depends on the type and number of ligands; ligand effects on the iron electrode potential are given in Table 10. Anionic ligands favor the oxidation of iron(II) by HOOH (at least thermodynamically) whereas pi-withdrawing ligands such as phenanthroline (phen) favor maintenance of the lower oxidation state.

Ligands inhibit the metal induced reactions of peroxide by means other than oxidation state stabilization. A multidentate ligand can block the peroxide reaction by occupying the metal binding sites; replacement of a water ligand by HOOH to form an inner sphere complex is considered important to radical generation. Also large organic chelate ligands such as EDTA (either bound or free) can act as HO• traps thus preventing either a substrate oxidation or a peroxide decomposition [40].

Table 10: Electrode potentials for iron couples[a].

Couple			E^0 (volts)
$Fe(OH)_{3(s)}$ + e^-	→	$Fe(OH)_{2(s)}$ + OH^-	- 0,56
$Fe(EDTA)^-$ + e^-	→	$Fe(EDTA)^{2-}$	-0.12
$Fe(C_2O_4)_3^{3-}$ + e^-	→	$Fe(C_2O_4)_2^{2-}$ + $C_2O_4^{2-}$	+0.02
$Fe(DTPA)^{2-}$ + e^-	→	$Fe(DTPA)^{3-}$	+0.17
$Fe(CN)_6^{3-}$ + e^-	→	$Fe(CN)_6^{4-}$	+0.36
$Fe(CN)_5(NH_3)^{2-}$ + e^-	→	$Fe(CN)_5(NH_3)^{3-}$	+0.37
$Fe(CN)_5(H_2O)^{2-}$ + e^-	→	$Fe(CN)_5(H_2O)^{3-}$	+0.49
$Fe(PO_4)_2^{3-}$ + e^-	→	$Fe(PO_4)^-$ + PO_4^{3-}	+0.61
$Fe(H_2O)_6^{3+}$ + e^-	→	$Fe(H_2O)_6^{2+}$	+0.77
$Fe(bipy)_3^{3+}$ + e^-	→	$Fe(bipy)_3^{2+}$	+1.10
$Fe(phen)_3^{3+}$ + e^-	→	$Fe(phen)_3^{2+}$	+1.14

a Data compiled by the authors.

Multidentate ligands can also play an entirely different role. Ferric ion, a catalyst for HOOH decomposition, forms a very insoluble hydroxide at the pH of cell fluid (7.4). The human body would find in the absence of strong ligands only a limited amount of available iron. The ligands hold the iron in homogeneous solution ready for transfer when needed. Iron(II) can also be held in solution at this pH by ligands such as EDTA, NTA and ATP, thus making it possible to have Fenton type reactions at physiological pH. An interesting discussion of catalytic activity for ferric-centered complexes over a wide range of pH values is available [15].

Depending on the conditions, EDTA can decrease the rate of Fe(III) decomposition of HOOH or it can form a catalytic complex [41, 42]. The decomposition of HOOH by $Fe(EDTA)^-$ does appear to proceed by a radical path similar to that by uncomplexed Fe^{3+}.

Rahhal and Richter [43] carried out a careful study of the reduction of HOOH by the DTPA (diethylenetriamine-N,N,N',N",N"-pentaacetate) complex of iron(II). The reaction is first order each in HOOH and $Fe(DTPA)^{3-}$. An oxidizing species not scavenged by *t* -BuOH is formed, and MeOH had essentially no effect on the reaction when HOOH was in excess. The authors concluded that an oxidant other than HO• was formed; this was postulated to be the iron-oxo species $Fe^{IV}(DTPA)(OH)^{2-}$.

5.4. Copper(I)

It appears probable that active oxidants other than (or along with) HO· are formed when Cu^{+}(aq) or $Cu(phen)_2^{+}$ react with HOOH [8, 27]. Johnson et al [8], on the basis of kinetics and spin trapping experiments, concluded that the active oxidant could not be HO·. The observed rate of formation of HCHO from MeOH by reaction of Cu^{+}(aq) and HOOH differed from the estimated rate using known rate constants.

Masarwa et al [27] came to the same conclusion. The ratio [$\cdot CH_2CH_2OH$] / [$\cdot CH_2CH(CH_3)OH$] formed from ethanol and 2-propanol in equal concentrations was not the same for the oxidant from Cu^{+}(aq) and HOOH as for the oxidant (presumably HO·) formed by radiation with copper present but HOOH absent.

There is no conclusive evidence as to the nature of the active oxidant formed from Cu^{+}(aq) and HOOH. A copper(I) peroxide $Cu^{+}\cdot OOH^{-}$ could be formed, and this could break down to form Cu(III)(aq). It is known that the equilibrium 49 is established in acid but not in neutral solution [25].

$$Cu^{III}(aq) \rightleftharpoons Cu^{II}(aq) + H^{+} + HO\cdot \qquad (49)$$

Similar difficulties in lack of agreement between observed rates and calculated rates (assuming HO· as intermediate) when $Cu(phen)_2^{+}$ and HOOH reacted [8].

Further details on the oxidant from copper(I) are to be found in Section VII.

The aqueous chemistry of copper is complicated. The aqueous copper(I) ion Cu^{+}(aq) tends to disproportionate so must be worked with in dilute solution. The +2 state is not a powerful oxidant but it reacts with organic radicals to form complexes which break down to oxidized products formed by ligand transfer from copper to the incipient carbonium ion concomitant with radical oxidation [44].

Copper(II) provides a useful test for perhydroxyl radical (HOO·). The metal ion oxidizes this radical [3, 22a] (eq.50) and this is followed by the oxidation of Cu^{+} by HOOH, which is followed in turn by reaction of HOOH with the oxidizing intermediate to regenerate HOO·. The HOO·, often a chain terminator, is converted to a chain propagator. The net effect of Cu^{2+} is to increase the chain length and therefore the overall rate.

$$Cu^{2+} + HOO\cdot \rightarrow Cu^{+} + H^{+} + O_2 \qquad (50)$$

5.5. Vanadium

Both +4 and +5 oxidation states of vanadium in the presence of HOOH appear to form HO•. In its aqueous acid form, vanadium(IV) is a cation having one oxide oxygen and is written here as VO^{2+}(aq). From studies with bidentate ligands such as acetylacetonate (acac$^-$) it is believed to have a tetragonal pyramid structure with oxide oxygen at the apex. The occasional presence of a weakly attached sixth ligand trans to oxide has been reported.

The reaction of VO^{2+} with HOOH has been studied by Brooks and Sicilio [45] and by Czapski et al [21] using flow-ESR techniques. It was found that both vanadium oxidation and dioxygen evolution occur. The reaction scheme proposed [45] is:

$$VO^{2+} + HOOH \underset{k_{1r}}{\overset{k_{1f}}{\rightleftharpoons}} VO(OOH)^{+} + H^{+} \qquad (50)$$

$$O{=}V(OOH)^{+} \xrightarrow{k_2} V(O)_2^{+} + HO\cdot \qquad (51)$$

$$V(O)_2^{+} + HOOH \rightleftharpoons VO(O_2)^{+} + H_2O \qquad (52)$$

$$VO^{2+} + HO\cdot \longrightarrow V(O)_2^{+} + H^{+} \qquad (53)$$

$$VO(O_2)^{+} + HO\cdot \longrightarrow I + \text{products} \qquad (54)$$

The peroxide complex $VO(O_2)^+$ is discussed below. The semi-stable product I is considered to be a superoxocomplex of vanadium(V) for which there is good ESR evidence [46]. The combined constant $k_{1f} k_2 / k_{1r}$ has a value of 0.21, and k_{1f} is 52 $M^{-1}s^{-1}$. On the basis of the fact that $VO(EDTA)^{2-}$ is oxidized more slowly than either VO^{2+} or $VO(IDA)^-$ where IDA is imido-diacetate, it was concluded that the oxidation involved an inner sphere intermediate.

In at least four reactions of vanadium(IV) with peroxides, evidence for complexes involving the two reactants formed prior to the redox step was reported [45, 47].

Reaction of VO^{2+} with HOOH was found to convert 2-propanol to the two radicals $(CH_3)_2(OH)C\cdot$ and $CH_3CHOHCH_2\cdot$ just as do Fe^{2+}, Ti^{3+} and Cr^{3+} [21]. It was noted that some VO^{2+} remained when the reaction mixture reached the ESR cell. This may result from reduction of $V(O)_2^+$ by HOOH, a reaction known to occur (see below). At pH 2.5, the second order constant for HO• formation is ~ 80 $M^{-1}s^{-1}$ according to Czapski et al [21].

In aqueous acid, vanadium(V) is a monovalent cation for which the composition has been written as either $V(O)_2^+$ or $V(OH)_4^+$; the former is more often used. This cation forms two complexes with HOOH. At moderate acidity ($[H^+] \geq 1$ M), a red monoperoxovanadate is produced by the equilibrium 52 which has a constant of 3×10^4 M^{-1} [48]. At lower acid concentrations and higher peroxide, a second peroxide is added to produce a yellow anion by the acid-dependent equilibrium 55 with a constant of ~ 1.

$$VO(O_2)^+ + HOOH \rightleftharpoons VO(O_2)_2^- + 2H^+ \qquad (55)$$

The monoperoxovanadate(V) cation decomposes slowly in aqueous solution to release dioxygen. If excess HOOH is present, that excess is used up prior to any intensity decrease in the $VO(O_2)^+$ color. The rate of decomposition, both with and without excess peroxide, shows sigmoid character suggestive of an intermediate building up in concentration prior to the regular reaction. A similar decomposition pattern in seen for the (pic)$VO(O_2)$ complex in acetonitrile and this peroxovanadate complex is known to hydroxylate benzene as well as to autodecompose to form O_2 [49]. These two reactions (autodecomposition and benzene oxidation) have kinetics consistent with (pic)$VO(O_2)$ and C_6H_6 competing for an intermediate. The intermediate may be HO•. The Hammett ρ value for substituted benzenes is -1.6 [49].

It is worth noting that in vanadium-peroxide systems, there is ESR evidence for regeneration of tetravalent vanadium even in the presence of excess HOOH [49].

5.6. Titanium(III)

Norman and his coworkers [26, 50] have advanced the use of the titanium(III) with HOOH reaction because of its rapid generation of HO•. The cation $Ti(H_2O)_6^{3+}$ has a d^1 electron configuration and a slightly distorted molecular structure. A minor complication that arises with this system is the formation of the monoperoxotitanium(IV) complex $Ti(O_2)^{2+}$. Many of the ideas that laid the foundation of our present understanding of the kinetics of Fenton's reaction arose in or because of the work of University of York group.

5.7. Chromium

The strong reductant $Cr^{2+}(aq)$ is a d^4 high spin complex with tetragonal distortion and extremely fast replacement reactions. Removal of the single antibonding electron is strongly favored by production of the d^3 chromium(III) state having high ligand field stabilization energy. The fast reaction rates for chromium(II) reactions are seen in Table IX; the rate constant given for reaction with HOOH is based on a radical concentration estimate rather than an experimental rate, but it is unquestionably large [21].

Recent data show that the reaction of HOOH with $Cr^{2+}(aq)$ forms HO• [43b].

The chromate ion is known to have carcinogenic properties; these can be traced to chromium(V) and presumably from this to HO•. This is discussed in the section on Fenton type reactions in biochemistry (see below).

5.8. Cobalt(II)

This cation is not generally considered to be an initiator of Fenton type reactions because its reaction with HOOH in acid solution is very slow. Nevertheless it has been found to generate HO• under physiological conditions (pH = 7.4 at 37°C.) [51].

Cobalt(II) is oxidized by peroxomonosulfate with a rate increasing with pH [52, 53] and the radical SO_4^{-}• is postulated. Therefore it seems probable that generation of HO• by the two steps 56, 57 can occur. It is to be noted that the strong oxidant cobalt(III) is

$$Co^{2+} + H_2O \rightleftharpoons Co(OH)^+ + H^+ \quad (56)$$

$$Co(OH)^+ + HOOH \longrightarrow CoO^+ + H_2O + HO\cdot \quad (57)$$

also formed when $HOOSO_3^-$ is the peroxide and is postulated in these steps with HOOH•. Cobalt(III) is known to oxidize HOOH. Further data on cobalt may be found in the section of biochemistry.

5.9. Other metal reductants

In Table 9, some metals not detailed are listed. These should be able to react with HOOH to form an oxidizing intermediate, but the amount known is not enough to warrant disucssion here.

5.10. Peroxonitrous acid

The ease of HO• formation directly from a peroxide of the class HOOX increases significantly with stability of the other radical XO• produced on homolysis

$$HOOX \rightarrow HO\bullet + XO\bullet \qquad (58)$$

because the bond dissociation energy is decreased. For peroxonitrous acid, the other radical is $\bullet NO_2$ which is known to be stable and poorly reactive. The homolysis of HOONO proceeds with a half-time of a few seconds at room temperature in the pH range from 2 to 5 [54]. That radicals are generated is confirmed by the ready oxidation of aromatics such as nitrobenzene [55]. The yield of HO• has been estimated as ~ 32% on the basis of HO• trapping by HOOH and by NO_2^- [56].

One can make solutions (pH 12) of the conjugate base $ONOO^-$ which are stable for hours, so a degree of control of mixing conditions is feasible. Satisfactory preparation methods are available.

At Brown we have found that a stabilized solution when introduced into a pH 7.4 solution of bovine serum albumin brings about the same degradation as does HO• produced by γ-ray radiation [57]. Further details concerning the biochemistry are given in Section VII.

6. Activation of Three Inorganic Peroxides

6.1. Extended Fenton types

There are three anionic peroxides with non-reducing groups attached to the peroxide oxygens. These three can undergo Fenton type activation yet each maintains a bit of individuality. With peroxodisulfate ion $S_2O_8^{2-}$, this individuality is facile formation of radicals by thermal homolysis; with peroxodiphosphates $H_nP_2O_8^{-4+n}$, it is the presence of two protonation steps in convenient pH ranges; with peroxomonosulfate $HOOSO_3^-$, it is the asymmetry of substitution at the peroxide oxygens with the consequent possibility of two different oxidizing radicals [9].

6.2. Peroxodisulfate

This is a long known and much studied peroxoanion [58-60], yet significant questions about chemical details remain. That sulfate radical-ion $SO_4^{-}\cdot$ is formed on photolysis and on reaction with reducing metals is not in question, nor is the conclusion that in many ways the reactions of $SO_4^{-}\cdot$ are akin to those of HO•. One of these ways is the initiation of olefin polymerization.

The anion undergoes thermal homolysis at convenient temperatures [61], however there is not complete agreement on the products of the first step in aqueous solution. Most chemists write this step as eq. 59 although a step of the type shown in eq. 60 has been proposed [62] and can not be ruled out. (It is certain that some $SO_4^{-}\cdot$ is formed in the first step because a product polymer whose olefin polymerization had been initiated by labeled $S_2O_8^{2-}$ contained a small amount of covalently-bound isotope label.)

$$S_2O_8^{2-} \rightarrow 2\,SO_4^{-}\cdot \qquad (59)$$

$$S_2O_8^{2-} + H_2O \rightarrow SO_4^{-}\cdot + HSO_4^{-} + HO\cdot \qquad (60)$$

The decomposition (net oxidation of water)

$$2\,S_2O_8^{2-} + 2\,H_2O \rightarrow O_2 + 4\,HSO_4^{-} \qquad (61)$$

has a rate constant at 60°C of 5.2×10^{-6} s^{-1}, and the activation parameters $\Delta H^{\ddagger}$ and $\Delta H^{\ddagger}$ are 133 kJ $mol^{-1}$and +47 J mol^{-1} K^{-1}, respectively [61].

Activation by Fe^{2+} is firmly established. The second order constant for the step

$$Fe^{2+} + S_2O_8^{2-} \rightarrow Fe^{3+} + SO_4^{2-} + SO_4^{-}\cdot \qquad (62)$$

is 141 $M^{-1}s^{-1}$ [63], and the photolysis is initiated presumably by simple homolysis with a quantum yield of ~ 0.6 [58, 64]. This radical can also be formed by pulse radiolysis [65].

The rate constants for reaction of $SO_4^{-}\cdot$ with both organic and inorganic substrates have been tabulated [66]. With organic compounds the constants for $SO_4^{-}\cdot$ are lower than those for HO•, often by two powers of ten.

Reactions with alcohols and ethers provide a test of the hypothesis that $SO_4^{-}\cdot$ reacts only by electron transfer. In Table 11, rate constants for protium and deuterium substrates are given. The k_H/k_D values range from 1.7 to 2.7, and this can be accepted as evidence for C-H bond breaking in the step where $SO_4^{-}\cdot$

and substrate react. These kinetic isotope effects are slightly larger than those reported by Burchill et al. [67] from careful studies on reaction of alcohols with HO•; these range from 1.24 to 2.28 and are expected to be lower because of the higher reactivity of HO• than of SO_4^{-}•.

Table 11: Constants for SO_4^{-}• with alcohols and ethers[a].

Substrate	k_H	k_D	k_H/k_D	k_α[b]
methanol	3.2×10^6	1.2×10^6	2.7	1.1×10^6
ethanol	1.6×10^7	6.7×10^6	2.4	7.9×10^6
2-propanol	3.2×10^7	1.2×10^7	2.7	3.2×10^7
dioxane	1.6×10^7	9.2×10^6	1.7	=
THF	1.0×10^8	5.1×10^7	2.0	=

[a] Data from ref. 65a. Units of k are $M^{-1}s^{-1}$. [b] Compare the ratios of change going from methanol to ethanol to 2-propanol with those for reactions of the same alcohols with HO• given in Table 5.

The study of Gilbert et al [65c] of the reaction of SO_4^{-}• with aliphatic carboxylic acids and their conjugate bases showed clearly that hydrogen abstraction and electron transfer compete. The order of SO_4^{-}• reactivity at different substrate sites is CO_2^{-} > β-H, γ-H > CO_2H > α-H. Thus, with propionic acid the dominant step would be (eq. 63):

$$SO_4^{-}\bullet + CH_3CH_2CO_2H \rightarrow HSO_4^{-} + \bullet CH_2CH_2CO_2H \qquad (63)$$

while with propionate ion it would be (eq. 64):

$$SO_4^{-}\bullet + CH_3CH_2CO_2^{-} \rightarrow SO_4^{2-} + CH_3CH_2CO_2\bullet \qquad (64)$$

to be followed by decarboxylation. Other organic oxidations by SO_4^{-}• are reported [65d,e].

It does appear correct that SO_4^{-}• is a strong electron transfer oxidant of both aromatic and aliphatic substrates. In attack at the α position of alcohols, the less reactive SO_4^{-}• is certainly more selective than HO• (see Table 11) [65b, 68-71].

Radiolysis of $S_2O_8^{2-}$ does produce the expected radical [72]; the results will be compared with those for radiolysis of peroxodiphosphates (see below).

In one reaction, $S_2O_8^{2-}$ is very different from HOOH; this is with Ag^+. The oxidation of Ag^+ proceeds by a two step mechanism

$$S_2O_8^{2-} + Ag^+ \rightarrow Ag^{2+} + SO_4^{2-} + SO_4^{-\bullet} \quad (65)$$

$$SO_4^{-\bullet} + Ag^+ \rightarrow Ag^{2+} + SO_4^{2-} \quad (66)$$

with eq. 65 being rate determining. Many oxidations by $S_2O_8^{2-}$ are carried out with Ag^+ present and the rate law with some substrates (Mn^{2+}, VO^{2+}, N_2H_4, etc.) is

$$\text{Rate} = k\,[S_2O_8^{2-}]\,[Ag^+].$$

The rate constant is essentially independent of the substrate [58]. Data with organic reductants suggest that Ag^{2+} is even more able than $SO_4^{-\bullet}$ to bring about electron transfer oxidations [68-70].

There is a most interesting difference in oxidation mechanisms for peroxodisulfate and peroxodiphosphate with tightly-chelated reducing metal ions; this is discussed below.

6.3. Peroxodiphosphates

The anion $P_2O_8^{4-}$ is isoelectronic with $S_2O_8^{2-}$, but there are fundamental differences in chemistry, Crutchfield [73] was the first to carry out experiments on proton equilibria, metal ion complexations, and hydrolysis rates for peroxodiphosphates. In recent years there have been a number of studies of the redox reactions [74] in spite of the difficulty of making and purifying this disubstituted peroxide.

The first two pK_a values of $H_4P_2O_8$ are less than 1.0; the second two are 5.2 and 7.7. From these it can be seen that pH can have a marked influence on peroxodiphosphate kinetics. Another new aspect is the ability of this peroxide to act as a chelate ligand even to metal ions such as Na^+ and Mg^{2+} [72]. In the peroxodiphosphate oxidation of VO^{2+}, evidence was obtained for a complex between the reactants [75a]. It is reasonable to suppose that electron transfer from vanadium to peroxide oxygen is facilited by inner sphere complexation. This supposition is well supported by a comparison of the reactions of $S_2O_8^{2-}$ and peroxodiphosphate with chelated metal ions. The propensity of the phosphate peroxide to form metal complexes was mentioned above. By way of comparison, the sulfate peroxide (to the best of our knowledge) does not do so.

In agreement with the argument, $S_2O_8^{2-}$ oxidizes VO^{2+} much more slowly [75b] than does the phosphate peroxide.

Towards inert complexes such as $Fe(phen)_3^{2+}$, the behavior of the two peroxides again differs. Towards the majority of inert complexes, $S_2O_8^{2-}$ acts as an outer sphere oxidant (an electron transfer oxidant). The second order rate is often faster than replacement in the metal coordination sphere. The rate step is presumed to be:

$$S_2O_8^{2-} + Fe(phen)_3^{2+} \rightarrow Fe(phen)_3^{3+} + SO_4^{2-} + SO_4^{\cdot-} \qquad \textbf{(67)}$$

Peroxodiphosphate reacts by an inner sphere mechanism, even for extremely inert complexes like $Os(terpy)(bipy)Cl^{+}$; this complex loses the Cl^- ligand and is also oxidized with the low same rate constant of $4.0 \times 10^6\ s^{-1}$ at 60°C. In this type of peroxodiphos-phate redox reaction, the rate is first order only in complex indicating that formation of an open position in the metal coordination sphere is the slow step. Monodentate attachment of peroxodiphosphate to metal and subsequent electron transfer proceed rapidly thereafter. The fact that an oxidizing radical is formed in these redox processes was confirmed by 2-propanol oxidation and by olefin polymerization which do not proceed with peroxodiphosphate alone.

Silver catalyzed oxidations by peroxodiphosphates proceed by the same type of mechanism as silver catalyzed oxidations by $S_2O_8^{2-}$.

The photolysis of $H_nP_2O_8^{-4+n}$ at 254 nm proceeds normally to give phosphate radicals which oxidize water. The quantum yield for peroxide loss is 0.80 ± 0.02 over the pH range from 3 to 13 even though protonation state changes [77]. On the other hand, thermal homolysis is very slow (rate is less than 1% of that for $S_2O_8^{2-}$).

Photochemical oxidation of ethanol and 2-propanol by peroxodiphosphate is observed but the chain lengths are short and pH dependent, in marked contrast to those for $S_2O_8^{2-}$ [77, 78]. This difference has been traced to the step

$$RC\cdot(H)(OH) + X_2O_8 \rightarrow O_3XOCH(R)OH + XO_4 \qquad \textbf{(68)}$$

which is over 100 more rapid for X = S than for X = P [79].

Pulse radiolysis and γ-ray radiolysis of peroxodiphosphate and peroxodisulfate ions showed similarities and differences akin to those seen in photolysis of the two peroxoanions. Radiolysis results for the two peroxides in the presence of alcohols also mirror the trends from photolysis [80]. In none of the comparative experiments was found any evidence that inorganic phosphate radical attack on alcohol was unusual. The short chain length for alcohol reaction with $H_nP_2O_8^{-4+n}$ is a consequence of the inefficiency of the step in which organic radical attacks the peroxodiphosphate. Therefore, chain termination must be the combination of two $(CH_3)(R)(OH)C\cdot$ to form nonradical products.

6.4. Peroxomonosulfate

The neutral acid H_2SO_5 is a strong acid ($pK_a1 << O$), but the second proton (on peroxide oxygen) is weak ($pK_a2 = 9.4$). Thus the monovalent anion is the predominant species over much of the convenient pH range. The anion does not undergo thermal homolysis, but there are a number of decomposition paths; some of these are metal catalyzed [81].

Most of the oxidation reactions of $HOOSO_3^-$ involve nucleophilic attack at the outer peroxide oxygen, but one electron reductants do react. Because of the asymmetry of attachment to peroxide oxygens, a reduction can occur in two ways:

$$HOOSO_3^- + e^- \rightarrow SO_4^{2-} + HO\cdot \qquad (69)$$

$$HOOSO_3^- + e^- \rightarrow OH^- + SO_4^{-}\cdot \qquad (70)$$

Gilbert and his coworkers [9, 82] looked at the reaction with three metal reductants; with Cu(I), $HO\cdot$ is the predominant radical, whereas with Fe(II) and Ti(III), the radical formed is $SO_4^-\cdot$. The complexes $Fe(EDTA)^{2-}$, $Fe(NTA)^-$, $Fe(DTPA)^{3-}$, $Ti(EDTA)^-$ and Ti(NTA) all gave $SO_4^-\cdot$ as the radical formed. It is worthy of note that monoperoxo-phthalate (also an unsymmetrical peroxide) gives different radicals with Cu^+ ($HO\cdot$ again) than with Fe^{2+} and Ti^{3+} ($ArCO_2\cdot$).

The photolysis of $HOOSO_3^-$ [81, 83] is first order in peroxide over a wide range of pH. The quantum yield increases from ~ 0.6 to 1.17 as the peroxide concentration increases. The reaction proceeds by a short chain (~ 4) and some $S_2O_8^{2-}$ is initially produced then photodecomposed in turn. Dioxygen is presumed to be the product; the source of the oxygen atoms has not been ascertained.

Radiolysis of $HOOSO_3^-$ does not give one radical; rather both $SO_4^{-}\cdot$ and $HO\cdot$ are produced in the ratio 1:4. The $G(HOOSO_3^-)$ value is large enough to indicate the presence of a chain mechanism [84].

The reduction of $HOOSO_3^-$ by VO^{2+} is first order in each reactant with a rate constant of ~ 13 $M^{-1}s^{-1}$. Trapping experiments showed the presence of a radical, presumably $SO_4^{-}\cdot$ [85].

A number of metals (Co^{2+}, VO^{2+}, Ce(IV), Ag^{2+}) interact with $HOOSO_3^-$ to produce mainly the peroxide radical $\cdot OOSO_3^-$ which has an interesting pattern of mechanism steps leading to dioxygen [81, 86]. These reactions have short chains, in part because the step:

$$SO_4^{-}\cdot + HOOSO_3^- \rightarrow HSO_4^- + \cdot OOSO_3^- \quad (71)$$

has an unexpectedly low rate constant ($k < 1 \times 10^5$ $M^{-1}s^{-1}$) for a radical-molecule process. The comparable process (eq. 72):

$$HO\cdot + HOOSO_3^- \rightarrow H_2O + \cdot OOSO_3^- \quad (72)$$

has a k value of 3.5×10^8 $M^{-1}s^{-1}$ [14].

7. Hydroxyl Radical in Biochemistry

7.1. Background

As a consequence of some reactions in the body, partial reductions of O_2 occur and $O_2^{-}\cdot$, HOOH and $HO\cdot$ are formed. Alkyl hydroperoxide is formed in lipid autoxidation.

Three enzymes act as body defense agents. There are *i.* superoxide dismutase (SOD) which catalyzes the dismutation (eq. 73):

$$2O_2^{-}\cdot + 2H^+ \xrightarrow{SOD} O_2 + HOOH \quad (73)$$

ii. catalase (CAT) which catalyzes the disproportionation (eq.74):

$$2HOOH \xrightarrow{CAT} O_2 + 2H_2O \quad (74)$$

and *iii.* peroxidase (POX) which catalyzes the redox process (eq.75):

$$HOOH + S \xrightarrow{POX} H_2O + SO \quad (75)$$

where S is a body substrate. For HO•, the most toxic, no specific defense agent has evolved; yet by enzymic destruction of O_2^- and HOOH, the amount of HO• formed in the body is minimized. This is necessary because HO• breaks DNA strands, oxidizes proteins, and is carcinogenic and mutagenic.

The conditions that lead to HO• formation and tissue damage are known. Peroxide alone (up to 100 μM) does not cause strand breaks in DNA, but they are observed in the presence of both iron and peroxide, particularly if iron is in the +2 oxidation state.

The evidence for toxicity of HO• is strong. Borg and Schaich [87] have said ".... potentially toxic hydroxyl radicals are now recognized to be the driving force initiating cascades of cytotoxic free radical reactions."

The biochemistry of O_2 and its reduction products is discussed in detail elsewhere [88-97] and much of that material is beyond the scope of this review. A general model for HO• production in the body is presented here. Then the oxidations of DNA (including deoxyribose and amine bases) and of proteins (including aminoacids and peptides) are discussed, as is the autoxidation of lipids. These are followed by descriptions of some unusual ways in which the defenses against HO• and its precursors are circumvented.

7.2. General model

Four steps, each presumably complicated, are involved in HO• generation. These are *i.* production of $O_2^{-}\cdot$ by one electron reduction of O_2, *ii.* reduction of ferric ion (enzyme bound) by $O_2^{-}\cdot$ or ascorbate or thiol, *iii.* formation of HOOH by O_2 reduction and/or $O_2^{-}\cdot$ disproportionation, and *iv.* reaction of iron(II) with HOOH as in the Fenton reagent. Because of the low concentrations of species involved ($O_2^{-}\cdot$, Fe^{2+}, HOOH), the overall HO• production rate will be slow even if the individual rate constants for the steps are large. When exogenous agents (radiation, smoke, chemicals such as CCl_4 and Paraquat) or excessive transition metals are present, the normal defenses against oxygenated intermediates may be overwhelmed.

7.3. Nucleic acids

The long chain molecule of DNA can be cleaved or otherwise damaged by HO• attack. A deoxyribose moiety of the backbone or an amine base can be lost. Minor damage may be corrected by the body's repair mechanisms but multiple damages close together can bring about permanent cleavages. The deoxyribose rings and the amine bases are the sites most susceptible to reaction with HO•. A selected group of biochemical compounds are listed in Table 12 along with their rate constants for reaction with the radical; the magnitude of k is awesome when the biological consequences are considered.

Table 12: Rates with biochemical compounds[a].

Compound	Charge	k ($M^{-1}s^{-1}$)
N-acetylalanine	(-)	$5x10^8$
adenine	(0)	$5x10^9$
alanine	(+)	$4x10^7$
alanine	(0)	$7x10^7$
alanine	(-)	$7x10^8$
ascorbate	(-)	$9x10^9$
cysteine	(0)	$2x10^{10}$
cystine	(0)	$2x10^9$
deoxyribose	(0)	$2x10^9$
formate	(-)	$2x10^9$
glutathione		$1x10^{10}$
glycine	(-)	$2x10^9$
guanine		$1x10^{10}$
histidine	(0)	$5x10^9$
methionine	(0)	$8x10^9$
phenylalanine	(0)	$6x10^9$
purine	(0)	$3x10^8$
thymine	(0)	$5x10^9$
tryptophan	(0)	$1x10^{10}$
tyrosine		$1x10^{10}$

[a] Selected by the authors.

The amine bases, being unsaturated compounds, can react either by hydrogen abstraction or by addition to double bonds; both types of reaction are observed with thymine [98]; see Scheme 3.

Scheme 3:

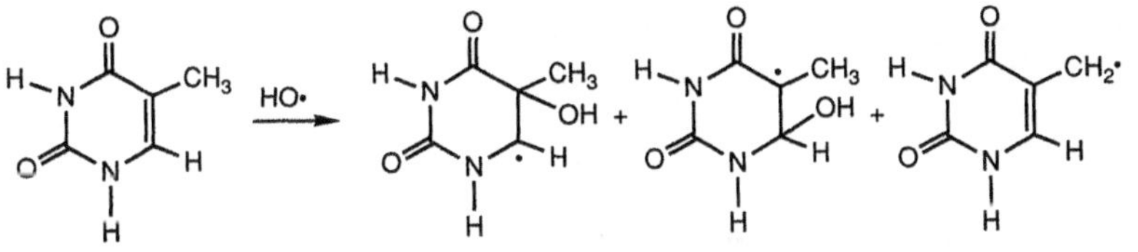

56% 35% 9%

Guanine generates four isomeric HO• adducts [96].

The CH bonds in deoxyribose are very reactive with HO•, and in nucleosides and polynucleotides about 20% of the HO• reacts wth the deoxyribose moiety. Five deoxyribose radicals are formed [96]. Damage to the sugar moiety of DNA not only leads to strand breaks but also to release of amine bases.

7.4. Proteins

Hydroxyl radicals attack aminoacids, peptides and proteins in the same rapid and nonselective way that they react with other organic compounds, and a large amount of relevant data (both stoichiometries and rates) have been accumulated, especially by Garrison et al [99].

A reactive site in both small and large peptides is the α-CH bond. With N-acetylglycine and N-acetylalanine, the reaction is:

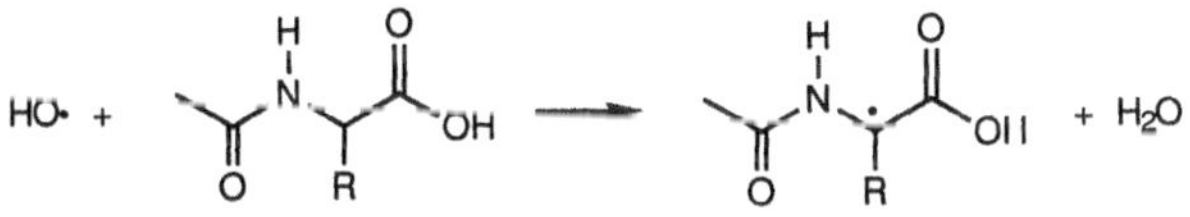

(76)

The resultant radical (in the absence of dioxygen) dimerizes to form α,α'-diamino-dicarboxylic acid derivatives

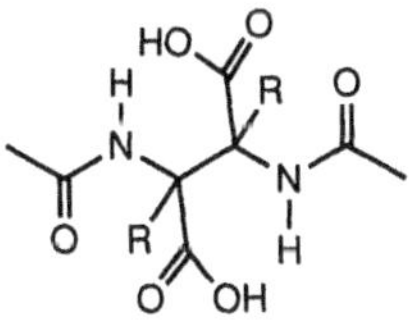

In the presence of O_2, the peroxoradical

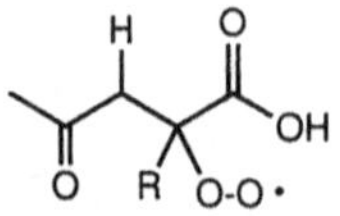

is formed by addition of O_2 to the aminoacid radical and this oxidation often leads to degraded and cleaved molecules.

With more complicated aminoacids, other types of reaction with HO• obtain. If the aminoacid has a hydrocarbon chain, attack at other CH bonds are noted. For example, valine is oxidized to 3-hydroxyvaline.

For those aminoacids that have an aromatic ring (phenylalanine, tyrosine, tryptophan and histidine), HO• attacks the ring rapidly ($k \sim 10^{10}\ M^{-1}s^{-1}$). Phenylalanine is first converted to a hydroxycyclohexadienyl radical and this goes on by oxidation to form three tyrosine isomers (ortho, meta and para) or by dimerization to give the biphenyl compound known as dityrosine. The appearance of ortho-tyrosine in a protein is considered to be evidence that the protein has encountered HO• at some point. Both tryptophan and histidine react by addition of HO• to the five membered ring double bond; in the presence of O_2, ring opened products are obtained.

Aminoacids having sulfur functions (cysteine, cystine, methionine and the tripeptide glutathione (GSH)) react rapidly at the sulfur.The mercaptide group transfers its hydrogen

$$RSH + HO\cdot \rightarrow RS\cdot + H_2O \qquad (77)$$

to form a thiyl radical at a rate which is essentially diffusion controlled. In the presence of O_2, a complicated set of steps obtains and products including RSSR are formed. Both thioethers and disulfides react with HO•, again to give a group of products; from the former, sulfoxides and sulfones are obtained.

The reaction of globular protein molecules with HO• was investigated both in the absence and in the presence of O_2 plus $O_2^{-}\cdot$ [100]. Multiple attacks on each protein molecule were observed. With HO• alone, aggregation (increased molecular weight, apparently by successive dimerizations) predominated; little fragmentation (decreased molecular weight) was found. On the other hand, when both O_2 and $O_2^{-}\cdot$ were also present little or no aggregation was found and fragmentation was usually extensive.

More detailed and specific studies of HO• and proteins have been initiated [101] and will continue. A matter of considerable current interest is the distance in cell fluid that HO• will travel from its site of formation before it reacts. If a reducing metal ion is bound near to the target area, the probability of reaction at this area is higher than at a similar molecular structure in solution. The matter arises from the collisional rate constants found for some biological molecules and from the inadequate protection afforded by HO• scavengers.

7.5. Lipids

Lipids, which are molecules made up of fatty acids esterified with glycerol, constitute a significant fraction of cell membrane material. When the fatty acid moiety is unsaturated, it undergoes the multistep process known as lipid autoxidation or as lipid peroxidation. This process, in essence an oxidation by O_2, leads to serious cell wall damage [90, 92].

In the absence of any peroxide (HOOH or ROOH), the autoxidation is extremely slow. However it is difficult to have lipids completely free of the related lipid hydroperoxide ROOH and of metal ions. For this reason it is difficult to pin-point the way in which the very first radical is formed, yet a likely step is the one often invoked i.e. eq. 10. The follow-up steps seem obvious: *i.* the radical HO• adds to a double bond or abstracts an allylic hydrogen, and by either path generates an organic radical R•, *ii.* R• adds dioxygen to give the peroxoradical ROO•, and *iii.* ROO• reacts slowly with RH to give the hydroperoxide ROOH and another R•. This set of steps constitutes the propagation part of a chain. Because of the relatively low reactivity of peroxoradicals a likely chain termination step is

$$2\,ROO\bullet \rightarrow ROOR + O_2 \qquad (78)$$

Chain branching is said to take place by the steps

$$ROOH + Fe^{2+} \rightarrow Fe^{3+} + OH^- + RO\bullet \qquad (79)$$

$$RO\bullet + RH \rightarrow ROH + R\bullet \qquad (80)$$

Direct implication of HO• in lipid autoxidation has not been made and may not soon be because the lipid hydroperoxide quickly becomes the dominant peroxide in initiating chains.

7.6. Cobalt and copper

Both of these metals are present in small quantity in the human body; under high stress conditions both can overwhelm the defense and repair apparati and initiate health problems. Cobalt salts are believed to be carcinogenic, to cause DNA strand breaks, and to stimulate lipid peroxidation. It is also known that Co^{2+} and HOOH can hydroxylate aromatic compounds and oxidize deoxyribose in the manner expected for HO• [102].

Cobalt(II) reacts with HOOH only very slowly whereas cobalt(III) oxidizes HOOH rapidly. Therefore the predominant state for the cobalt in homogeneous solution (even with HOOH present) is +2. On the other hand, $Co(EDTA)^{2-}$ can be oxidized to $Co(EDTA)^{-}$ at which state it tends to remain. Provided a reducing agent is available, the $Co(EDTA)^{-}$ can be recycled to $Co(EDTA)^{2-}$. Both the predominant state of the cobalt atom and the rates of $+2 \rightarrow +3$ interconversion are strongly dependent on ligand nature.

Cobalt(II) salts and HOOH degrade deoxyribose slowly and the rate is not accelerated by ascorbate, whereas iron(II) salts with HOOH are more reactive and are stimulated by ascorbate. The degradation by cobalt is inhibited by mannitol, ethanol and formate ion but not by urea; all these effects are consistent with their intermediacy of HO• [102]. Also consistent is the observation that greater amounts of inhibitor are needed when EDTA is absent than when it is present, presumably [102] because EDTA itself is a known trap for HO•.

The data of Moorhouse et al [102] can be interpreted in terms of the postulation that shuttling between the +2 and +3 states with concomitant formation of HO• takes place for both cobalt and iron. The rates for $Co^{2+}(aq)$, for $Co(EDTA)^{2-}$, for $Fe^{2+}(aq)$ and for $Fe(EDTA)^{2-}$ along with the +3 compounds in their redox shuttle processes must be considered in explanations of the extents of HO• formation under physiological conditions.

It is worth noting here that manganese(II) with HOOH is expected to behave in a manner similar to that seen with cobalt(II), but the rate of HO• formation should be still slower

Hydroxyl radical has also been invoked in the $Cu(phen)_2^{+}$ degradation and cleavage of DNA [103,104]. It seems certain that an oxidizing intermediate is formed, but doubt has arisen as to it being HO• (see Section 5 above). Besides metal and ligand, HOOH is required [104]. Acetate, benzoate and formate ions all act as inhibitors of the degradation, and as in other cases the inhibition is attributed to trapping. The particular ligand used (phen) is crucial, as indicated by the fact that eight other strong ligands including neocuproine, bipy, and EDTA do not form catalytic complexes with copper. The cleavage occurs for double-stranded DNA but not for single stranded so attachment of $Cu(phen)_2^+$ to DNA by intercalation prior to HO• formation appears to be an important aspect of the cleavage process [104]. A proximity effect seems probable here; the very high rate constants for the reaction of HO• with organic molecules suggests that this radical must be formed near to the target DNA in order to be an effective strand breaker.

7.7. Peroxonitrites

Up to this point, we have steered clear of the heated discussion in the literature on the toxicity level and danger of superoxide in the body. It is mentioned now only in that some recent results bear both on the toxicity question and on the formation of HO• in certain body tissues.

Superoxide is certainly not as agressive a radical as HO•, yet the tissue of both plants and animals contains SOD in at least three forms differing in part by the metal at the active site. On this basis and others, it has been the contention of Fridovich [105] that O_2^{-}• plays a prime toxic role. On the other hand, others [106] have felt that O_2^{-}• is too poorly reactive a radical to play a direct role.

The recent work of Beckman et al [107] indicates that at least in one reaction O_2^{-}• plays both a primary and a secondary role. The radical ON• (nitric oxide) or a derived compound has been found to be a blood vessel dilating substance. The time period of action for this necessary dilator is consistent with lifetimes in the cell fluid ranging from 4 to 50 seconds, and the lifetime is approximately doubled by SOD. It was suggested by Beckman et al [107] that superoxide can destroy the ON• by the known reaction

$$ON\cdot + O_2^{-}\cdot \rightarrow ONOO^- \quad (81)$$

which produces peroxonitrite ion. The enzyme SOD acts, of course, by destroying O_2^{-}• which action in turn decreases the loss of ON•, thereby extending the active lifetime of this nitrogen radical. A prime danger of O_2^{-}• thus is seen in its destruction of ON• by a combination step.

Further the SOD prevention of $ONOO^-$ formation eliminates the subsequent generation of HO• by the conjugate acid of $ONOO^-$. This step (eq. 82) is known [54] to take place

$$HOONO \rightarrow HO\cdot + O_2N\cdot \quad (82)$$

rapidly (about 2 seconds at body temperature 37°C and pH 7.4) after very rapid protonation of the anion.

Superoxide is doubly dangerous. In its prime role, it destroys the needed ON• by combining with this nitrogen radical. In its secondary role, it produces peroxonitrous acid which is a known HO• generator.

Peroxonitrite has been pretty much of a laboratory curosity over the years since first made in 1901 [54]. The anion form $ONOO^-$ is fairly stable at pH 12 in the absence of adventitous metals. In acid solution (pH 2 to 5) the predominant form is HOONO (the pKa is about 6) and this compound decomposes with a half-time of seconds at 25°C. Halfpenny and Robinson [55] and Mahoney [56] gave clear evidence for homogeneous scission of the peroxide link. Radical escape from cage occurs in about 32% of homolysis events [56].

One of the methods for preparation of stable samples of $ONOO^-$ is by photolysis of solid KNO_3 [54]. Professor V. Anderson and Mr. Peter King of the Brown University Chemistry Department have recently shown that samples of KNO_3 irradiated by 254 nm ultraviolet, on dissolution in an aqueous solution of bovine serum albumin, cleave the protein in the same way as does γ-irradiation.

7.8. Chromate

Chromium(VI), known to be an oxidant, has been found to be carcinogenic, toxic and mutagenic [108-111]. By way of comparison, chromium(III), which forms inert complexes and reacts slowly with DNA, does not bring on these health problems. It was early recognized that the anion CrO_4^{2-} could be readily transported across cell membrance barriers by nonspecific anion carries [112]. The anion in the absence of reductants does not damage DNA; once in the cell, it has to be activated.

The mode of activation of chromium(VI) and the chemistry of the activated form of this element then became the focus of intensive investigation by several groups including those of Kawanishi [113], of O'Brien and Kortenkamp [114], of Wetterhahn and Floyd [110, 115] and of Shi and Dalal [116]. Some of the questions asked of their experiments were *i.* What is the importance of chromium(V) and how is it formed from CrO_4^{2-}?, *ii.* Is HO• formed and is it the DNA damaging intermediate?, *iii.* Are the known chromium(V) and chromium(VI) species that have been identified and/or isolated on the direct track to the active intermediate?, and *iv*. How important are the tripeptide glutathione GSH and its chromate complexes?

Some answers to those questions and other are available. Cell fluid contains significant amounts of thiols (especially GSH), plus some ascorbate ion, and a little HOOH; all of these can reduce Cr(VI). Indeed the ESR of chromium(V) species has been seen by several of the investigators. The thiol GSH does form complexes with chromium (see below). The formation of chromate/peroxide complexes is well established [15, 117] although detection of them under physiological conditions does not appear to have been reported in the literature. The fact that no chromium(V)/HOOH complexes have been observed under physiological conditions is not surprising since the reaction

$$Cr^{V} + HOOH \rightarrow Cr^{VI} + HO^{-} + HO\cdot \qquad (83)$$

is expected to proceed readily.

Some attention has been paid to the peroxoanion $Cr(O_2)_4^{3-}$ as a source of HO• in the body. This is unwarranted due to the low [HOOH] and mild pH of cell fluid; as can be seen by inspection of the formation stoichio-metry

$$2\,CrO_4^{2-} + 9\,HOOH + 2\,OH^{-} \rightarrow 2\,Cr(O_2)_4^{3-} + 10\,H_2O \qquad (84)$$

an enormous dependence on concentration of HOOH is demanded. A definitive article on decomposition of $Cr(O_2)_4^{3-}$ which elaborates come complexities of this species has gone rather unrecognized [118].

Questions about formation of HO• and its role as the active particle to attack DNA deserve a positive but not yet certain response in view of the spin trap and ESR studies.

Particularly interesting is the recent work of Jones et al. [114a] who showed the formation of methyl radical when the solvent for a Cr(V)/GSH complex was 50% DMSO and 50% H_2O. It is known that HO• reacts with this sulfoxide molecule rapidly in accordance with the step

$$HO\cdot + CH_3SOCH_3 \rightarrow CH_3SO_2H + \cdot CH_3 \qquad (85)$$

Found in mM quantities in the cell, GSH can do a number of things. Besides the chromium(V) complex, it can also form an unstable chromium(VI) complex [119]. Also GSH can carry out chemical repair of organic radicals [120] by the step

$$R\cdot + GSH \rightarrow RH + GS\cdot \qquad (86)$$

In the HO• generation by hypoxanthine/xanthine oxidase system, thiols including GSH can either increase or decrease the amount of HO• depending on concentration [121].

8. Closing Remarks

Since the pioneering studies by Fenton, the chemistry of the ferrous-ion-catalyzed decomposition of hydrogen peroxide has been explored in a considerable number of variants and continues to be the topic of much concern. This is so because of the obvious biological implications, but also because this system can find important applications. In fact, the high reactivity and lack of selectivity that is characteristic of Fenton-type systems make these suitable when dealing with waste-water treatment and with pollution problems. Detailed reviews are available concerning the oxidation of the various classes of organic compounds by Fenton systems [122]. In this chapter, the authors choose to present a critical overview of the key features of Fenton-type activation of hydrogen peroxide, combining various aspects of the chemistry of H_2O_2 in the presence of relevant transition-metal ions, and the reactivity of the hydroxyl radical. Because of space and time limitations, a comprehensive coverage of the literature dealing with the Fenton reaction and the chemistry of hydroxyl radical was not feasible. Our regrets to those Authors whose work was omitted.

Acknowledgments

We thank Drs. V. Anderson, G. R. A. Johnson, P. Jones, P. O'Brien, R. C. Plumb, and Mr. P. King for suggestions and references. R.Curci. wishes to thank the Directorate for Science, R.&D. of the European Communities for generous funding under a research grant (CEC-STEP Programme CT 90-0068) administered by Montefluos S.p.A. (Milan, Italy). Thanks are also due to Dr. Fulvio Burzio (Montefluos S.p.A.) for many helpful discussions.

References

1. H. J. H. Fenton, *J. Chem. Soc.*, **65**, 899 (1984).
2. (a) F. Haber and J. J. Weiss, *Proc. Roy. Soc. London*, Ser. A, **147**, 332 (1934); (b) N. Uri, *Chem. Rev.*, **50**, 375 (1952).
3. (a) W. G. Barb, J. H. Baxendale, P. George and K. R. Hargrave, *Trans. Faraday Soc.*, **47**, 462 (1951); (b) *ibid.*, **47**, 591 (1951).
4. (a) J. H. Merz and W. A. Waters, *Disc. Faraday Soc.*, **2**, 179 (1947); (b) ibid, *J. Chem. Soc.*, **S15**, 2427 (1949); (c) see also: I. M. Kolthoff and A. I. Medalia, *J. Am. Chem. Soc.*, **71**, 3777, 3784, 3789 (1949).
5. (a) C. Walling and S. Kato, *J. Am. Chem. Soc.*, **93**, 4275 (1971); (b) C. Walling and G. M. El-Taliawi, *J. Am. Chem. Soc.*, **95**, 844 (1973); (c) *ibid*, 95, 848 (1973); (d) C. Walling, G. M. El-Taliawi and R. A. Johnson, *J. Am. Chem. Soc.*, **96**, 133 (1974); (e) C. Walling, *Acc. Chem. Res.*, **8**, 125 (1975); (f) C. Walling and R. A. Johnson, *J. Am. Chem. Soc.*, **97**, 363 (1975).
6. (a) W. C. Bray and M. H. Gorin, *J. Am. Chem. Soc.*, **54**, 2124 (1932); (b) see also A. E. Cahill and H. Taube, *J. Am. Chem. Soc.*, **74**, 2312 (1952).
7. J. T. Groves and M. Van Der Puy, *J. Am. Chem. Soc.*, **98**, 5290 (1976) and references therein.
8. (a) G. R. A. Johnson and N. B. Nazhat, *J. Am. Chem. Soc.*, **109**, 1990 (1987); (b) G. R. A. Johnson, N. B. Nazhat and R. A. Saadalla-Nazhat, *J. Chem. Soc. Faraday Trans.*, **84**, 501 (1988); (c) G. R. A. Johnson and N. B. Nazhat in '*Free Radicals, Metal Ions and Biopolymers* ', Eds. P. C. Beaumont, D. J. Deeble, B. J. Parsons and C. Rice-Evans, Richelieu Press, London 1989, p. 55.
9. (a) B. C. Gilbert and J. K. Stell, *J. Chem. Soc. Perkin II*, 1281 (1990); (b) B. C. Gilbert and J. K. Stell, *J. Chem. Soc. Faraday Trans.*, **86**, 3261 (1991).
10. A. J. Bard, R. Parsons and J. Jordan, '*Standard Potentials in Aqueous Solution* ', IUPAC, M. Dekker, New York 1985.
11. W. M. Latimer, '*Oxidation Potentials* ', 2nd Ed., Prentice-Hall, Englewood Cliffs, N.J. 1961.
12. R. J. P. Williams, in '*Oxygen and Life* ', Roy. Soc. Chem., Spec. Publ. no 39 (1981), p. 18-44.
13. T. J. Hardwick, *Can. J. Chem.*, **35**, 428, 437 (1957), and references therein.
14. Farhataziz and A. B. Ross, NSRDS-NBS no 59, U. S. Dept. of Commerce, U. S. Gov. Printing Office, Washington, D. C. 1977.
15. S. B. Brown, P. Jones and A. Suggett, *Prog. Inorg. Chem.*, **13**, 170-204 (1970).
16. (a) M. G. Evans, P. George and N. Uri, *Trans. Faraday Soc.*, **45**, 236 (1949); (b) M. L. Haggett, P. Jones and W. F. K. Wynne-Jones, *Disc. Faraday Soc.*, **29**, 153 (1960).
17. (a) M. L. Kremer and G. Stein, *Trans. Faraday Soc.*, **55**, 959 (1959); (b) M. L. Kremer, *Trans. Faraday Soc.*, **59**, 2535 (1963).
18. C. Walling and A. Goosen, *J. Am. Chem. Soc.*, **95**, 2987 (1973).
19. C. F. Wells and M. A. Salam, *J. Chem. Soc. (A)*, 308 (1968) and references therein.
20. P. Jones, M. L. Tobe and W. F. K. Wynne-Jones, *Trans. Faraday Soc.*, **55**, 91 (1959).
21. For review see: (a) M. C. R. Symons, in '*Peroxide Reaction Mechanisms* ', Ed. J. O. Edwards, Wiley-Interscience, New York 1962, p. 137-151; (b) Y. Ogata, K. Tomizawa and K. Furuta, in '*The Chemistry of Peroxides* ', edited by S. Patai, Wiley, New York 1983, p. 711-775.

22. (a) J. H. Baxendale and J. A. Wilson, *Trans. Faraday Soc.*, **53**, 344 (1957); (b) J. P. Hunt and H. Taube, *J. Am. Chem. Soc.*, **74**, 5999 (1952); D. H. Volman and J. C. Chen, *J. Am. Chem. Soc.*, **81**, 4141 (1959).
23. (a) R. Livingston and H. Zeldes, *J. Chem. Phys.*, **44**, 1245 (1966); (b) R. Livingston and H. Zeldes, *J. Magn. Reson.*, **1**, 169 (1969) and references therein.
24. For review see: (a) G. Stein, '*Radiation Chemistry in Aqueous Systems* ', Wiley-Interscience, New York 1968; (b) J. W. T. Spinks and R. J. Woods, '*An Introduction to Radiation Chemistry* ', 3rd edition, Wiley, New York 1990; (c) M. Matheson and L. M. Dorfman, '*Pulse Radiolysis* ', MIT Press, Cambridge 1969; (d) K. Eiben and R. W. Fessenden, *J. Phys. Chem.*, **75**, 1186 (1971).
25. D. Meyerstein, *Inorg. Chem.*, **10**, 638 (1971).
26. For general review see: R. O. C. Norman and B. C. Gilbert, *Adv. Phys. Org. Chem.*, **5**, 53 -119 (1967).
27. M. Masarwa, H. Cohen, D. Meyerstein, D. L. Hickman, A. Bakac and J. H. Espenson, *J. Am. Chem. Soc.*, **110**, 4293 (1988).
28. G. Czapski, A. Samuni and D. Meisel, *J. Phys. Chem.*, **75**, 3271 (1971).
29. (a) J. K. Kochi, *Tetrahedron*, **18**, 482 (1962); (b) H. F. Delamare, J. K. Kochi and F. F. Rust, *J. Am. Chem. Soc.*, **85**, 1437 (1963).
30. (a) P. Maggioni and F. Minisci, *Chim. Ind. (Milan)*, **59**, 239 (1977); (b) H. Pfitzner and H. Baumann, *Angew. Chem.*, **70**, 232 (1958); (c) M. Kamel and S. A. Amin, *J. Prakt. Chem.*, **36**, 230 (1967); (d) M. Bonchio and F. DiFuria, personal communication.
31. G. Sosnovsky in '*Organic Peroxides* ', Ed. D. Swern, Wiley-Interscience, New York 1971, vol. 2, p. 269.
32. D. D. Coffmann and coworkers, *J. Am. Chem. Soc.*, **80**, 2864, 2872, 2877, 2880, 2882 (1958).
33. B. Bertilsson, B. Gustafsson, I. Kuhn and K. Torsell, *Acta Chem. Scand.*, **24**, 3590 (1970).
34. (a) J. K. Kochi in '*Free Radicals* ', Ed. J. K. Kochi, Wiley, New York 1973, chapter 11; (b) J. K. Kochi, *Acc. Chem. Res.*, **7**, 352 (1974).
35. (a) A. R. Gallopo and J. O. Edwards, *J. Org. Chem.*, **36**, 4089 (1971); (b) B. C. Gilbert, J. K. Stell and M. Jeff, *J. Chem. Soc. Perkin II*, 1867 (1988).
36. D. H. Ellison, G. A. Salmon and F. Wilkinson, *Proc. Roy. Soc. London Ser. A.*, 328 (1572), 23 (1972).
37. J. H. Baxendale, M. G. Evans and G. S. Park, *Trans. Faraday Soc.*, **42**, 155 (1946).
38. For review and references see C. Walling, '*Free Radicals in Solution* ', Wiley, New York 1957.
39. K. Kustin and J. Swinehart, *Prog. Inorg. Chem.*, **13**, 114 (1970).
40. (a) D. L. Ball and J. O. Edwards, *J. Am. Chem. Soc.*, **78**, 1125 (1956); (b) E. Koubek, M. L. Haggett, C. J. Battaglia, K. M. Ibne-Rasa, H. Y. Pyun and J.O. Edwards, *J. Am. Chem. Soc.*, **85**, 2263 (1964); (c) D. F. Evans and M. W. Upton, *J. Chem. Soc. Dalton Trans.*, 1141, 1151, 2525 (1985); (d) D. M. Davies, D. Dunn, M. Haydarali, R. M. Jones and J. M. Lawther, *J. Chem. Soc. Chem. Commun.*, 987 (1986).
41. (a) C. Walling, R. E. Partch and T. Weil, *Proc. Natl. Acad. Sci. USA*, **72**, 140 (1975).
42. K. C. Francis, D. Cummins and J. Oakes, *J. Chem. Soc. Dalton Trans.*, 493 (1985), and references therein.
43. S. Rahhal and H. W. Richter, *J. Am. Chem. Soc.*, **110**, 3126 (1988).
44. C. L. Jenkins and J. K. Kochi, *J. Am. Chem. Soc.*, **94**, 843, 856 (1972).
45. H. B. Brooks and F. Sicilio, *Inorg. Chem.*, **10**, 2530 (1971).
46. G. Czapski, *Ann. Rev. Phys. Chem.*, **22**, 171 (1971) and references therein.

47. (a) M. Andersen, J. O. Edwards, A. A. Green and M. D. Wiswell, *Inorg. Chim. Acta*, **3**, 655 (1969); (b) C. C. Su, J. W. Reed and E. S. Gould, *Inorg. Chem.*, **12**, 337 (1973), (c) N. A. Johnson and E. S. Gould, *J. Am. Chem. Soc.*, **95**, 5198 (1973).
48. (a) M. Orhanovic and R. G. Wilkins, *J. Am. Chem. Soc.*, **89**, 278 (1967); (b) F. Secco, *Inorg. Chem.*, **19**, 2722 (1980).
49. M. Bonchio, personal communication.
50. (a) R. O. C. Norman and P. R. West, *J. Chem. Soc. B*, 389 (1969); (b) W. T. Dixon and R. O. C. Norman, *Nature*, **196**, 891 (1962); (c) W. T. Dixon and R.O. C. Norman, *J. Chem. Soc.*, 3119 (1963); d) R. O. C. Norman and P. M. Storey, *J. Chem. Soc. B*, 1099 (1970) and references therein.
51. C. P. Moorhouse, B. Halliwell, M. Grootveld and J. M. C. Gutteridge, *Biochim. Biophys. Acta*, **843**, 261 (1985).
52. R. C. Thompson, *Inorg. Chem.*, **20**, 1005 (1981).
53. C. Marsh and J. O. Edwards, *Prog. Reaction Kinetics*, **15**, 35-75 (1989).
54. A review on the chemistry of peroxonitrites has been prepared (J. O. Edwards and R. C. Plumb) and will soon be submitted. Inquires about this may be directed to J.O.E.
55. E. Halfpenny and P. L. Robinson, *J. Chem.Soc. A*, 939, 946 (1952).
56. L. R. Mahoney, *J. Am. Chem. Soc.*, **92**, 5262 (1970).
57. V. Anderson, J. O. Edwards, P. King and R. C. Plumb, unpublished experiments.
58. W. K. Wilmarth and A. Haim, in 'Peroxide Reaction Mechanisms ', Ed. J. O. Edwards, Wiley-Interscience, New York 1962, p. 175-225.
59. D. A. House, *Chem. Rev.*, **62**, 185 (1962).
60. E. J. Behrman and J. E. McIsaac Jr. in '*Mechanisms of Reactions of Sulfur Compounds* ', 2nd edition, Ed. N. Kharasch, Intrascience Pub., Los Angeles 1968, p. 193-218.
61. For review see: E. J. Behrman and J. O. Edwards, *Rev. Inorg. Chem.*, **2**, 179 (1980); see particularly: I. M. Kolthoff and I. K. Miller, *J. Am. Chem. Soc.*, **73**, 3055 (1951).
62. S. Fronaeus, *Acta Chem. Scand.*, **A40**, 572 (1986).
63. J. W. L. Fordham and H. L. Williams, *J. Am. Chem. Soc.*, **73**, 4855 (1951).
64. (a) M. Tsao and W. K. Wilmarth, *J. Phys. Chem.*, **63**, 346 (1959); (b) O. P. Chawla and P. W. Fessenden, *J. Phys. Chem.*, **79**, 2693 (1975) and references therein.
65. (a) H. Eltenberger, P. Steenken, P. O'Neill and D. Schulte-Frohlinde, *J. Phys. Chem.*, **82**, 749 (1978); (b) P. Neta, V. Madhavan, H. Zemel and R. W. Fessenden, *J. Am. Chem. Soc.*, **99**, 163 (1977); (c) M. J. Davies, B. C. Gilbert, C. B. Thomas and J. Young, *J. Chem. Soc. Perkin Trans. II*, 1199 (1985); M. J. Davies and B. C. Gilbert, *J. Chem. Soc. Perkin Trans. II*, 1809 (1984); e) B. C. Gilbert, J. K. Stell and M. Jeff, *J. Chem Soc. Perkin Trans. II*, 1867 (1988).
66. A. B. Ross and P. Neta, NSRDS - NBS no 65, U. S. Goverment Printing Office, Washington, D. C. 1979.
67. (a) C. E. Burchill and G. F. Thompson, *Can. J. Chem.*, **49**, 1305 (1971); (b) C. E. Burchill and P. W. Jones, *Can. J. Chem.*, **49**, 4005 (1971), and references therein.
68. A. Ledwith, P. J. Russell and L. H. Sutcliffe, *Proc. Roy. Soc. London A*, **332**, 151-166 (1973).
69. M. E. Snook and G. A. Hamilton, *J. Am. Chem. Soc.*, **96**, 860 (1974).
70. (a) C. Arnoldi, A. Citterio and F. Minisci, *J. Chem Soc. Perkin Trans. II*, 531 (1983); and (b) F. Minisci, *Acc. Chem. Res.*, **16**, 27 (1983).
71. (a) C. Walling, D. M. Camaioni and S. S. Kim, *J. Am. Chem. Soc.*, **100**, 4814 (1978); (b) C. Walling and K. Amarnath, *J. Am. Chem. Soc.*, **104**, 1185 (1982).
72. G. Levey and E. J. Hart, *J. Phys. Chem.*, **79**, 1642 (1975) and references therein.

73. (a) M. M. Crutchfield, in '*Peroxide Reaction Mechanisms* ', Ed. J. O. Edwards, Wiley-Interscience, N.Y., p. 41-65; (b) M. M. Crutchfield and J. O. Edwards, *J. Am. Chem. Soc.*, **82**, 3533 (1960).
74. For a review on peroxophosphate chemistry see I. I. Creaser and J. O. Edwards, *Top. Phosphorus Chem.*, **7**, 379-432 (1972).
75. (a) M. R. Hoffman, P. A. Stern, P. H. Rieger and J. O. Edwards, *Inorg. Chim. Acta*, **19**, 181 (1976); (b) S. A. Frennesson and S. Fronaeus, *Acta Chem. Scand.*, **26**, 3917 (1972).
76. J. O. Edwards, *Coord. Chem. Rev.*, **8**, 87 (1972) and references therein.
77. R. J. Lussier, W. M. Risen Jr. and J. O. Edwards, *J. Phys. Chem.*, **74**, 4039 (1970).
78. G. Bida, R. Curci and J. O. Edwards, *IntlJ. Chem. Kinetics*, **5**, 859 (1973).
79. G. Levey, G. Bida and J. O. Edwards, *Intl. J. Chem. Kinetics*, **13**, 497 (1981).
80. G. Levey and E. J. Hart, *J. Phys. Chem.*, **79**, 1642 (1975).
81. C. Marsh and J. O. Edwards, *Prog. Reaction Kinetics*, 15, 35-75 (1989).
82. B. C. Gilbert, J. K. Stell, A. C. Whitwood, C. Halliwell and W. R. Sanderson, *J. Chem. Soc. Perkin II*, 629 (1991).
83. P. Kanakaraj and P. Maruthamuthu, *Intl. J. Chem. Kinetics*, **15**, 1301 (1983).
84. P. Maruthamuthu and P. Neta, *J. Phys. Chem.*, **81**, 937 (1977) and references therein.
85. R. C. Thompson, *Inorg. Chem.*, **20**, 3745 (1981).
86. Z. Zhang, Ph. D. thesis, Brown University 1990.
87. D. C. Borg and K. M. Schaich, in ref. 90, p. 20.
88. '*Metal-Ion Activation of Dioxygen* ', Ed. T. G. Spiro, Wiley-Interscience, New York 1980.
89. '*Chemical and Biochemical Aspects of Superoxide and Superoxide Dismutase* ', Eds. J. V. Bannister and H. A. O. Hill, Elsevier, N.Y. (1980).
90. 'Oxygen Radicals and Tissue Injury ', Ed. B. Halliwell, published by FASEB, Bethesda, MD 1988.
91. '*Oxygen and Life* ', Second BOC Priestley Conference, Proc. Roy. Soc. Spec. Pub. no 39 (1981).
92. '*Free Radicals: Recent Developments in Lipid Chemistry, Experimental Pathology and Medicine* ', Eds. F. Dormandy and C. Rice-Evans, Richelieu Press, 1988.
93. (a) Series '*Free Radicals in Biology* ', Ed. W. A. Pryor, Academic Press, New York; (b) For articles bearing on oxygenation mechanisms: *Isr. J. Chem.*, vol. 23 no 4 (1983) and vol. 24 no 1 (1984).
94. (a) W. A. Pryor, *Ann. Rev. Physiol.*, **48**, 657-67 (1986); (b) J. A. Imlay and S. Linn, *Science*, **240**, 1302 (1988).
95. M. Vuillaume, *Mutation Res.*, **186**, 43 (1987).
96. K. Kim, S. G. Rhee and E. R. Stadtiman, *J. Biol. Chem.*, **260**, 15394 (1985).
97. M. G. Simic, D. S. Bergtold and L. R. Karam, *Mutation Res.*, **214**, 3 (1989).
98. (a) S. V. Jovanovic and M. G. Simic, *J. Am. Chem. Soc.*, **108**, 5968 (1986); (b) G. Scholes, J. F. Ward and J. Weiss, *J. Mol. Biol.*, **1**, 379 (1960).
99. For review see: W. M. Garrison, *Chem. Rev.*, **87**, 381 (1987).
100. (a) K. J. A. Davies, *J. Biol. Chem.*, **262**, 9895 (1987); (b) K. J. A. Davies, M. E. Delsignore and S. W. Lin, *J. Biol. Chem.*, **262**, 9902 (1987); (c) K. J. A. Davies and M. E. Delsignore, *J. Biol. Chem.*, **262**, 9908 (1987); (d) K. J. A. Davies, S. Wilin and R. E. Pacifici, *J. Biol. Chem.*, **262**, 9914 (1987).
101. (a) I. Z. Nagy and R. A. Floyd, *Biochim. Biophys. Acta*, **790**, 238 (1984); (b) T. M. Rana and C. F. Meares, *J. Am. Chem. Soc.*, **112**, 2457 (1990); (c) C. Czapski, *Isr. J. Chem.*, **24**, 29 (1984).
102. C. P. Moorhouse, B. Halliwell, M. Grootveld, and J. M. C. Gutteridge, *Biochim. Biophys. Acta*, **843**, 261 (1985).

103. B. G. Que, K. M. Downey and A. G. So, *Biochem.*, **19**, 5987 (1980).
104. (a) D. S. Sigman, D. R. Graham, V. D'Aurora and A. M. Stern, *J. Biol. Chem.*, **254**, 12269 (1979); (b) D. R. Graham and D. S. Sigman, *J. Biol. Chem.*, **257**, 12121 (1982); (c) D. S. Sigman, *Acc. Chem. Res.*, **19**, 180 (1986).
105. (a) I. Fridovich, Science, 201, 875 (1978); (b) I. Fridovich, in ref 90; (c) E. Fridovich in 'Advances in Inorganic Chemistry ', Vol. 1, Eds. G. L. Eichhorn and L. G. Marzilli, Elsevier/North Holland, New York 1979.
106. (a) J. A. Fee, in ref. 91, page 77; (b) D. T. Sawyer and J. S. Valentine, *Acc. Chem. Res.*, **14**, 393 (1981).
107. J. S. Beckman, T. W. Beckman, J. Chen, P. A. Marshall, and B. A. Freeman, *Proc. Natl. Acad. Sci. USA*, **87**, 1620 (1990).
108. '*Chromium: Metabolism and Toxicity* ', Ed. D. Burrows, CRC Press, Boca Raton, FL 1983.
109. '*Biological and Environmental Aspects of Chromium* ', Ed. S. Langard, Elsevier Biomedical Press, Amsterdam 1982.
110. A. S. Standeven and K. E. Wetterhahn, *J. Am. Coll. Toxicol.*, **8**, 1275 (1989).
111. 'IARC Monographs on the Evaluation of Carcinogenic Risk of Chemicals to Humans, Some Metals and Metallic Compounds ', vol. 23, p. 205, International Agency for Research on Cancer, Lyon, France 1980.
112. (a) G. Tamino, L. Peretta and A. G. Levis, *Chem.-Biol. Interactions*, **37**, 309 (1981); (b) A. G. Levis, V. Bianchi, G. Tamino and B. Pegoraro, *Br. J. Cancer*, **37**, 386 (1978).
113. S. Kawanishi, S. Inoue and S. Sano, *J. Biol. Chem.*, **261**, 5952 (1986) and references therein.
114. (a) P. Jones, A. Kortenkamp, P. O'Brien, G. Wang, and G. Yang, *Arch. Biochem. Biophys.*, **286**, 652 (1991); (b) A. Kortenkamp, G. Oetken, and D. Beyersman, *Mutat. Res.*, **232**, 155 (1990), (c) P. O'Brien, J. Pratt, F. Swanson, P. Thornton and G. Wang, *Inorg. Chim. Acta*, 169, (1990), and references therein. **??**
115. (a) J. Aiyar, K. M. Borges, R. A. Floyd and K. E. Wetterhahn, *Toxicol. Environ. Chem.*, **22**, 135 (1989); (b) R. A. Floyd, M. S. West, K. L. Eneff, W. E. Hogsett, and D. T. Tingey, *Arch. Biochem. Biophys.*, **262**, 266 (1988), and references therein.
116. X. Shi and N. S. Dalal, *Arch. Biochem. Biophys*, **281**, 90 (1991) and references therein.
117. J. A. Connor and E. A. V. Ebsworth, *Adv. Inorg. Chem. Radiochem.*, **6**, 279 (1964).
118. S. B. Brown, J. O. Edwards, I. J. Herman, P. Jones, J. R. Mills and J. E. Earley, *Inorg. Chim. Acta*, **3**, 351 (1969).
119. S. L. Brauer and K. E. Wetterhahn, *J. Am. Chem. Soc.*, **113**, 3001 (1991).
120. M. S. Akhlaq, H. P. Schuchmann and C. von Sonntag, *Int. J. Radiat. Biol.*, **51**, 91 (1987).
121. D. A. Rowley and B. Halliwell, *FEBS Lett.*, **138**, 33 (1982).

CHAPTER 5

Activation of Hydrogen Peroxide with Biomimetic Systems

Bernard MEUNIER

Laboratoire de Chimie de Coordination du CNRS, 205 route de Narbonne, 31077 Toulouse cedex, France

Introduction

Hydrogen peroxide is a natural oxidant formed and used in biological systems, since Evolution provided molecular oxygen on earth. Mitochondria and peroxisomes are the essential sites of the molecular oxygen uptake by cells. Proteins such as D-aminoacid oxidase, urate oxidase, galactose oxidase, amine oxidases are able to produce hydrogen peroxide in O_2-mediated oxidation of the corresponding substrate [1]. In order to control the hydrogen peroxide level in these sub-cellular compartments, catalase is present and catalyzes the H_2O_2 dismutation in water and dioxygen at a rate of 10^5 cycles/second [2].

Peroxidases are also using hydrogen peroxide as co-factor [2b, 3]. Unexpected overproduction of hydrogen peroxide by cells during respiration cycle or under irradiation would lead to mutagenic or lethal events to biological systems, through the production of hydroxyl radicals HO• [4]. Because of the very low dissociation energy of the peroxide O-O bond (only 49 Kcal/mol, compared to 89 Kcal/mol for the dissociation of an O-H bond in H_2O_2 [5]), the homolytic cleavage of the O-O bond resulting from an electron transfer to H_2O_2 is the easiest reaction at ambient temperature. When the electron is provided by an iron(II) salt or complex, this generation of HO• is known as the Fenton reaction [6].

For hemoproteins such as catalase or peroxidases which are handling hydrogen peroxide as a single oxygen atom donor (*i.e.* a heterolytic cleavage of the O-O bond), their distal pocket is equipped with several acid and basic aminoacid residues in order to facilitate the heterolytic, *vs* . homolytic, cleavage of hydrogen peroxide (for details on the structure of catalase or peroxidase and the role of the distal aminoacids in the push-pull mechanism of the O-O cleavage, see ref. 7). We will see that one of the key points in the use of hydrogen peroxide in biomimetic catalyst-mediated oxidations is to avoid

G. Strukul (ed.), Catalytic Oxidations with Hydrogen Peroxide as Oxidant, 153–175.

the homolytic cleavage of the peroxide O-O bond.

This account describes the recent H_2O_2 oxidation reactions catalyzed by biomimetic catalysts. Since many hemoproteins are using hydrogen peroxide, nearly all the biomimetic catalysts are based on synthetic tetraarylporphyrins metallated by iron or manganese. As admitted by enzymologists, a distinction will be made throughout this review between *oxygenation* and *oxidation* reactions (namely oxygen atom insertion in the substrate *vs.* electron abstraction from the substrate).

For general recent review articles on metalloporphyrin-catalyzed oxygenation or oxidation reactions, see ref. 8 to 13.

1. Catalase Activity of Metalloporphyrins

All aerobic organisms have a high content of catalases in order to protect themselves from any excess of hydrogen peroxide [2]. Most of the known catalases are tetramers of identical subunits containing a ferriprotoporphyrin as prosthetic group (a non-heme Mn-catalase has recently been isolated and identified [14]). Each heme has a tyrosinato ligand as proximal residue and is buried 20 Å below the protein surface [7b]. There is a channel (length = 30 Å; width = 15 Å) for the access to each heme site. Only small molecules are able to enter this channel: hydrogen peroxide, methanol or ethanol.

The dismutation of hydrogen peroxide by heme-catalase (EC 1.11.1.6) occurs *via* the formation of an intermediate iron-oxo species generated by the first H_2O_2 molecule (Scheme 1, pathway *a*). This green intermediate, Compound I, is two redox equivalents above the resting state of the native enzyme and is considered as an Fe(IV)=O species with a delocalized π-cation radical on the porphyrin ligand [15]. A second molecule of hydrogen peroxide interacts with catalase Compound I leading to the evolution of molecular oxygen and formation of a water molecule. Catalase is also able to oxidize ethanol by a stereospecific removal of the pro-R-hydrogen [16]. One of the most impressive features of catalase is its high catalytic activity: 10^5 to 10^7 catalytic cycles *per* second [17]. Since H_2O_2 can easily be cleaved by homolysis of the O-O bond, this means that the protein is able to handle an average of 10^6 molecules of H_2O_2 *per* second without generating HO• radicals.

Scheme 1:

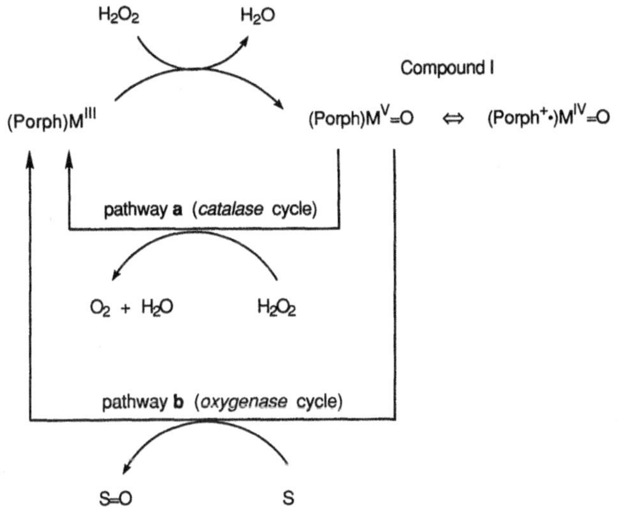

Any formation of HO• during the dismutation reaction should give rise to a denaturation of catalase, *i.e.* to an inhibition of the H_2O_2 dismutation. If we assume that one homolytic cleavage event corresponds to the loss of the catalase activity, one can calculate that for a continuous H_2O_2 dismutation by catalase over 10 minutes, this corresponds to a selectivity of 6 x 10^8 to 1 between the dismutation and the homolysis of hydrogen peroxide! Up to now, no chemical catalysts have been able to reach such chemoselectivity in reactions involving hydrogen peroxide.

The catalase activity of iron-porphyrin complexes based on natural porphyrin ligands has long been recognized by Von Euler [18] and Jones [19]. The activity of catalase is pH-independent in the range 5-10. In contrast, heme complexes exhibit a pH-dependent dismutase activity. At pH 13, the activity of iron protoporphyrin is close to the activity of catalase [19a]. However, the stability of these flat metalloporphyrins is poor and they are quickly degraded [20].

Two different approaches have been recently used in order to increase the catalyst stability. One method is to adsorb ionic metalloporphyrins on a ion-exchange resin. The second one is to take advantage of the cage-effect created by sterically hindered tetrarylporphyrin ligands with substituents at the *ortho* and *ortho'* positions of the *meso* -phenyl groups. In both cases, the oxidative degradation due to an intermolecular oxygen atom transfer from high-valent iron-oxo species to the ligand of another catalyst molecule is considerably reduced.

The first approach has been developed by Saito *et al.* using cobalt, iron or manganese *meso*-tetrakis(4-sulfonatophenyl)porphyrin complexes adsorbed by strong electrostatic interactions on an ion-exchange resin such as Amberlite® IRA 900 [21]. After 10 repeated uses of 30 min each, both manganese- and cobalt-supported sulfonated porphyrin complexes exhibit a high catalase activity (50-70% of the initial activity) [21b]. The catalase activity of this supported manganese porphyrin catalyst was evidenced using the hydrogen peroxide oxidation of methanol [21c]. The formaldehyde formed was quantified by its reaction with 4-amino-3-hydrazino-5-mercapto-1,2,4-triazole (AHMT), the resulting dye showing a strong absorption band at 550 nm. The highest catalytic activities have been obtained with the supported manganese catalyst. The activity is divided by a factor of two using the corresponding iron or cobalt complexes. The authors suggest that these catalase models might have a future for the determination of hydrogen peroxide concentrations in clinical analyses, since they are more stable than proteins during a long term storage.

The corresponding iron-supported complex is less stable compared to manganese and cobalt catalysts, suggesting that an intramolecular oxidative degradation pathway might be predominant in the case of iron porphyrin complexes. Metalloporphyrins can be linked to polymers only by the axial ligand, *i.e.* the basic nitrogen of polylysine [22]. In this case, *i.e.* the poly-L-lysine bound heme, the catalase activity is similar to that of the free heme. However, the metalloporphyrin binding on polylysine is not irreversible like in the case of sulfonated metalloporphyrin on Amberlite® resins: the free heme is bleached, so the equilibrium is driven in favor of the free heme leading to the displacement of the bonded heme from polylysine.

The second approach to obtain efficient catalase models is to create a cage around the metal site to avoid intermolecular oxygen atom transfer during the catalytic cycle. If the substrate S in the pathway b of Scheme I is a second catalyst molecule, this reactions leads to the oxidative intermolecular degradation of the metalloporphyrin.

In catalytic hydrocarbon oxygenation reactions, hindered metalloporphyrins with substituents at the *ortho* and *ortho'* positions of the *meso*-phenyl groups have been successfully used as catalysts [23]. The presence of alkyl or halogen substituents on such positions creates a cage around the highly reactive metal-oxo species and consequently avoids the transfer of the oxygen atom to another catalyst molecule. This approach was successfully developed by Bruice *et al.* in the case of catalase models using the iron derivative of the *meso*-tetrakis(2,6-dimethyl-3-sulfonatophenyl)porphyrin [24]. The ligand was obtained by the sulfonation of the corresponding porphyrin ligand in sulfuric acid at 100°C for 4 hours. This water-soluble iron-porphyrin complex exists as two possible forms depending on the pH value of the solution: with a hydroxo ligand at basic pH (Soret band at 413 nm) or with two water molecules in axial positions at acidic pH (hexacoordinated complex, Soret band at 393 nm) [24a]. The rate constants of the reaction of this water-soluble iron-porphyrin with hydrogen peroxide were determined by using a water-soluble substrate used for clinical peroxidase assays: ABTS or 2,2'-azinobis(3-ethylbenzothiazoline-6-sulfonic acid) [25]. The oxidized form of ABTS, namely its radical-cation form, exhibits a strong absorption band at 660 nm.

Both hydrogen peroxide and ABTS are competing in the reaction with the intermediate high-valent iron-oxo species formed by cleavage of the hydrogen peroxide molecule. The oxidation of hydrogen peroxide to molecular oxygen by the iron-oxo complex is favored at basic pH [24]. The rate constant for the formation of the iron-oxo species by the proton-catalyzed O-O bond scission is estimated to be close to 10^9 M^{-1} s^{-1}. The rate constants for the formation of the corresponding manganese-oxo complex have also been determined and are in the range 10^4-10^5 M^{-1} s^{-1} [24b]. Traylor and Xu have also evidenced that high-valent iron-oxo complexes are able to abstract H• from alkylhydroperoxide [26]. However it is not clear whether such reaction may occur with hydrogen peroxide itself in catalase-type reactions.

Up to recently, no attention was paid to the role of the axial ligand *trans* to the metal-oxo species in the catalytic dismutation of hydrogen peroxide by metalloporphyrins. Among all heme-enzyme, catalase is the only one with an oxgyen atom (from a tyrosine residue) in proximal position [7b]. Cytochrome P-450 and peroxidases have a sulfur-containing and a nitrogen-containing proximal ligand, respectively. So, why an oxygen atom in a proximal position in catalase? In order to answer this question, we have performed a comparative study using super-structured metalloporphyrins having an oxygen proximal ligand or a nitrogen proximal ligand (see Scheme 2 for structures and Table 1 for data on the catalytic oxygenation and H_2O_2 dismutation) [27]. Three different
key points should be noted: *i* the proximal ligand being the same, manganese catalysts have a higher catalase activity than the corresponding iron complexes,

ii complexes having one imidazole proximal ligand are more efficient catalysts for the H_2O_2 dismutation compared to complexes having one alkoxo or phenoxo proximal ligand, but *iii* whereas the imidazole-containing catalysts are able to epoxidize olefins, those having an oxygen atom in proximal position are not able to do so.

Table 1: Competitive oxygenation (olefin epoxidation) and hydrogen peroxide dismutation catalyzed by various manganese or iron porphyrin complexes having a oxygen or a nitrogen proximal ligand (see Scheme 2).

Catalyst	M	Conversion (%)	
		Catalase (oxygen yield)[a]	Oxygenase (epoxide yield)[a]
1	Mn	26% (89)	0%
2	Fe	18% (62)	0%
3	Mn	27% (93)	1% (2)
4	Fe	21% (72)	0%
5	Mn	31% (106)	4% (6)

[a] The numbers in parentheses are the turnover numbers based on the product yields (epoxide or molecular oxygen, calculated from the olefin or the initial amount of H_2O_2 respectively) at 5 minutes.

It can be noticed that, among the manganese complexes, the one (**1**) with a phenoxo proximal ligand like in catalase itself, is the only manganese complex without any oxygenase activity. Thus, one of the main role of the tyrosine proximal ligand in catalase might be to inhibit completely the oxygenase cycle (pathway b in Scheme I) in order to avoid a possible protein degradation by an oxygen atom transfer reaction from the high valent metal-oxo species (Compound I) to a distal protein residue.

2. H_2O_2 Oxygenation Reaction Catalyzed by Metalloporphyrin Complexes

Two main oxygenation reactions have been reported using hydrogen peroxide as oxygen atom source and a metalloporphyrin as catalyst: olefin epoxidation and saturated alkane hydroxylation. However, it should be noted

Scheme 2:

M = Mn (**1**); Fe (**2**)

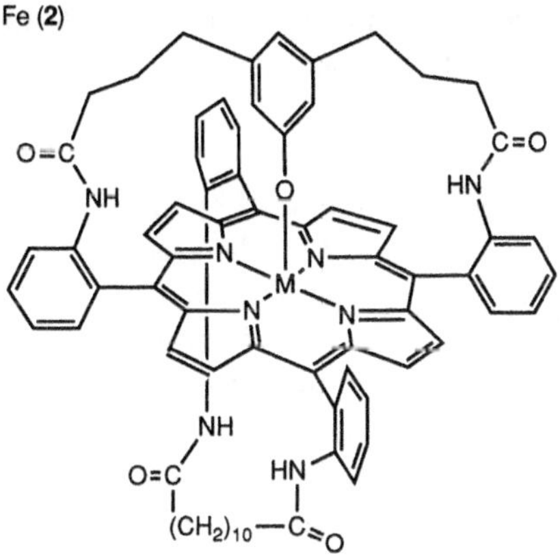

M = Mn (**3**); Fe (**4**)

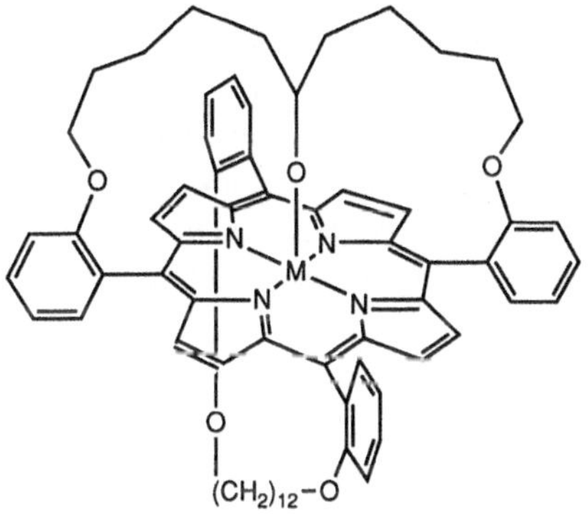

M = Mn (**5**); Fe (**6**)

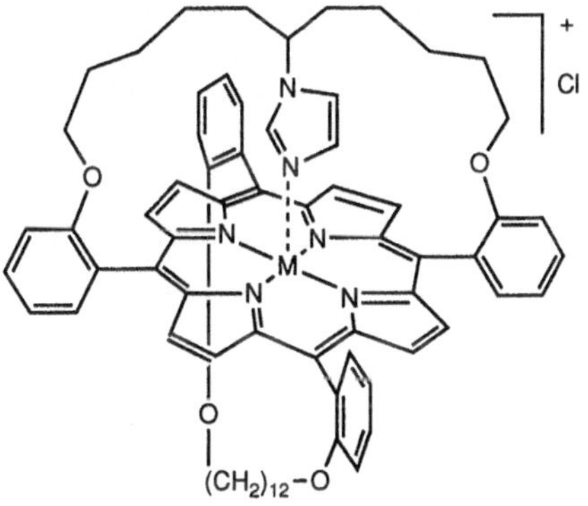

that the first metalloporphyrin-catalyzed oxygen transfer reaction from H_2O_2 to a substrate was reported by Oae *et al.* on the S-oxygenation of organic sulfides [28].

$$\text{Ph-S-CH}_3 \xrightarrow[\text{imidazole}]{\text{Fe(TPP)Cl / H}_2\text{O}_2} \text{Ph-SO-CH}_3 + \text{Ph-S-S-Ph} + \text{CH}_2\text{O}$$

Besides the sulfoxide Ar-SO-CH_3, the authors also observed the corresponding disulfide and formaldehyde resulting from the S-dealkylation reaction. Both reactions are catalyzed by cytochrome P-450. Imidazole played a key role in these H_2O_2 dependent oxygenation reactions (see next paragraph).

2.1. Olefin Epoxidation

The same catalytic system has been used by Mansuy *et al.* for the olefin epoxidation by hydrogen peroxide [29, 30]. Manganese porphyrin derivatives are more efficient than the corresponding iron complexes. As observed for other oxygen atom donors, catalysts derived from the regular tetraphenylporphyrin ligand are rapidly destroyed before the complete olefin conversion [29]. Other robust porphyrins, TMP or TDCPP (respectively the dianion of tetramesitylporphyrin or tetrakis(2,6-dichlorophenyl)porphyrin, see Scheme 3 for structures) are suitable ligands for manganese porphyrin catalysts.

In the case of epoxidations catalyzed by Mn(TDCPP)Cl, the compound is found intact at the end of the olefin conversion. With this catalyst 90% epoxide yields are reached, even for a terminal olefin. The catalytic olefin epoxidation by H_2O_2 is stereospecific for aliphatic olefin and only a little loss of stereoselectivity is detected in the case of *cis* -stilbene (*cis/trans* epoxide ratio: 97/3), an olefin leading usually to a mixture of both stereoisomers in metalloporphyrin-catalyzed epoxidation, depending on the nature of the oxygen atom donor, the metalloporphyrin and the proximal ligand [23a, 31].

The exact role of imidazole was elucidated by Mansuy *et al.* using basket-handle type porphyrin ligands [30]. Since these ligands have handles from one *ortho* -phenyl position to the corresponding *trans* position, the approach to the central metal is restricted. Such properties were successfully used in metalloporphyrin-mediated alkane hydroxylation by alkylhydroperoxides in order to distinguish between a metal-oxo pathway (resulting from the heterolytic cleavage of the peroxide bond by the metalloporphyrin catalyst) or a Fenton-like mechanism (resulting from the homolytic cleavage of the O-O bond, without formation of metal-oxo species) [32]. In fact, these highly hindered complexes are unable to bind imidazole and do not catalyze the olefin epoxidation by H_2O_2 [30]. Still, the coordination of imidazole in one of the axial positions of the manganese porphyrin is not sufficient: in addition, imidazole must interact with hydrogen peroxide on the distal site in order to facilitate the heterolytic cleavage of the O-O bond (Scheme 4).

Scheme 3:

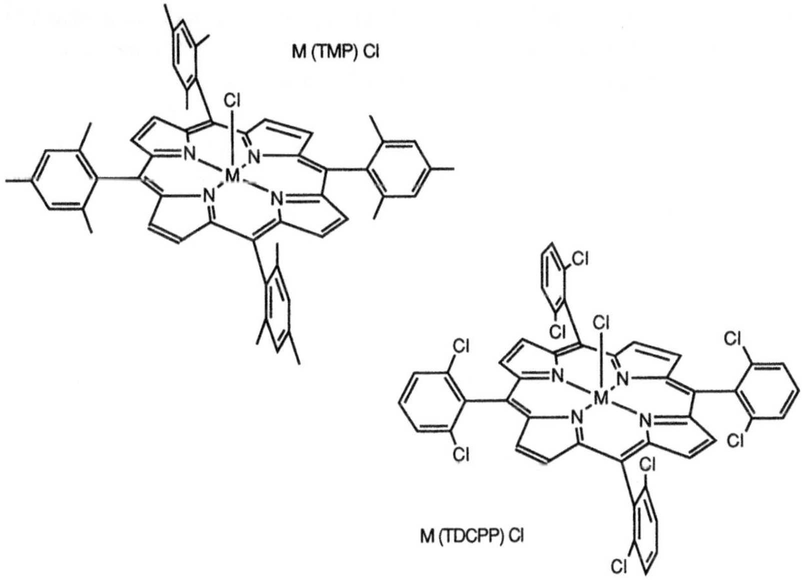

M = Mn, Fe

Scheme 4:

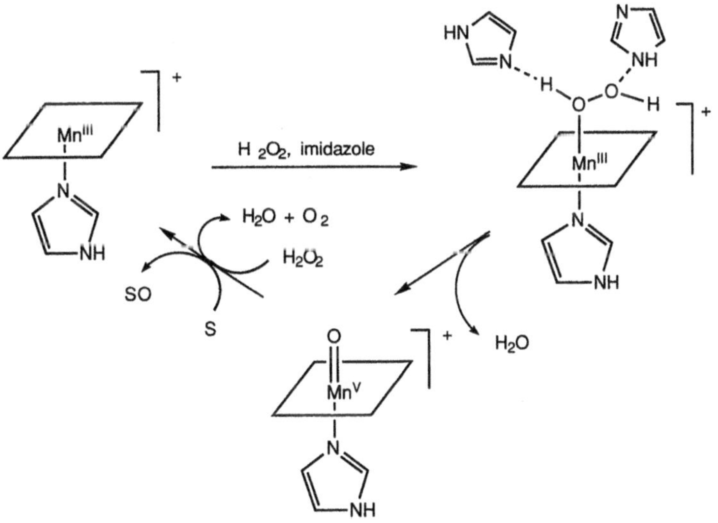

This was demonstrated using another basket-handle manganese porphyrin complex containing an imidazole ligand coming from one handle, in proximal position (this complex was previously used, as well as the pyridine analogue, in hypochlorite olefin epoxidation in order to compare the proximal effect of these two nitrogen-containing ligands [33]). In the absence of extra free imidazole in the reaction mixture, the manganese complex having a proximal imidazole is unable to catalyze the H_2O_2 dependent epoxidation [30]. This fact supports the role of imidazole on the distal side of the catalyst in order to make possible the heterolytic cleavage of the peroxide O-O bond. Two distal imidazole molecules have been indicated in Scheme 4, one acting as base to remove a proton from H_2O_2, the other being the proton source for the formation of the water molecule. This drawing is made by analogy of the push-pull mechanism reported for peroxidases (see ref. 7c for a review article on this topic) and can also be extended to catalase. In the original proposed catalytic cycle for these metalloporphyrin-mediated oxidations, only one distal imidazole was indicated [30].

Besides all these mechanistic aspects, it should be noted that diluted hydrogen peroxide (30% in H_2O) is suitable as oxygen atom source in these metalloporphyrin-mediated epoxidations. This safety aspect is not negligible and will be probably a major factor for the possible application of this epoxidation method, at least at the laboratory scale. Recently, Sawyer *et al.* emphasized the use of anhydrous hydrogen peroxide with iron salts in dry acetonitrile [34]. But one should keep in mind any possible explosive decomposition of anhydrous hydrogen peroxide (for some comments on how to safely handle hydrogen peroxide, see ref. 35).

A new trend in the development of diluted hydrogen peroxide solutions has been developed recently by Montanari *et al.* [36]. Using Mn(TDCPP)Cl as catalyst, a biphasic system (CH_2Cl_2/H_2O) and a 30% H_2O_2 water solution (the pH being adjusted to 4.5-5.0), they found that high catalytic activities are obtained in the presence of 1 to 8 equivalents of carboxylic acids (acetic or benzoic acid) and 1 equivalent of lipophilic imidazole or pyridine (n-hexyl imidazole or 4-t-butylpyridine) with respect to the metalloporphyrin.

$$\text{olefin} + H_2O_2 \xrightarrow[\text{axial ligand/carboxylic acid}]{\text{Mn(TDCPP)Cl}} \text{epoxide} + H_2O$$

With a [cyclooctene]/[H_2O_2]/[*N* -hexylimidazole]/[benzoic acid]/

[Mn(TDCPP)Cl] molar ratio of 200:400:1:1:1, the corresponding epoxide is obtained within 10 min at 0°C with a nearly quantitative yield. The catalytic activity, 20 cycles/min, is above that observed for a cytochrome P-450 monooxygenase. In this case, the carboxylic acid favors the generation of the high-valent manganese-oxo species.

However it is not clear yet, whether this is due to an acid-catalyzed heterolytic cleavage of the peroxide O-O bond or to the intermediate formation of a peracid. The latter compounds are known to be very efficient oxygen atom donors, even at very low temperature [37]. In addition, it must be noted that only a 2-fold molar excess of H_2O_2 is required to achieve complete olefin epoxidation. This suggests that the catalase activity is highly minimized in the Montanari-Banfi system. Prior to this report, a group of the Degussa company in collaboration with Buchler have reported attempts to use molybdenum and tungsten porphyrin catalysts in order to reduce the catalase activity and thus to enhance the hydrogen peroxide consumption with respect to the epoxide formation [38]. Epoxidation of cyclohexene in a mixture of propylacetate and methyl trimethylacetate at 60°C with 85% aqueous H_2O_2 is performed in 24 h with 74% of H_2O_2 consumption. The 90% selectivity observed for the hydrogen consumption *versus* epoxide formation indicates that the side-catalase reaction is very low compared to the epoxide reaction. However these molybdenum or tungsten porphyrin catalyzed epoxidations are rather slow compared to the highly efficient manganese-porphyrin based systems.

2.2. Alkane and Arene Hydroxylation

Hydroxyl radicals are very efficient agents for the hydroxylation of aromatic compounds including benzene itself [6b, 39]. The rate of the HO• addition to aromatic substrates is close to that of diffusion. These hydroxyl radicals are easily generated from H_2O_2 by different methods: in the presence of iron(II) salts [6b, 39] (Fenton's reagent) or by UV photolysis [40]. Even in aprotic solvents, it has been proposed that the aromatic hydroxylation by H_2O_2/Fe(II) or Fe(III) salts occurred *via* hydroxyl radicals [41], whereas using also acetonitrile as solvent it has been suggested that aliphatic hydroxylation by H_2O_2/Fe(II) salts might occur *via* Fe=O species [41]. Considering these remarks, it is not easy to say whether some reported aromatic hydroxylations by H_2O_2/metalloporphyrin are due to a Fenton-type or a metal-oxo chemistry, especially when the catalytic activity is rather low.

However, since Fenton's reagent is known to lead to a low phenol selectivity and a wide variety of quinones (monomers, dimers...), one can assume that any metalloporphyrin-catalyzed hydroxylation of aromatic compounds giving rise to a good selectivity in phenol derivatives may occur via a monooxygenase-type reaction (oxygen atom transfer from a metal-oxo species opposed to hydroxyl radical addition).

This is the case of the hydroxylation of phenylalanine by aqueous H_2O_2 in the presence of $[Fe^{III}(TMPyP)](OTs)_5$ as catalyst (see Scheme 5) [43]. The tyrosine/DOPA ratio decreased from 10 to 5 during the first hour of the hydroxylation reaction which occurred at a rate of 7 cycles/hour.

An unambiguous monooxygenase-type hydroxylation system based on manganese porphyrin complexes has been developed by Mansuy *et al.* [44]. In the presence of imidazole, hydroxylation of alkanes by H_2O_2 is catalyzed by Mn(TDCPP)Cl (2.5% *versus* substrate) in a mixture of CH_3OH/H_2O (9/1). Within one hour, 54% of cyclohexane is converted to cyclohexanol (yield 30%) and cyclohexane (10%).

Scheme 5:

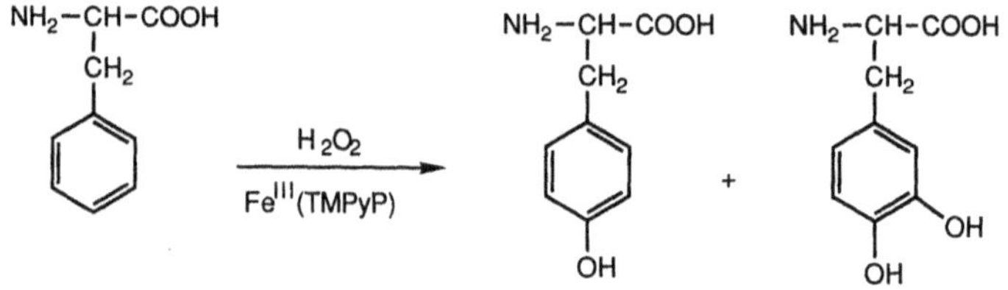

Phenylalanine　　Tyrosine　　DOPA

Adamantane gives both alcohols with a predominance for the tertiary alcohol as expected for a mechanism involving hydrogen atom abstraction (adamantan-1-ol, 63% and adamantan-2-ol, 19%), whereas the ketone (adamantan-2-one) yield is only 3%. No isotope effects have been reported so far with the $Mn(TDCPP)Cl/H_2O_2$/imidazole system. The k_H/k_D values are expected to be as high as 8-11 as in the case of cytochrome P-450 itself or the models using iodosylbenzene as single oxygen atom donor [8], but lower values have been found when potassium monopersulfate is used as oxidant in metalloporphyrin catalyzed hydroxylations [45].

It has been shown that iron complexes are less efficient oxygenation catalysts compared to the corresponding manganese complexes because of their tendency to catalyze the H_2O_2 dismutation reaction [30].

In addition to the catalase activity of iron porphyrins, these complexes also give quite often an unstable derivative which is unable to catalyze an oxygenation reaction [46]. To reinforce the robustness of metalloporphyrin catalysts, one approach is to protect the β-pyrrole positions from hydrogen atom abstraction by halogenation. This approach has been chosen in the case of the TDCPP ligand [47], TMP [48], or TFPFP [49]. The iron derivative of the latter perfluorinated porphyrin TFPFP (the so-called "teflon porphyrin") is able to catalyze the hydroxylation of benzene to phenol (55 turnover numbers within 2 h) [49]. 1,4-Benzoquinone is also observed in the reaction mixture. This complex is intact at the end of the catalytic reaction. The same complex also catalyzes the H_2O_2 epoxidation of cyclooctene at a rate of 10 cycles/min.

Another approach to reduce metalloporphyrin destruction in H_2O_2 hydroxylations is to immobilize these catalysts onto imidazole-modified silica gel [50] (for the use of immobilized metalloporphyrin as catalase models, see the previous main paragraph and as peroxidase models, see the next one). The manganese porphyrin complex, Mn(TPP)Cl, has been linked to a silica gel containing Si-O-$(CH_2)_3$-imidazole groups by coordination of the nitrogen base in axial position of the complex itself [50]. Cyclohexane is converted to cyclohexanol (32 turnovers) and cyclohexanone (10 turnovers). Adamantane is mainly hydroxylated at the tertiary positions, but the normalized [tertiary C-H]/[secondary C-H] hydroxylation ratio is 5.4 and similar to those of other cytochrome P-450 models and far below that of a Fenton-type free radical oxidation [51].

3. Peroxidase Models or Electron Abstraction Reactions Catalyzed by Metalloporphyrins in the Presence of H_2O_2

3.1. Early Peroxidase Models

Most of the known peroxidases are heme proteins with a histidine residue as proximal ligand of the iron protoporphyrin IX [2b, 3]. These enzymes make use of hydrogen peroxide (the so-called "co-factor") as electron acceptor, the overall stoichiometry of the reaction being the abstraction of two electrons from a substrate molecule and the reduction of H_2O_2 to H_2O.

$$AH_2 + H_2O_2 \xrightarrow{\text{peroxidase}} A + 2\,H_2O$$

Jones [52] and Traylor [53] have studied the use of synthetic iron(III) porphyrins as models of peroxidases. Some difficulties related to non-robust metalloporphyrins (catalyst destruction, dimerization in basic media) prevented investigators from carrying out thorough studies. More recently, Saito *et al.* found that anionic manganese porphyrin complexes immobilized onto an anion-exchange resin were efficient peroxidase models [54]. Intermolecular destruction of catalyst molecule is prevented by isolation of active sites on the resin. Consequently, high catalytic activities are observed, even after repeated uses of the same modified resin [54, 55].

$$\text{4-aminoantipyrine} + \text{phenol} \xrightarrow[2\,H_2O_2]{Mn^{III}TPPS/Amberlite} \text{quinonedye} + 4\,H_2O$$

Manganese tetrakis(4-sulfonatophenyl)porphyrin was immobilized onto Amberlite IRA 900 by strong electrostatic interactions. This supported catalyst was used for the determination of H_2O_2 concentration by titration of a quinone dye resulting from the oxidative condensation of 4-aminoantipyrine with phenol. However, a large excess of phenol is required to obtain efficient dye formation. But, in place of phenol as chromogen, *N,N* -diethylaniline can be advantageously used [56]. This approach is an alternative method in clinical analyses of glucose or cholesterol. These compounds are routinely monitored by measurement of hydrogen peroxide produced during peroxidase-mediated oxidations [57].

Another method to obtain a reactive polymer exhibiting a peroxidase activity is to attach covalently hemin, which has two carboxyl groups, to monomethoxypolyethylene glycol, PEG, through ester bonds [58]. This hemin-modified PEG is soluble in neutral aqueous solutions as well as in organic solvents such as 1,1,1-trichloroethane. In the latter solvent, a strong Soret band is observed at 398 nm [58].

The peroxidase characteristics of this reactive polymer were determined by measuring the oxidation of *o* -phenylenediamine in the presence of hydrogen peroxide. The initial rate of the reaction is linearly dependent on the concentration of hydrogen peroxide or on that of PEG-hemin as expected for a peroxidase-type reaction. The rate constant of the first-order reaction is 2.3 x 10^3 M^{-1} sec^{-1} in trichloroethane and only 3.0 x 10 M^{-1} sec^{-1} in aqueous solution buffered at pH 7.0. Such behavior, *i.e.* a lower activity in water compared to organic solvents, was also observed in the catalytic H_2O_2 dismutation by catalase models [27b]. All in all, despite a lower catalytic activity compared to peroxidases, these functional polymers have probably a future as substitutes for the corresponding enzymes in analytical chemistry because of their stability and facile storage at room temperature [59].

The same approach, *i.e.* immobilization of metalloporphyrins, was also successfully used in the modelling of lignin, a peroxidase involved in delignification of wood.

3.2. Ligninase Models Based on Supported Metalloporphyrins

A heme-containing H_2O_2-dependent enzyme has been recently purified from the extra-cellular medium of the *Phanerochaete Chrysosporium.* [60] This peroxidase, *i.e.* lignin peroxidase (ligninase or LiP), is involved, as well as manganese peroxidase, in the enzyme-catalyzed degradation of lignin in wood (for a recent review article on delignification of wood, see ref. 61). The search for new clean delignification procedures has stimulated all studies related to biodegradation of lignin by H_2O_2-dependent enzymes or by enzyme models. Recent reports have shown that synthetic iron porphyrin complexes associated to alkylhydroperoxides [62], sodium hypochlorite [63], or molecular oxygen and a reducing agent [64] are able to oxidize lignin models and also to delignify wood chips [62b]. Because of our own interest in demethoxylation reactions catalyzed by peroxidase [65], we have recently developed efficient ligninase models based on the association of a H_2O_2 surrogate, potassium monopersulfate with manganese or iron porphyrin either free or immobilized on an ion-exchange resin [66] (see Scheme 6 for structures of different sulfonated metalloporphyrin complexes and for a representation of a supported catalyst on a cationic exchange resin). The activity of these different metalloporphyrin complexes were assayed using two benzylic alcohol derivatives which are generally used in lignin degradation model studies, *i.e.* veratryl alcohol (**1**) and 1-(3,4-dimethoxyphenyl)-2-(2-methoxyphenoxy)propane-1,3-diol (**2**). The latter dimer molecule is used to evidence the oxidative cleavage of the C_α-C_β bond, a key step in delignification (see Scheme 7).

Scheme 6:

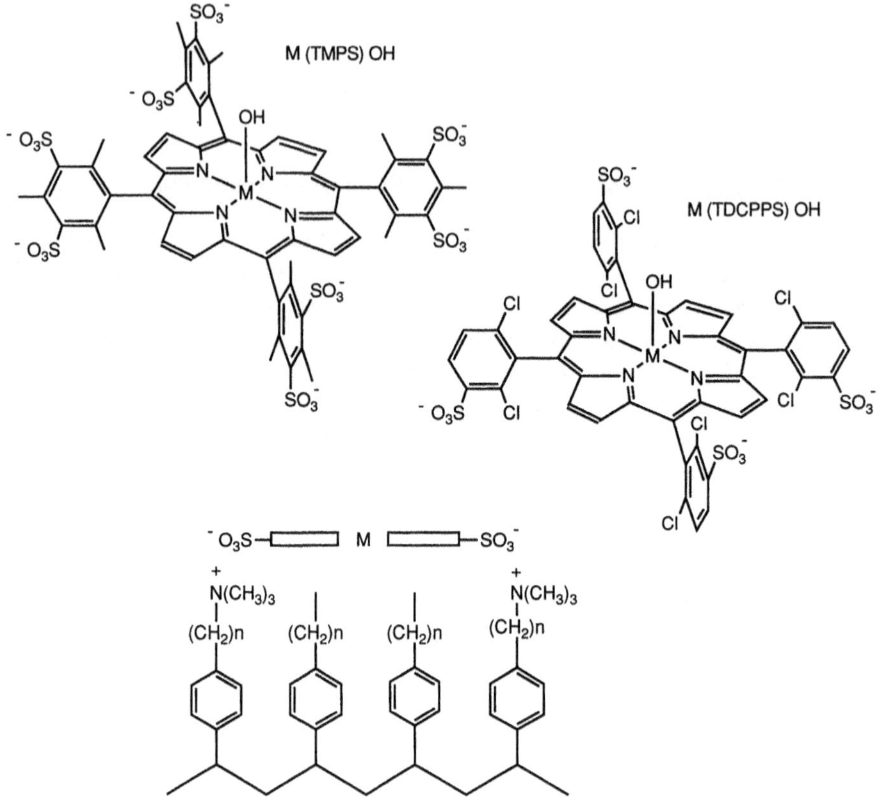

Potassium monopersulfate is more efficient than hydrogen peroxide in the oxidation of veratryl alcohol, **1**, catalyzed by FeTPPS or MnTPPS, both used as soluble or supported catalysts. This lower activity observed with H_2O_2 is essentially due to a rapid bleaching of the metalloporphyrin complexes (the corresponding data are reported in Table 2). Iron complexes are highly active at pH 3.0: only 0.2% of catalyst *vs.* substrate is required to obtain a complete oxidation of veratryl alcohol within a few minutes. The two main oxidation products are the corresponding substituted benzaldehyde (**3**) and 2-methoxy-5-hydroxymethyl-1,4-benzoquinone (**4**) a quinone resulting from the oxidation of the aromatic

Scheme 7:

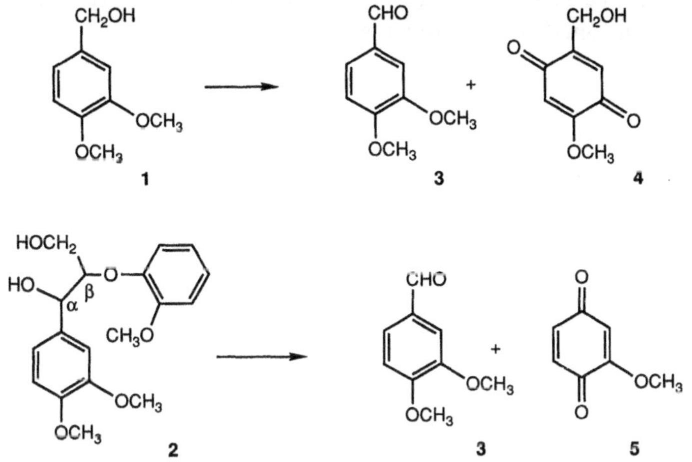

cycle of the substrate (see Table 3). The manganese catalysts are active at pH 6.0 (Table 2) and the two same oxidation products, **3** and **4**, are observed (Table 3). Yet, in the case of manganese complexes, the major oxidation product is the aldehyde and not the quinone like in the case of iron catalysts.

Table 2: Comparison between $KHSO_5$ and H_2O_2 in the oxidation of veratryl alcohol (**1**) by Fe^{III}(TPPS) and Mn^{III}(TPPS) in solution and supported on Amberlite (-Ad).

catalyst	%[a]	oxidant	imidazole/cat	conversion[b] (%)
Fe(TPPS)	0.2	H_2O_2	-	5
	0.2	H_2O_2	100	7
	0.2	$KHSO_5$	-	67
	0.2	$KHSO_5$	100	65
Mn(TPPS)	3	H_2O_2	100	7
	3	$KHSO_5$	100	67
Fe(TPPS)-Ad	10	H_2O_2	-	6
	10	$KHSO_5$	-	50
Mn(TPPS)-Ad	10	H_2O_2	100	7
	10	$KHSO_5$	100	61

a % of catalyst with respect to substrate. *b* reaction time *one* min.

Table 3: Product distribution in the $KHSO_5$ oxidation of veratryl alcohol (**1**) or the dimer (**2**) by Fe^{III}(TPPS) and Mn^{III}(TPPS) in solution and supported on Amberlite (-Ad).

Catalyst	%[a]	pH	Substrate	Product yields (%)[b] 3	4	5	Total
Fe(TPPS)	0.3	3.0	**1**	20	70		90
			2	10		30	40
Fe(TPPS)-Ad	10	3.0	**1**	35	50		85
			2	10		35	35
Mn (TPPS)[c]	5	6.0	**1**	40	25		65
			2	15		20	35
Mn(TPPS)-Ad[c]	10	6.0	**1**	40	20		60
			2	15		15	30

a % of catalyst with respect to substrate. *b* reaction time 15 min (complete conversion). *c* 100 equiv. 4-tert-butylpyridine/catalyst added.

Both manganese and iron complexes are able to oxidize the dimer **2** into two compounds, **4** and **5**, which result from the oxidative cleavage of the C_α-C_β bond of this lignin model (see Table 3 for the yields of these products in MTPPS catalyzed reactions).

In order to improve the efficiency of these ligninase models, we decided to synthesize more "robust" catalysts by sulfonation of two porphyrin ligands having substituents in *ortho* and *ortho'* positions of the phenyl rings, $TMPH_2$ and $TDCPPH_2$. Tetramesithylporphyrin was easily sulfonated at both *meta* position of the mesityl groups, whereas $TDCPPH_2$ was only sulfonated at one *meta* positions [67]. The corresponding water-soluble iron and manganese complexes, M(TMPS) and M(TDCPPS) (Scheme 6), are obtained in good yields from $TPMH_2$ or $TDCPPH_2$ [67]. The capacity of these metalloporphyrin complexes to resist to the bleaching by potassium monopersulfate was checked by monitoring the disappearance of their Soret band during the incubation without any substrate. Mn(TMPS) and Mn(TDCPPS) were intact after a 10 min incubation while 10% and 80% of Fe(TDCPPS) and Fe(TMPS) were respectively destroyed under the same conditions. These metalloporphyrin complexes are highly efficient catalysts in the oxidation of lignin molecule models and can be re-used with only a small decrease in their initial catalytic activity (see Table 4 for some data on potassium monopersulfate oxidation of veratryl alcohol **1** by robust sulfonated metalloporphyrin complexes). In this field of ligninase modeling, synthetic metalloporphyrin complexes seem to have a chance to be considered as efficient catalysts compared to ligninase which is a rather fragile enzyme [68].

Table 4: Product distribution in the $KHSO_5$ oxidation of veratryl alcohol (**1**) by robust metalloporphyrins M(TMPS) and M(TDCPPS) (M = Fe, Mn) in solution and supported on Amberlite (-Ad).

					Product yields (%)			
Catalyst	%[a]	pH	Time	Conv.(%)	3	4	6	Total
Fe(TMPS)	1	3.0	1 min	100	35	60		95
Fe(TDCPPS)	1	3.0	1 min	100	10	70		80
Fe(TMPS)-Ad	10	3.0	1 min	100 (90)[b]	40	30		70
Mn(TMPS)[c]	10	6.0	15 min	100	5	12	71	88
Mn(TDCPPS)[c]	10	6.0	15 min	85	7	18	57	82
Mn(TMPS)-Ad[c]	10	6.0	15 min	82 (75)[b]	30	10		40

[a] % of catalyst with respect to substrate. [b] numbers in parentheses refer to the conversion after a second run with the same catalyst. [c] 100 equiv. 4-tert-butylpyridine/ catalyst added.

4. Perspectives

During the last decade, a general trend in biomimetic oxidation has been to use hydrogen peroxide as oxygen source. Catalytic oxidations are likely to evolve in this direction for two main reasons: (i) hydrogen peroxide is readily available through the anthraquinone autoxidation process and (ii) this chemical is a non-polluting oxidizing agent. In addition, the content of active oxygen is high: even for a 30 wt.% solution, the content of active oxygen is 14%, much more than many other available oxygen sources, besides molecular oxygen itself.

Within the last decade, the knowledge on how metalloenzymes are able to handle hydrogen peroxide without producing hydroxyl radicals has largely increased. During the same period, the first uses of H_2O_2 in biomimetic catalytic oxidations have been described. Thus, we can hope that industrial applications of such biomimetic oxidations based on hydrogen peroxide will be started before the end of the century.

Acknowledgments

The contributions of many co-workers and colleagues have been cited in the references. All of them are gratefully acknowledged. The corresponding work was made possible with the financial support of the Centre National de la Recherche Scientifique and of Elf-Aquitaine for the part describing lignin modelling.

Abbreviations

Abbreviations for porphyrin derivatives: TPP stands for the *meso* -tetraphenylporphyrinato ligand, TMP stands for *meso* -tetramesitylporphyrinato, TDCPP for *meso* -tetrakis(2,6-dichlorophenyl)porphyrinato, TMPyP for *meso* -tetrakis(4-*N* -methylpyridiniumyl)porphyrinato, TFPFP for *meso* -tetrakis(pentafluorophenyl)-β-octafluoroporphyrinato, TMPS for *meso* -tetrakis(3,5-disulfonatomesityl)porphyrinato, TPPS for *meso* -tetrakis(4-sulfonatophenyl)porphyrinato and TDCPPS for *meso* -tetrakis(2,6-dichloro-3-sulfonatophenyl)porphyrinato.

References

1. B. G. Malmström, L. E. Andreasson and B. Reinhammar in '*The Enzymes* ', Ed. P. D. Boyer, Academic Press, New York 1975, vol XII, p. 507 .
2. (a) G. R. Schonbaum and B. Chance, in '*The Enzymes* ', Ed. P. D.Boyer, Academic Press, New York, 1976, vol. XIII, p. 365; (b) J. E. Frew and P. Jones, *Adv. Inorg. Bioinorg. Mech.*, **3**, 175 (1984).
3. (a) H. B. Dunford, *Adv. Inorg. Biochem.*, **4**, 41 (1982); (b) '*Peroxidases* ', Ed. J. Everse, CRC Press, Boca Raton 1990, vol. I and II.
4. (a) B. Demple and J. Halbrook, *Nature* , **304**, 466 (1983); (b) J. A. Imlay, S. C. Chin and S. Linn, *Science* , **240**, 640 (1988).
5. '*Comprehensive Inorganic Chemistry* ' Eds. J. C. Bailar, H. J. Emeleus, R. S. Nyholm and A. F. Trotman-Dickenson, Pergamon, New York 1973, vol. 2, p.774.
6 (a) E. Graf, J. R. Mahoney, R. G. Bryant and J. W. Eaton, *J. Biol. Chem.*, **259**, 3620 (1984); (b) C. Walling, *Acc. Chem. Res.*, **8**, 125 (1975).

7. (a) B. C. Finzel, T. L. Poulos and J. Kraut, *J. Biol. Chem.*, **259**, 13027 (1984); (b) I. Fita and M. G. Rossman, *J. Mol. Biol.*, **185**, 21 (1985); (c) J. H. Dawson, *Science* , **240**, 433 (1988).
8. T. J. McMurry and J. T. Groves, in '*Cytochrome P-450; Structure, Mechanism and Biochemistry* ' Ed. P. Ortiz de Montellano, Plenum Press, New York, 1986, chapter I.
9. B. Meunier, *Bull. Soc. Chim. Fr.*, 578 (1986).
10. D. Mansuy, *Pure Appl. Chem.*, **59**, 579 (1987).
11. B. Meunier, *Gazz. Chim. Ital.* , **118**, 485 (1988).
12. T. C. Bruice, *Aldrichim. Acta* , **21**, 87 (1988).
13. I. Tabushi, *Coord. Chem. Rev.*, **86**, 1 (1988).
14. (a) Y. Kono and I. Fridovich, *J. Biol. Chem.*, **258**, 6015 (1983); (b) R. Fronko, J. E. Penner-Hahn and C. J. Bender, *J. Am. Chem. Soc.*, **110**, 7554 (1988).
15. D. Dolphin, A. Forman, D. C. Borg, J. Fajer and R. H. Felton, *Proc. Natl. Acad. Sci. USA* , 68, 614 (1971).
16. R. J. M. Corrall, H. M. Rodman, J. Margolis and B. R. Landau, *J. Biol. Chem.*, **249**, 3181 (1974).
17. T. E. Barman, '*Enzyme Handbook* ', Springer Verlag, Berlin 1969, vol. I, p. 232.
18. H. Von Euler and K. Josephson, *Justus Liebigs Ann. Chem.*, **456**, 111 (1927).
19. (a) P. Jones, T. Robson and S. B. Brown, *Biochem. J.*, 135, 353 (1973) and references therein, (b) H. C. Kelly, D. M. Davies, M. J. King and P. Jones, *Biochemistry* , 16, 3543 (1977).
20. I. Wilson, K. R. Bretscher, C. K. Chea and H. C. Kelly, *J. Inorg. Biochem.*, **19**, 345 (1983).
21. (a) Y. Saito, M. Mifume, J. Odo, Y. Tanaka, M. Chikuma and H. Tanaka, *Reactive Polymers* , **4**, 243 (1986); (b) Y. Saito, M. Mifume, T. Kawaguchi, J. Odo, Y. Tanaka, M. Chikuma and H. Tanaka, *Chem. Pharm. Bull.*, **34**, 2885 (1986); (c) Y. Saito, M. Satouchi, M. Mifume, T. Tai, J. Odo, Y. Tanaka, M. Chikuma and H. Tanaka, *Bull. Chem. Soc. Japan* , **60**, 2227 (1987).
22. M. Barteri, P. Jones and O. Mantovani, *J. Chem. Soc. Dalton Trans.*, 333 (1986).
23. (a) J. T. Groves and T. E. Nemo, *J. Am. Chem. Soc.*, **105**, 5786 (1983); (b) P. S. Traylor, D. Dolphin and T. G. Traylor, *J. Chem. Soc. Chem. Commun.*, 279 (1984); (c) B. de Poorter and B. Meunier, *Tetrahedron Lett.*, **25**, 1895 (1984); (d) O. Bortolini and B. Meunier, *J. Chem. Soc. Perkin II* , 1967 (1984); (e) K. S. Suslick and B. Cook, *J. Chem. Soc. Chem. Commun.*, 200 (1987).
24. (a) M. F. Zipplies, W. A. Lee and T. C. Bruice, *J. Am. Chem. Soc.*, **108**, 4433 (1986); (b) P. N Balasubramanian, E.S. Schmidt and T. C. Bruice, *J. Am. Chem. Soc.*, **109**, 7865 (1987).
25. H. Gallati, *J. Clin. Biochem.*, **17**, 1 (1979).
26. T. G. Traylor and F. Xu, *J. Am. Chem. Soc.*, **109**, 6201 (1987).
27. (a) R. Belal, M. Momenteau and B. Meunier, *J. Chem. Soc. Chem. Commun.*, 412 (1989); (b) R. Belal, M. Momenteau and B. Meunier, *New J. Chem.*, **13**, 856 (1989); (c) A. Robert, B. Look, M. Momenteau and B. Meunier, *Inorg. Chem.*, **30**, 706 (1991).
28. S. Oae, Y. Watanabe and K. Fujimori, *Tetrahedron Lett.*, **23**, 1189 (1982).
29. J. P. Renaud, P. Battioni, J. F. Bartoli and D. Mansuy, *J. Chem. Soc. Chem. Commun.*, 888 (1985).
30. P. Battioni, J. P. Renaud, J. F. Bartoli, M. Reina-Artiles, M. Fort and D. Mansuy, *J. Am. Chem. Soc.*, **110**, 8462 (1988).

31. (a) B. Meunier, E. Guilmet, M. E. de Carvalho and R. Poilblanc, *J. Am. Chem. Soc.*, 106, 6668 (1984); (b) A. J. Castellino and T. C. Bruice, *J. Am. Chem. Soc.*, 110, 156 (1988).
32. D. Mansuy, J. F. Bartoli, M. Momenteau, *Tetrahedron Lett.*, **23**, 2781 (1982).
33. B. Meunier, M. E. de Carvalho, O. Bortolini and M. Momenteau, *Inorg. Chem.*, **27**, 161 (1988).
34. (a) H. Sugimoto and D. T. Sawyer, *J. Org. Chem.*, **50**, 1786 (1985); (b) H. Sugimoto and D. T. Sawyer, *J. Am. Chem. Soc.*, **107**, 5712 (1985).
35. J. -P. Schirmann and S. Y. Delavarenne, '*Hydrogen Peroxide in Organic Chemistry* ', Edition et Documentation Industrielle, Paris 1979, p. 181.
36. (a) L. Anelli, S. Banfi, F. Montanari and S. Quici, *J. Chem. Soc. Chem. Commun.*, 779 (1989); (b) F. Montanari, S. Banfi and S. Quici, *Pure Appl. Chem.*, **61**, 1631 (1989).
37. (a) J. T. Groves, Y. Watanabe and T. J. McMurry, *J. Am. Chem. Soc.*, **105**, 4489 (1983); (b) J. T. Groves and Y. Watanabe, *J. Am. Chem. Soc.*, **110**, 8443 (1988).
38. (a) M. Schmidt and G. Prescher, Ger. DE 3,704,532, *Chem. Abs.*, **110**, 57494h (1989); (b) J. W. Buchler, G. Herget, M. Schmidt and G. Prescher, Ger. DE 3,800,973, *Chem. Abs.*, **111**, 23297b (1989).
39. (a) S. Udenfriend, C. T. Clark, J. Axelrod and B. B. Brodie, *J. Biol. Chem.*, **208**, 731 (1954); (b) A. Kunai, S. Hata, S. Ito and K. Sasaki, *J. Am. Chem. Soc.*, **108**, 6012 (1986).
40. C. R. E. Jefcoate, J. R. Lindsay-Smith and R. O. C. Norman, *J. Chem. Soc. B*, 1013 (1969).
41. M. A. Brook, L. Castle, J. R. Lindsay-Smith, R. Higgins and K. P. Morris, *J. Chem. Soc. Perkin Trans. II* , 687 (1982).
42. J. T. Groves and Van der Puy, *J. Am. Chem. Soc.*, **98**, 5290 (1976).
43. T. Schimidzu, T. Iyoda and N. Kanda, *J. Chem. Soc. Chem. Commun.*, 1206 (1981).
44. P. Battioni, J. P. Renaud, J. F. Bartoli and D. Mansuy, *J. Chem. Soc. Chem. Commun.*, 341 (1986).
45. A. Robert and B. Meunier, *New J. Chem.*, **12**, 885 (1988).
46. P. Hoffmann, A. Robert and B. Meunier, unpublished results.
47. T. G. Traylor and S. Tsuchiya, *Inorg. Chem.*, **26**, 1338 (1987).
48. P. Hoffmann, G. Labat, A. Robert and B. Meunier, *Tetrahedron Lett.*, **31**, 1991 (1990).
49. S. Tsuchiya and M. Seno, *Chem. Lett.*, 263 (1989).
50. T. Tatsumi, H. Nakamura and H. Tominaga, *Chem. Lett.*, 419 (1989).
51. N. Kitajima, H. Fukui and Y. Moro-oka, *J. Chem. Soc. Chem. Commun.*, 485 (1988).
52. P. Jones, D. Mantle and I. Wilson, *J. Chem. Soc. Dalton Trans.*, 161 (1983) and references therein.
53. T. G. Traylor, W. A. Lee and D. V. Stynes, *J. Am. Chem. Soc.*, **106**, 755 (1984).
54. Y. Saito, S. Nakashima, M. Mifume, J. Odo, Y. Tanaka, M. Chikuma and H. Tanaka *Analytica Chim. Acta* , **172**, 285 (1985).
55. Y. Saito, M. Mifume, S. Nakashima, Y. Tanaka, M. Chikuma and H. Tanaka, *Chem. Pharm. Bull.*, **34**, 5016 (1986).
56. Y. Saito, M. Mifume, S. Nakashima, H. Nakayama, J. Odo, Y. Tanaka, M. Chikuma and H. Tanaka, *Chem. Pharm. Bull.*, **35**, 869 (1987).
57. P. Sharp, *Clin. Chim. Acta* , **40**, 115 (1972).
58. K. Takakashi, A. Matsushima, Y. Saito and Y. Inada, *Biochem. Biophys. Res. Commun.*, **138**, 283 (1986).
59. (a) H. Tanaka, T. Nakagawa, Y. Okabayashi, H. Aoyama, K. Itoh, M. Chikuma, Y.

Saito, H. Sakurai and M. Nakayame, *Pure Appl. Chem.*, **59**, 573 (1987); (b) Y. Saito, M. Mifume, S. Nakashima, J. Odo, Y. Tanaka, M. Chikuma and H. Tanaka, *Analytical Sciences* , **3**, 171 (1987).
60. (a) M. Tien and T. K. Kirk, *Science* , **221**, 661 (1983); (b) J. K. Glenn, M. A. Morgan, M. B. Mayfield, M. Kuwahara and M. H. Gold, *Biochem. Biophys. Res. Commun.*, **114**, 1077 (1983); (c) V. Renganathan and M. H. Gold, *Biochemistry* , **25**, 1626 (1986); (d) M. S. A. Leisola, B. Kozulic, F. Meussdoerffer and A. J. Fiechter, *J. Biol. Chem.*, **262**, 419 (1987).
61. (a) '*Methods in Enzymology* ', Section I, **161**, 3 (1988); (b) G. Labat and B. Meunier, *Bull. Soc. Chim. Fr.*, **127**, 553 (1990).
62. (a) T. Habe, M. Shimada, T. Okamoto, B. Panijpan and T. Higuchi, *J. Chem. Soc. Chem. Commun.*, 1323 (1985); (b) A. Paszczynski, R. L. Crawford and R. A. Blanchette, *Appl. Environmental Microbiol.*, **54**, 62 (1988).
63. D. Dolphin, T. Nakano, T. E. Maione, T. K. Kirk and R. Farell, *Chem. Abs.*, **109**, 69402 (1988).
64. T. Okamoto, K. Sasaki and S. Oka, *J. Am. Chem. Soc.*, **110**, 1187 (1988).
65. (a) G. Meunier and B. Meunier, *J. Am. Chem. Soc.*, **107**, 2558 (1985); (b) G. Meunier and B. Meunier, *J. Biol. Chem.*, **260**, 10576 (1985).
66. (a) G. Labat and B. Meunier, *New J. Chem.*, **13**, 801 (1989); (b) G. Labat and B. Meunier, *J. Org. Chem.*, **54**, 5008 (1989).
67. G. Labat, PhD Thesis, University Paul Sabatier, Toulouse 1989.
68. H. Wariishi and M. H. Gold, *FEBS Letters* , **243**, 165 (1989).

CHAPTER 6

Nucleophilic and Electrophilic Catalysis with Transition Metal Complexes

Giorgio STRUKUL

Dipartimento di Chimica, Università di Venezia, Dorsoduro 2137, 30123 Venezia, Italy

1. Introduction

As outlined in chapter 1 the activation of H_2O_2 in the presence of transition metal complexes to give metal peroxo species is one of the major pathways through which hydrogen peroxide can be employed as primary oxidant for the synthesis of oxygenated compounds. In this chapter it will be discussed how this activation will lead to peroxo complexes having either electrophilic or nucleophilic characteristics and how this guideline may be helpful in suggesting the appropriate catalyst for a given reaction. However, before doing so it seems appropriate to summarize some general physico-chemical features of hydrogen peroxide.

Crystal structure determination of pure H_2O_2 by neutron diffraction data has shown that the corrected H-O and O-O bond distances can be indicated as 1.01 and 1.453 Å respectively [1] , while the corresponding H-O-O-H dihedral angle is close to 90° (actually 90.2° from neutron diffraction data [1] and 93.5° from X-ray [2]). Indeed, the latter seems to vary up to 180° thereby giving rise to virtually flat H_2O_2 molecules when associated with other molecules in crystalline complexes [3] . Examples are $Na_2C_2O_4{\cdot}H_2O_2$ (C_2O_4 = oxalate) [4], $NH_4F{\cdot}H_2O_2$ [5], or $Na_2CO_3{\cdot}3/2H_2O_2$ [6], the latter being particularly interesting since two different H-O-O-H dihedral angles (125° and 170°) were found within the same unit cell, the difference being ascribed to interaction of each peroxy oxygen with two and one Na atoms respectively indicating a very low rotational barrier. Another interesting structural feature of H_2O_2 both pure and associated with other molecules is the presence, where possible, of hydrogen bonding through the peroxy oxygens indicating as expected a significant Lewis basicity. The H-O bond dissociation energy (ΔH^0) was found

G. Strukul (ed.), Catalytic Oxidations with Hydrogen Peroxide as Oxidant, 177–221.

to be 88 Kcal/mol compared to 119 Kcal/mol for H_2O [7] and this trend is parallelled by the acidity constant (pK_a) that for H_2O_2 was calculated as being 11.6 [8], considerably higher than H_2O.

The overall picture that emerges from the above data is that of a molecule with amphoteric character capable of reacting easily both with acids and with bases and hence changing its original properties either in an electrophilic or in a nucleophilic sense. This formalism is in fact widely accepted in organic chemistry and is implied in the observation that polar reaction mechanisms requiring the prior transformation of hydrogen peroxide into a more reactive form constitute the basis for the vast majority of organic oxidation reactions involving the ultimate use of H_2O_2.

Hydrogen peroxide is the key chemical for the preparation of organic peroxy acids [9] that is generally accomplished by reaction with carboxylic acids or the corresponding acyl halides (eq. 1). Indeed, peroxy acids are the best characterized of all organic reactants capable of delivering an electrophilic oxygen [10, 11] and this feature is testified by the whole body of their reactivity which involves reactions with electron-rich molecules like olefins, acetilenes, amines, sulfides, etc. [9, 12] . In eq. 2 the epoxidation of olefins

$$R\text{-}COOH + H_2O_2 \xrightleftharpoons{H^+} R\text{-}COO_2H + H_2O \qquad (1)$$

is reported as an example.

(2)

On the other hand, alkaline hydrogen peroxide is known to add to electron-poor carbon atoms like in aldehydes (Dakin reaction) [13] and in substituted cyclobutanones (Baeyer Villiger reaction) [14] or nitriles [15-17] or α,β-unsaturated ketones and esters [18, 19] , the driving force for this class of reactions having been recognized as being the nucleophilic attack of the hydroperoxy anion onto the electron-poor end of the substrate (eq. 3).

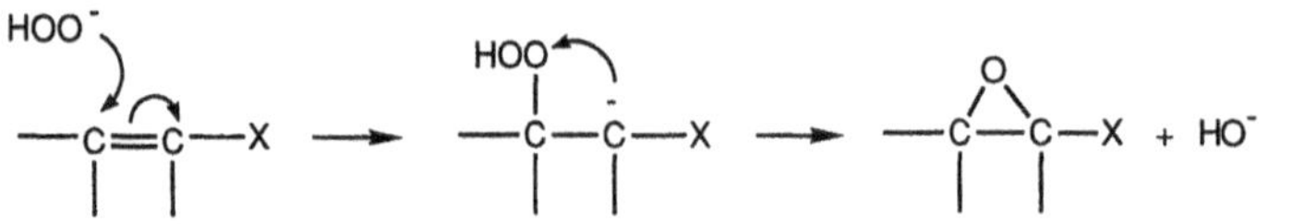

(3)

2. Interaction of Hydrogen Peroxide with Transition Metal Centers

By analogy with the different behavior that H_2O_2 can display in purely organic systems, it seems appropriate to apply a similar formalism to the case of reactions with transition metals. This choice seems to be justified by the different nature of the metal centers involved, i.e. either early transition metals in their highest oxidation states or group VIII metals in low oxidation state. In other words and to continue with the above analogy, the two different types of metal centers can be viewed as "acidic" or "basic" respectively, at least in a Lewis sense. Therefore, it is conceivable that by reaction with H_2O_2 these centers will produce peroxidic complexes where the character of the peroxy oxygens is modified in an opposite sense.

2.1. Synthesis of metal-peroxo complexes.

The synthesis of peroxo metal complexes starting from H_2O_2 can be accomplished in essentially two different ways:

1. With early transition metals peroxo complexes are generally synthesized from the corresponding oxides by direct dissolution in aqueous H_2O_2 in the presence of the appropriate ancillary ligands (eq. 4):

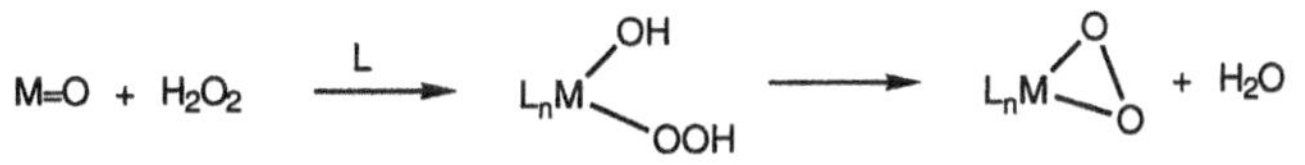

(4)

M = Ti, V, Nb, Ta, Cr, Mo, W, U, Th
L = py, HMPA, pic, pico, Cl, C_2O_4, etc.

The procedure is essentially similar for a variety of complexes and is based on the recipe reported some 20 years ago by Mimoun and coworkers for the complex $MoO(O_2)_2(HMPA)(H_2O)$ [20, 21]. Usually, eq. 4 is considered a stepwise process, although only in very few cases could the intermediate hydroperoxy complex be observed. An example relevant to oxygen transfer reactions is $[MoO(O_2)_2(OOH)]_2\cdot 2(pyH)$ [22]. Generally, ring closure as in eq. 4 does not take place when ROOH is used instead of H_2O_2 giving rise to alkylperoxy species [23], where even the second oxygen of the peroxo moiety interacts with the metal to produce a sort of triangular geometry that is generally considered very important to the purpose of oxygen transfer [24]. As indicated above, to the coordination sphere of the metal ancillary ligands may be added during the preparation; these include: neutral nitrogen donors like pyridine (py) or hexamethylphosphortriamide (HMPA) as for example in $CrO(O_2)(py)$ [25], $MoO(O_2)_2(HMPA)(H_2O)$ [20] or $WO(O_2)_2(HMPA)(H_2O)$ [20] ; bidentate carboxylic acids anions like oxalate (C_2O_4), picolinate (pic) and picolinate-N-oxide (pico) as for example in $Ti(O_2)(pic)_2(HMPA)$ [26], $VO(O_2)(pic)(H_2O)_2$ [27], $MoO(O_2)(pico)$ [28], $[Nb(O_2)_2(C_2O_4)_2]\ (NH_4)_3$ [29], $MoO(O_2)(pic)(HMPA)Cl$ [30] or $Th(O_2)(pic)_2$ [31]. A complementary procedure to the above is the coordination of the ancillary ligands prior to H_2O_2 interaction as is the case with cyclopentadienyl deri-vatives like in $Nb(O_2)(Cp)_2Cl$ [32, 33], or with metalloporphyrins like in $Ti(O_2)(OEP)$ (OEP = octaethylporphyrin) [34] or $Mo(O_2)_2(porph)$ [35].

2. With late transition metals the synthetic method generally exploits the mild acidic properties of hydrogen peroxide. The reaction is regarded as an acid-base exchange

$$L_nM\text{-}X + H_2O_2 \rightleftharpoons L_nM\text{-}OOH + HX \qquad (5)$$

M = Pd, Pt, Cu

$X = OOCCF_3, OOCCH_3, OH$

starting from a suitable precursor and producing in most cases hydroperoxy complexes (eq. 5). It has to be pointed out that this procedure is far less general than the previous case having been applied successfully only to a limited number of complexes such as $[(RCOO)Pd(OOH)]_2$ ($R = CF_3, CH_3$) [36], $(P\text{-}P)Pt(CF_3)(OOH)$ (P-P = various diphosphines) [37-39], $[(CH_3COO)Cu_2(OH)_2(OOH)]_n$ [40].

The limited ability of eq. 5 to generate hydroperoxy species by direct interaction with H_2O_2 is mainly due to other possible competing reactions. In the case of some Co complexes the occurrence of ring closure leading to side-on peroxy complexes [41, 42] similar to those obtained through reaction 4 has been reported. Another alternative possibility is related to the low oxidation state of the late transition metals involved, which can make Haber-Weiss decomposition of H_2O_2 the dominant process. In fact, even if we do not consider the classical Fe(II)/Fe(III), Co(II)/Co(III) or Cu(I)/Cu(II) systems, also with Ir(I) or Rh(I) complexes, like *trans* -$(PPh_3)_2Ir(CO)(OH)$ [43] ,

[(COD)Ir(μ-OH)]$_2$ [44] and *trans* -$(PPh_3)_2Rh(CO)(OH)$ [45], eq. 5 was found to lead only to radical-type H_2O_2 decomposition [46] with no transfer to external substrates. This problem appears to affect also some reducible d^0 transition metals centers despite their being in their highest oxidation states and in an oxidizing environment. Finally, in one case, i.e. with $(CH_3COO)_2Pd$, eq. 5 was reported to produce also a superoxo complex [47] depending on the experimental conditions.

Alternative synthetic procedures leading to group VIII metals hydroperoxy complexes involve either dioxygen insertion into a metal-hydride bond (eq. 6)

$$L_nM\text{-}H + O_2 \longrightarrow L_nM\text{-}OOH \qquad (6)$$

M = Co, Rh, Ir, Pt

or protonation of easily accessible metal-dioxygen complexes with strong acids (eq. 7).

(7)

M = Pd, Pt

Examples of eq. 6 include some older studies on $[RhH(NH_3)_5]^{2+}$ [48], $[RhH(CN)_4]^{2-}$ [49] or $[CoH(CN)_5]^{3-}$ [50] and more recently on $[(COD)IrHCl_2]_2$ [51, 52], P_2PtHCl [53] and (diphoe)Pt(CF_3)H [54] . Equation 7 has been successfully applied in the case of $(PPh_3)_2Pd(O_2)$ [55] and $(PPh_3)_2Pt(O_2)$ [56, 57].

2.2. Structural considerations and bonding description

Many peroxo complexes of transition metals have been structurally characterized, although the situation is quite different in the two classes of complexes synthesized through eq. 4 and 5 respectively. In fact, while a wide variety of d^0 peroxo complexes have been structurally characterized, in the case of late transition metals no crystal structures of hydroperoxo complexes have been reported to date and hence a direct comparison of the structural features of the metal peroxo moiety in the two cases is impossible. In order to get some indications (if any) of the different character of the peroxygens from structural data we will consider for comparison some dioxygen complexes and some alkylperoxo complexes of group VIII metals. In Table 1 a list of representative examples of structurally characterized complexes is reported, some of which are also of historical importance. A more thorough compilation of structural data can be found elsewhere (see for example references 24 and 58).

The vast majority of the d^0 complexes synthesized via reaction 4 exhibit pentagonal-bypyramidal geometry around the metal center with the peroxo group coordinated side-on although the two M-O bond distances are usually slightly different. This triangular geometry for the metal peroxo moiety is preserved also in M-OOR complexes as was

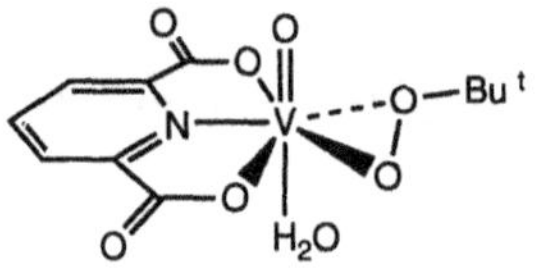

ref. 70

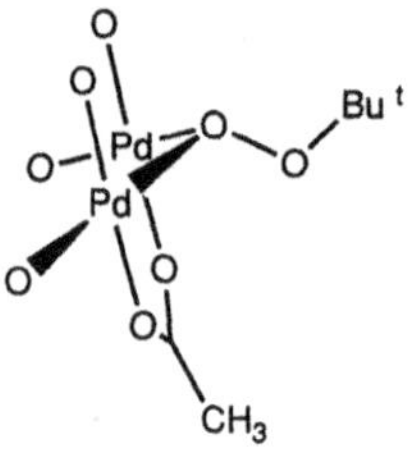

ref. 65

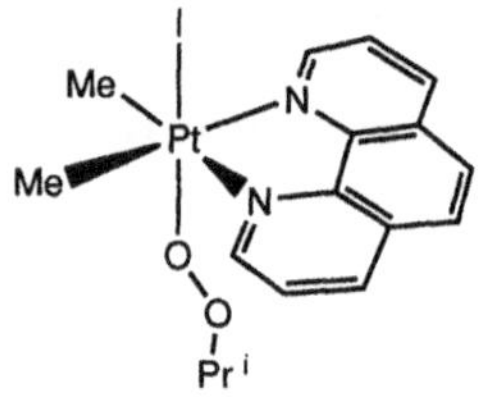

ref. 69

shown for example by the X-ray structure of $VO(dipic)(OOBu^t)(H_2O)$ [70]. Conversely, a variety of geometries have been observed with late transition metals mostly depending on the oxidation state of the metal. In the case of alkylperoxo complexes the peroxo ligand is coordinated either in a linear or in a bridging fashion, but always through one oxygen only.

If we consider the bond distances, no particular trend seems to be evident, independently of the metal involved, its oxidation state, the coordination geometry, or the type of ancillary ligands completing the coordination sphere. The O-O bond distance falls invariably between 1.4-1.5 Å very similar to that of free hydrogen peroxide (see above), while the M-O distances seem to reflect mostly the ionic radius of the metal as is suggested by grouping together first row, second row and third row transition metals independently of their position in the periodic table (Table 1).

Table 1: Selected bond distances (in Å) in a series of metal peroxy complexes.

Complex	d(M-O)	d(O-O)	Ref.
First row transition metals			
$Ti(O_2)(OEP)$	1.827 1.822	1.445	34
$VO(O_2)(pic)(H_2O)_2$	1.878 1.867	1.425	27
$[Cr(O_2)_4]\ K_3$	1.87 1.97 (av)	1.47 (av)	59
$[Co(O_2)(diphoe)_2)]\ BF_4$	1.871 1.902	1.420	60
$Co(DMG)_2(py)(OOCum)$	1.909	1.455	61
$Ni(O_2)(CNBu^t)_2$	1.808	1.45	62
Second row transition metals			
$Nb(O_2)_2(C_2O_4)_2$	1.961 1.981 1.982 1.969	1.480 1.485	29
$MoO(O_2)(DML)$	1.930 1.935 1.901 1.915	1.46 1.45	63
$[MoO(O_2)_2(OOH)]_2(pyH)_2$	1.948 1.925 1.920 1.954 2.047	1.473 1.466 4.458	22
$[Rh(O_2)(AsPhMe_2)_4]\ ClO_4$	2.03, 2.03	1.46	64
$[Pd(OOCCCl_3)(OOBu^t)]_4$	1.994	1.49	65
Third row transition metals			
$[(WO(O_2)_2)_2O]\ K_2$	1.93 1.96 1.96 2.00	1.50 1.54	66
$Ir(O_2)I(CO)(PPh_3)_2$	2.035, 2.082	1.509	67
$Pt(O_2)(PPh_3)_2$	2.01, 2.01	1.45	68
$Pt(Me)_2I(OOPr^i)(phen)_2$	2.032	1.465	69

OEP = octaethylporphyrin; pic = picolinate; diphoe = *cis* -1,2-bis-diphenylphosphino- ethylene; DMG = dimethylglyoximate; Cum = cumyl; C_2O_4 = oxalate; DML = dimethyl-lactamide.

While structural data seem to give no clues as to the nature of the peroxygens in the two classes of complexes, the bonding description that has been given is quite different. In general, it can be observed that possessing d^0 transition metals no valence electrons, bond formation with the surrounding ligands (including the peroxygens) will occur only through a net donation from the latter. The situation of the metal peroxo moiety has been analyzed theoretically [71-74] by considering the nature and the interaction of the frontier orbitals in the case of the species shown below. These studies have shown that the most important bonding characteristic of these species is the bonding of a filled in-plane orbital with π*-type symmetry on the peroxide with an empty d-orbital on the metal. The conclusion was that the direction of the electron flow is the opposite of that of the Dewar-Chatt-Duncanson model for the π-complexation of alkenes with metals [71] supporting the idea that the oxygen proximal to the metal is electrophilic in nature.

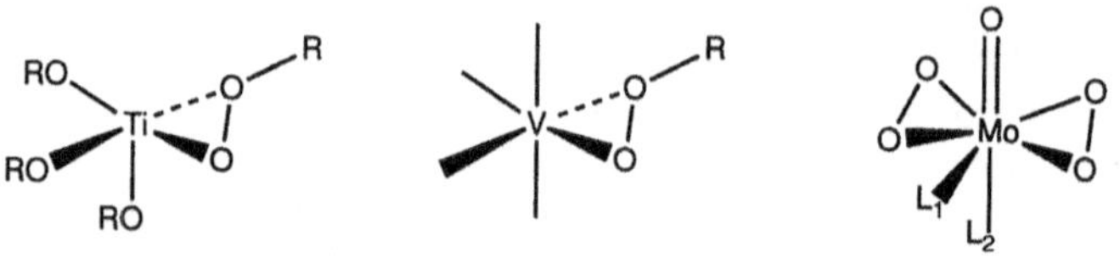

In group VIII transition metals the bonding model that has been suggested is quite different [75]. Similarly to olefins, a significant back-donation (predominantly π-bonding) from the filled d orbitals of the metal to the π* orbitals of dioxygen has been suggested. Since the initial discovery by Vaska of the reversible dioxygen binding of *trans* -$(PPh_3)_2Ir(CO)Cl$ to give $(PPh_3)_2Ir(O_2)(CO)Cl$ [76], this reaction on similar d^8 complexes has been often described as an oxidative addition, thus emphasizing the "peroxide ion" character (and hence the nucleophilic nature) of coordinated dioxygen.

However, despite their historical and theoretical significance group VIII metals dioxygen complexes are of little practical value in the context of catalytic oxidation. More relevant to the use of hydrogen peroxide and effective in oxidation reactions are hydroperoxo complexes as stated above. Although a comparison between MOOH and the model MO_2 systems may be acceptable as far as the nature of the peroxidic oxygen is concerned, a more direct determination of the polar nature of the metal-oxygen in MOOH complexes would be desirable. This was performed spectroscopically through NMR techniques in the case of PtOOH complexes that are the best characterized species of this class of compounds.

Some years ago, Appleton and Bennett [77] determined a *trans* -influence series on an homologous class of square planar complexes of Pt(II) of the type (dppe)Pt(Me)X through the analysis of the ^1J(P-Pt) coupling constant for the phosphorus atom trans to X. This series parallels also the extent of the polarization of the Pt-X bond as is indicated by some representative data reported in Table 2. As shown the -OH ligand falls between strong σ donors giving covalent bonds like the -Me group and and very weak, mostly anionic ligands like -ONO_2. Peroxo ligands like -OOH or -$OOBu^t$ behave very similarly (Table 2), indicating a significant polarization of the corresponding Pt-O bond [78].

On the other hand, evidence on the even strong basicity of Pt-OH complexes from which the corresponding hydroperoxo complexes can be prepared is rather abundant; this include conductivity measurements [79] and initial pH determinations [80, 81] and these properties have been exploited in the catalytic hydration of nitriles to carboxamides

Table 2: P-Pt coupling constants for a series of homologous complexes of the type[a]

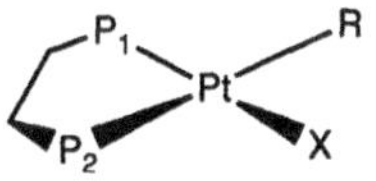

$^{1}J(P_1\text{-}Pt)$ (Hz)	X	R	Ref.
1794	CH_3	CH_3	75
2580	CH_2NO_2	CH_3	75
3546	OH	CH_3	75
3263	OH	CF_3	76
3139	OOH	CF_3	76
3075	$OOBu^t$	CF_3	76
3840	O-Ph-OMe	CH_3	75
4122	OAc	CH_3	75
4510	ONO_2	CH_3	75

a $^{31}P\{^{1}H\}$ NMR spectroscopy data.

[80-90] and the catalytic hydration of olefins to alcohols [88-90]. It was also shown by NMR that the polarization of the Pt-O bond in this class of complexes can be significantly increased (probably by solvation) in polar media, particularly in the presence of water [79]. In fact by analyzing the $^{1}J(P_1\text{-}Pt)$ of (diphoe)Pt(Ph)(OH) in THF (Table 3) on addition of an increasing amount of water, a corresponding increase of the coupling constant toward the ionic end of the series reported by Appleton and Bennett [77] was observed.

3. Reactivity in Oxidation Reactions

The above considerations on the different nature of the peroxo oxygen when bound to d^0 or late transition metal centers (especially d^8) is supported by the oxygen transfer properties of these species toward organic substrates. In this respect, the distinction between nucleophilic peroxo metal complexes and electrophilic peroxo metal complexes in dioxygen activation has been already suggested by Sheldon and Kochi some years ago in their book on metal catalyzed oxidations [91] . Indeed, as far as hydrogen peroxide is concerned, it can be said that in reactions with organic substrates (in Scheme 1 the case of olefins is represented as an example) depending on the type of transition metal center involved to promote the oxygen transfer, the role of the electrophile and that of the nucleophile in the system will be inverted.

Table 3: Effect of the addition of H_2O on the $^1J(P_1\text{-}Pt)^a$ of the complex

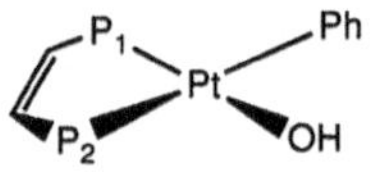

$^1J(P_1\text{-}Pt)$ (Hz)	%H_2O in THF
3317	0
3480	5
3566	10
3612	25

a $^{31}P\{^1H\}$ NMR spectroscopy data taken from ref. 79.

Scheme 1:

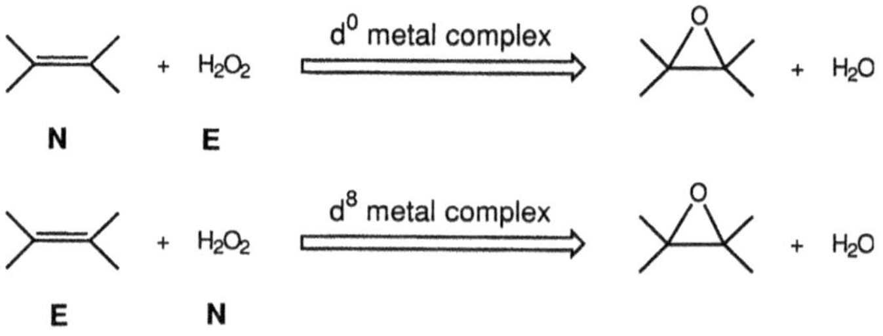

E = electrophile
N = nucleophile

As a consequence of this observation it is predictable that depending on the characteristics of the substrate to be oxidized one of the two classes of catalysts (d^0 or d^8 transition metals) will be better suited to bring about the reaction. Like most generalized, schematic views Scheme 1 is in many cases a simplified version of reality and indeed the examples where the reactivity of the two classes of catalysts overlap or escape heterolytic reaction mechanisms are numerous. Nevertheless, as will be seen in the foregoing, where the reactivity of d^0, d^6 and d^8 transition metal catalysts toward a variety of substrates will be examined, Scheme 1 can indeed constitute a useful guideline in the selection of the most appropriate catalyst for a given reaction.

3.1. Molybdenum and Tungsten Complexes

The most typical complexes showing clear electrophilic behavior in the oxidation of a variety of organic substrates are those of molybdenum and tungsten. The oxidation of nucleophiles like sulfides to sulfoxides with complexes like $[MoO(O_2)_2(pic)]$ NBu_4 or $[MoO(O_2)_2(pic)]$ NBu_4 has been analyzed in a careful kinetic study by Di Furia, Modena, et al. [92] and has shown second order behavior. The effect of substituents (see Table 4) indicates that the reaction rate increases with the nucleophilicity of sulfur.
The coordinatively saturated nature of the complex, the observed lack of solvolytic processes in a variety of solvents, the absence of effects on the reaction rate by addition of external ligands like HMPT and the relative insensitivity to the different atom donors in the chelating ligand (N donor = pic, O donor = pico), rule out the possible coordination of the substrate through ligand displacement prior to oxygen transfer. All these data strongly support a mechanism involving a simple bimolecular interaction similar to that established for the oxidations with organic peroxy acids [9a, 12, 93].

Other species that have been found to be active in the oxidation of sulfides with similar reactivity include $Mo(CO)_6$ [94], $MoO_2(acac)_2$ [94, 95], $MoO(O_2)_2(HMPT)$ [94] which catalyze the reaction in water miscible solvents, or Na_2MoO_4 and Na_2WO_4 [96] which are also catalysts under phase transfer conditions. The latter species are employed in the presence of a slight excess of sulfuric acid and an organic base (generally a variety of tri- and hexaalkylphosphorictriamides) that increases the solubility of the catalyst in the organic phase. It is believed that the oxidizing species involved is of the type $MO(O_2)_2(base)$ (M = Mo, W). All these compounds generally oxidize more easily the more nucleophilic sulfides to sulfoxides than the corresponding sulfoxides to sulfones, thereby emphasizing the nature of the metal peroxide/substrate interaction.

Table 4: Second order kinetic data for the reaction:

(8)

O—X	R	R'	10^3 x k
pico	Ph	Ph	0.26
	p -NO$_2$-Ph	Me	0.76
	p -Cl-Ph	Me	1.8
	Ph	Me	2.4
	p -Me-Ph	Me	3.3
	Bu	Bu	11.8
pic	Ph	Me	3.1

Data taken from ref. 92.

Another example of clear electrophilic behavior of Mo and W peroxo species is in the oxidation of amines. Amines give rise to different types of products depending on substitution at nitrogen. Thus, primary amines will give oximes, secondary amines will give hydroxylamines and tertiary amines will give amine oxides (eq. 9-11).

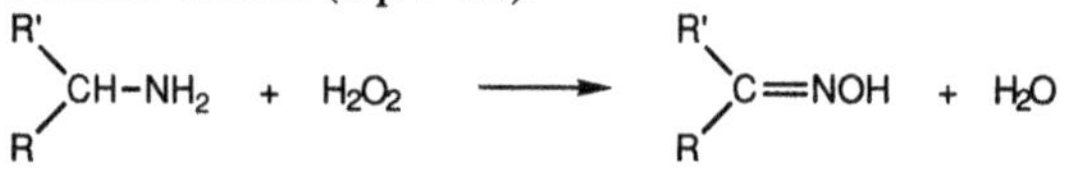

(9)

(10)

(11)

The above reactions have been found to be catalyzed by Na molybdates and tungstates [97-99].

The study of the interaction of hydrogen peroxide and olefins is historically credited to the pioneering work of Milas who reported [100-102] over fifty years ago the hydroxylation of olefins to glycols catalyzed by a variety of simple metal oxides like MoO_3, WO_3, V_2O_5, CrO_3 and OsO_4. The osmium tetroxide can produce cis-glycols as a stoichiometric oxidant even in the absence of H_2O_2 involving a mechanism that is substantially different from that of the other oxides [103] since metal peroxo species may not be involved (eq. 12). Osmium tetroxide based reagents are still the oxidants of choice for the cis-hydroxylation of olefins on a laboratory scale [104]. The simple oxides of Mo, W, V, etc. need to react with H_2O_2 (see eq. 4 in the absence of added ligands) to form species frequently called "inorganic peracids" [105]. The trans nature of the glycols produced in the oxidation of olefins and later studies have suggested that the original reaction reported by Milas proceeds via an epoxide intermediate that is hydrolyzed under the acidic reaction

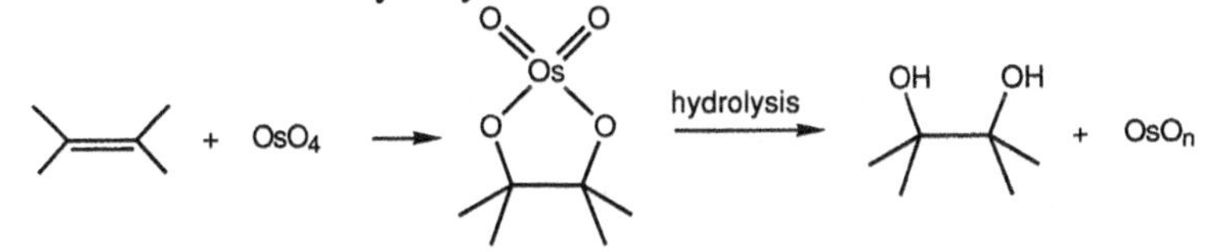

(12)

conditions [106]. The reaction can be stopped at the epoxide stage by using neutral or basic conditions [106-111].

Transition metal catalyzed epoxidations have been recently reviewed in an excellent and comprehensive work by Jørgensen [74]. Within this large body of data it must be said that the use of hydrogen peroxide as oxidant still occupies a minor role, although Mo and W peroxo species are central in this field.

The use of W(VI) species as catalysts for the epoxidation of olefins dates back to the late 40's [106] and since that time has proved useful for a variety of carbon-carbon double bonds like in simple olefins [110, 113-115], allylic and homoallylic alcohols [108, 116] and α,β-unsaturated acids [107, 117]. The behavior of Mo complexes is similar and with both metals the epoxidation has been found to be stereospecific [74] *cis* and *trans* olefins giving respectively *cis* and *trans* epoxides. However, W is superior as catalyst since tungstate is more stable in aqueous solutions, it decomposes H_2O_2 to a limited extent and allows the use of a broader pH range [118, 119].

Molybdenum complexes modified with chiral ligands like (S)-N,N-dimethyllactamide have been employed as stoichiometric oxidants for the asymmetric epoxidation of simple olefins giving a maximum 35% e.e. in the case of *trans* -2-butene [62, 120] or *trans* -2-octene [121].

Indeed, for years the use of Mo complexes for the epoxidation of olefins has been strongly limited in scope because of the difficulty to regenerate the peroxo oxidant under the conditions employed. The use of water miscible solvents necessary to dissolve commercial H_2O_2 retards the approach of hydrophobic olefins to the metal center and these problems have been solved only relatively recently with the use of phase-transfer techniques (see chapter 7). On the other hand, being Mo peroxo complexes more soluble in organic solvents they have been more extensively studied.

As a general trend these Mo and W complexes are more effective oxidants toward electron-rich olefins rather than toward electron-poor ones.

The mechanism of oxygen transfer from this class of peroxo species has been the subject of some debate over the last two decades. The reaction has been thoroughly investigated in the case of $MoO(O_2)_2(HMPA)$ [20, 21] and has shown the following mechanistic features:
i. a prior reversible complexation of the olefin, followed by irreversible oxygen transfer [20, 21] ; this point has been questioned by some authors since in some cases Michaelis-Menten type kinetic behavior was not observed [122, 123] ; *ii.* a slowing effect by σ-donor ligands that retard the reaction with a complete inhibition in complexes having anionic ligands or bidentate or tridentate ligands in the equatorial positions adjacent to the peroxo group [124] ; *iii.* the oxygen transferred to the olefin comes exclusively from the peroxo group [123] ; *iv* the reactivity increases with the nucleophilicity of the olefin (in Table 5 some representative data are reported); *v* peroxo anions being less reactive than the corresponding neutral species [125, 126] .

Table 5: Pseudo-first order rate constants for the reaction:

R	R_1	R_2	R_3	k x 10^4 (s^{-1})
H	H	H	C_6H_{13}	0.49
H	C_2H_5	H	CH_3	2.76
H	CH_3	CH_3	CH_3	6.65
CH_3	CH_3	CH_3	CH_3	9.5

Data taken from ref. 21.

The two possible oxygen transfer processes that have been suggested are outlined in Scheme 2 and involve opposite characters for the peroxo ligand and the substrate. The former (path A), first suggested by Sharpless [123] , is quite similar to the behavior of organic peroxy acids and involves the nucleophilic attack of the free olefin on the electrophilic peroxo ligand via a "butterfly" transition state. The latter (path B) suggested by Mimoun [20, 21, 127] requires the prior coordination of the olefin with consequent loss of nucleophilic character, followed by internal nucleophilic attack of the peroxygen to the olefin to give a peroxymetallacyclic intermediate which then decomposes to products.

This long lasting controversy has been eventually solved by a theoretical study [72, 74] that on the basis of extended Hückel calculations has proved the electrophilic character of the peroxygens in these complexes and has shown that the total energy of the "butterfly" transition state (Scheme 2, path A) is lower than that of the peroxymetallacycle. The mechanistic consequence, based also on the analysis of the frontier orbitals, has been that the coordination of the olefin simply helps the slipping of the substrate toward the peroxygens with the oxygen transfer arising from the interaction between a virtually free (nucleophilic) olefin and the (electrophilic) peroxygens.

Scheme 2:

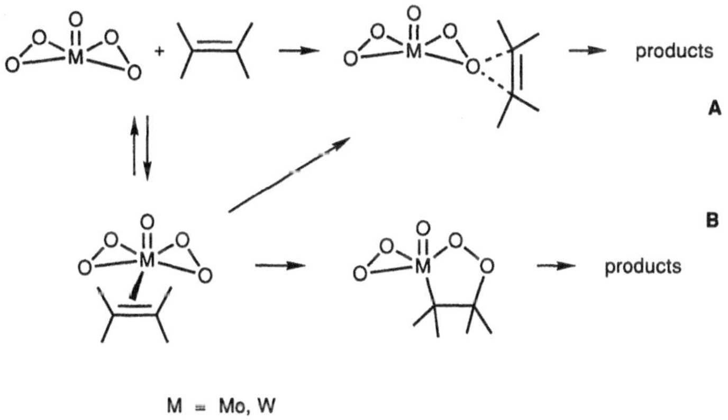

It has long been known that lithium alkyls [128] or aryl Grignard reagents [129] can be converted into the corresponding alkoxy derivatives by reaction with $MoO(O_2)_2$(HMPA) derivatives. The related complex $MoO(O_2)_2$(py)(HMPA) was also used for the hydroxylation of esters, lactones and ketones having enolizable methine or methylene groups after their conversion into the corresponding enolate anions by reaction with lithium amides [130, 131] . This reaction is normally accomplished at temperatures below 0°C and has found interesting applications in organic synthesis (chapter 7). Although mechanistic details are lacking, these reactions have been interpreted [131] as resulting from the nucleophilic attack of a carbanion on one of the electrophilic peroxygens of the complex (eq. 13).

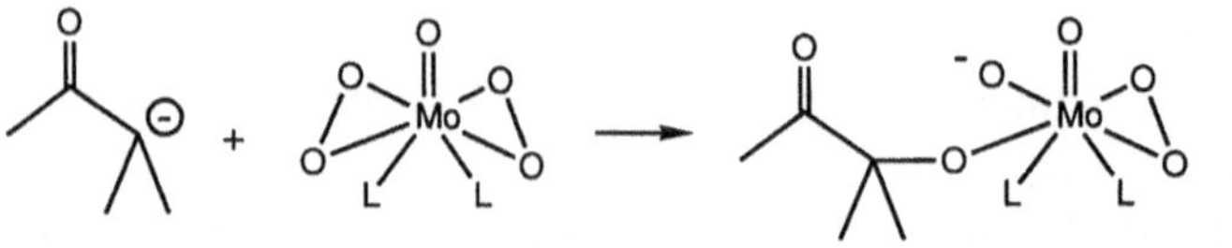

(13)

It is not without significance that, in contrast to the easy reaction with enolizable ketones, these Mo complexes interact with the carbonyl functional group of ketones to give Baeyer-Villiger type oxidations only under more forcing conditions. In fact, in the only example reported to date for this reaction [132], $[MoO(O_2)_2(pic)]^-$ or $[MoO(O_2)(dipic)]$ were found to catalyze the lactonization of a variety of simple cyclic ketones at 60°C using 90% H_2O_2 as primary oxidant. Some data are reported in Table 16 chapter 7. Surprisingly, better results were observed with the neutral complex which should be a poorer electrophile. The moderate yields obtained and the conditions used are in line with Scheme 1 and seem to reflect the difficulty to bring together two putative electrophiles i.e. the carbonyl carbon and one of the peroxygens.

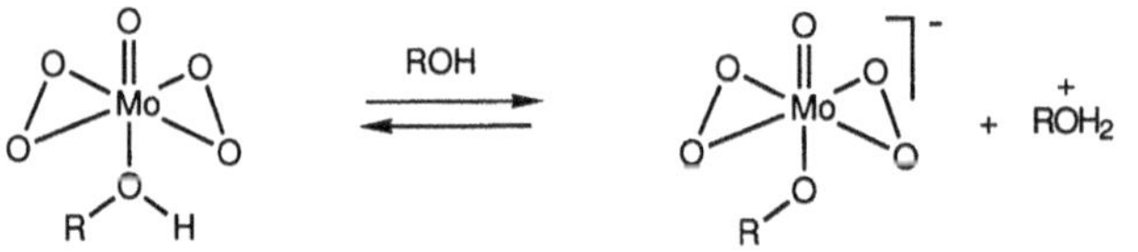

(14)

The oxidation of alcohols to carbonyl compounds escapes the simple view shown in Scheme 1. Simple oxo, diperoxo complexes of Mo(VI) and W(VI) are known to behave as strong acids [125, 126, 133-135] in alcohol solvents (reaction 14), although the localyzation of the negative charge cannot be safely attributed [136-139]. In the anionic species the electrophilic character of the peroxygens would be partly lost as is their reactivity toward typical nucleophiles, even if a complete inversion of character is not observed, as should be expected if the negative charge were located on the peroxygens (see above the Baeyer-Villiger oxidation of ketones).

(15)

Interestingly, peroxo molybdates and peroxo tungstates are good catalysts for the selective oxidation of primary alcohols to aldehydes and secondary alcohols to ketones under phase transfer conditions [140] and, as stoichiometric oxidants, even in the presence of carbon carbon double bonds (eq. 15) [141]. Some synthetic applications of this reaction are reported in chapter 7.

3.2. Vanadium and Chromium Complexes

Peroxo complexes of V(V) and Cr(VI) are easily formed from suitable metal oxo precursors and hydrogen peroxide according to eq 4. However, at variance with similar Mo and W peroxo species, they easily decompose H_2O_2 in aprotic media [27] , where they are also reactive as oxidants.

The utility of V peroxo complexes like $VO(O_2)(pic)$ and that of the related triangular *t*-butylperoxo complexes in the epoxidation of simple olefins is rather limited and has been recently reviewed [74] . In both cases oxygen transfer occurs in a non stereo-selective manner and this is associated to the relatively easy reducibility of the V(V) centers. It is suggested [27, 74] that V(IV) peroxo radicals of the type shown below

are involved in the oxygen transfer to olefins and this behavior contrasts sharply with the related V(V) catalyzed stereoselective epoxidation of allylic alcohols with *t*-BuOOH as oxidant [142], where the complexes display typical electrophilic character. Interestingly, the above behavior is at the origin of the ability of these complexes to yield the direct hydroxylation of aromatics and alkanes [24, 27]. Scheme 3 has been suggested and, as shown, the radical peroxo species is involved directly in the oxygen transfer to olefins and aromatic hydrocarbons. Conversely, in the case of alkanes, to account for the extensive epimerization at the hydroxylated carbon atom, a simple bimolecular electro-philic attack at the hydrocarbon to give hydrogen abstraction has been suggested [27].

Scheme 3:

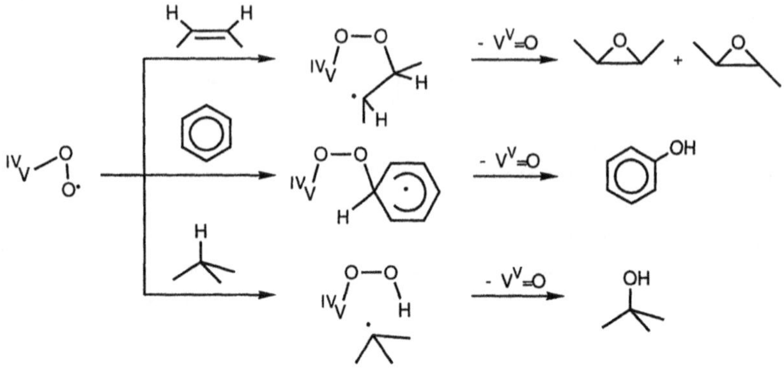

Some controversy has arisen as to the nature of the reactive intermediate [143]. To account for the promoting effect of one electron donors such as Co(II) on the rate of hydroxylation of benzene, it has been suggested that a species like that shown in eq 16 formed from one electron addition to the peroxo moiety without changing the formal

 (16)

oxidation state of the metal is more likely to be responsible of the overall reactivity of $VO(O_2)(pic)$. The reaction scheme suggested for the hydroxylation of benzene [143] is considerably more complicated than Scheme 3, but accounts for the autocatalytic reaction profiles observed that cannot be explained by the simple pathway shown in Scheme 3.

Chromium complexes behave somewhat differently. For example $CrO(O_2)_2(py)$ oxidizes alcohols to carbonyl compounds [25b] , while $CrO(O_2)_2(OPPh_3)$ oxidizes alkanes to alcohols and carbonyl compounds [144], but is rather inefficient in the oxidation of olefins and aromatics. Here again, it is believed that peroxo radicals are responsible for the reactivity observed [144] (eq 17).

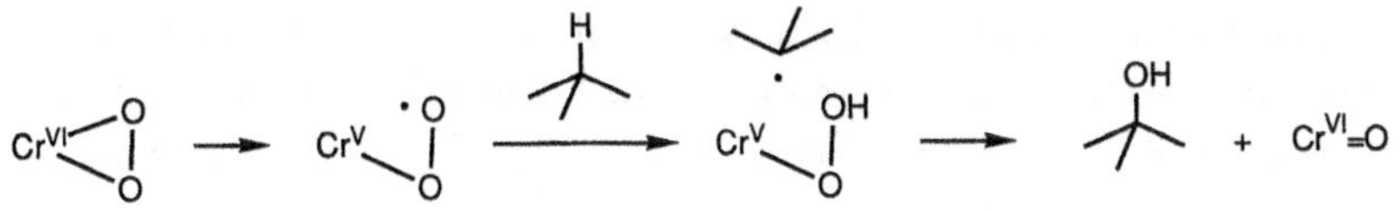

(17)

3.3. Titanium Species

As already stated in chapter 1 Ti is the base metal catalyst of the Shell heterogeneous process for the production of propylene oxide [145]. Indeed, soluble Titanium complexes have been widely employed with alkylhydroperoxides as oxidants, the most prominent system being still the well known "Sharpless reagent" for the asymmetric epoxidation of allylic alcohols [146], that has proved useful even in the synthesis of natural products [147]. This wealth of results with the Ti(IV)/ROOH combination contrasts sharply with the paucity of data obtained with H_2O_2. This is due to the easy reducibility of Ti(IV) to Ti(III) that invariably leads to Haber-Weiss decomposition of hydrogen peroxide.

In the past few years a class of new materials, Ti silicalites, has emerged as the major breakthrough in the use of H_2O_2 as oxidant in industrial organic chemistry [148] . One of these materials, the so called TS1, has proved useful also for large scale operations, as a 10,000 tons/year new plant for the production of diphenols has been built in Ravenna, Italy since 1986 [149]. To date, TS1 has been successful in a variety of oxidation reactions, namely the selective epoxidation of olefins and diolefins, the selective oxidation of primary alcohols to aldehydes and secondary alcohols to ketones, the ammoximation of ketones to oximes, the hydroxylation of aromatic and aliphatic compounds. Some representative data are reported in Table 6.

Table 6: TS-1 catalyzed oxidation with hydrogen peroxide.

Reagents	Products (selectivity %)	H_2O_2 conv.	Ref.
propylene	propylene oxide (98)	99	150
allyl chloride	epichloridrine (93)	98	151
benzene	phenol (76) + p-benzoquinone (24)	100	152
phenol	cathecol (70) + hydroquinone (30)	98	153
cyclohexanone + NH_3	cyclohexanone oxime (98)	93	154
n-butane	2-butanol (55) + 2-butanone (45)	69	155

Silicalites are crystalline silica molecular sieves, isostructural with the ZSM-5 zeolites family, that possess an internal tubular channel system of fixed diameter, but at variance with ZSM-5 that are strongly hydrophilic because of the presence of cations and anions, silicalites being neutral are mildly hydrophobic. They are generally prepared by hydrolysis from alkoxides followed by crystallyzation at high temperature under autogenous pressure. In the case of TS-1 the synthetic procedure is quite critical and depends on a number of factors [149] like the purity of the chemicals involved, the adventitious presence of alkali, the crystallite dimensions, etc. A structural unit is shown in A;

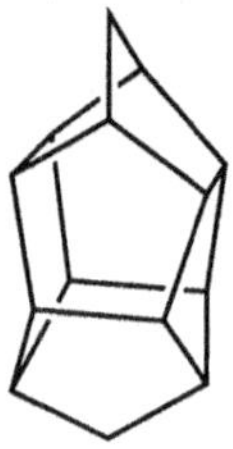

A

B

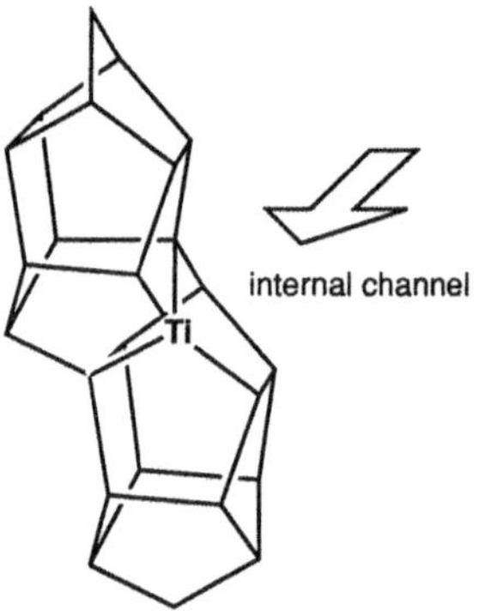

silicon atoms are located at the vertexes while each segment represents an oxygen bridge. In TS-1 some Si centers are statistically substituted with Ti atoms as in B that are therefore open to the internal tubular channels of the structure. This seems to be at the origin of the shape selectivity displayed by these catalysts in a number of instances like in the oxidation of primary alcohols to aldehydes [156] or in the oxidation of saturated hydrocarbons [155, 157]. On the other hand, the fixed dimensions of the channels determine a drop in reactivity toward long-chain aliphatic hydrocarbons [157b].

The ability of these materials to activate H_2O_2 avoiding extensive radical-type decomposition seems to be associated mainly to the mildly hydrophobic nature of the internal channel system and the lack of interaction between contiguous Ti centers, that is prevented by the fact that Ti is rigidly held in the lattice and diluted in the structure [149]. On the basis of IR and UV-VIS spectra of these materials it has been proposed [158] that the surface is partly hydrated, forming surface titanols and silanols that can reversibly dehydrate (eq. 18).

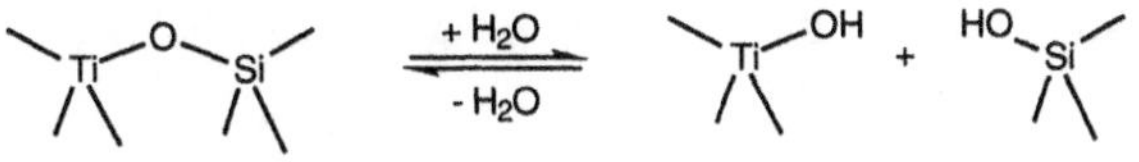

(18)

By reaction with H_2O_2 these species are believed [149, 158] to form peroxo species and these are suggested to account for the observed ability to epoxidize olefins similarly to Mo and W complexes and in agreement to the usual electrophilic behavior displayed by the peroxo ligand in d^0 transition metal centers. However, Ti side-bonded peroxo species were never observed to be active in oxygen transfer to olefins [26, 159], on the other hand, triangular type hydroperoxo derivatives, similar to the alkylperoxo complexes involved in ROOH oxidations, have never been observed. The reactivity reported for TS-1 toward a series of butenes [156] i.e. cis-2-butene > 1-butene > iso-butene > trans-2-butene does not follow the order for the simple electrophilic/nucleophilic interaction shown in Scheme 1 (although the order could be altered by shape selectivity), neither does the oxidation of alcohols, aromatics and alkanes as previously shown for Mo or V. It has been recently suggested [155] that in the reaction with H_2O_2 surface hydroperoxy species may be trapped by stabilization in the lattice from silanols, titanols or methanol which would explain also the unique role of methanol as solvent in reactions with hydrogen peroxide catalyzed by TS-1.

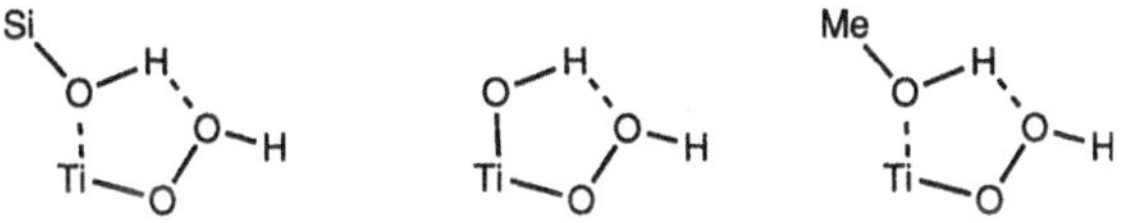

This is reminiscent of eq. 14 for Mo and W complexes in the oxidation of alcohols and the above species are believed [155] to be in equilibrium with the corresponding anionic forms, the position of the equilibrium being critical in promoting either alkene/alkane or alcohol oxidation. Since alkenes and alkanes oxidation occurs under acidic conditions, it is believed [155] that the hydroperoxo form is highly electrophilic and able to interact also with poor nucleophiles like simple saturated hydrocarbons.

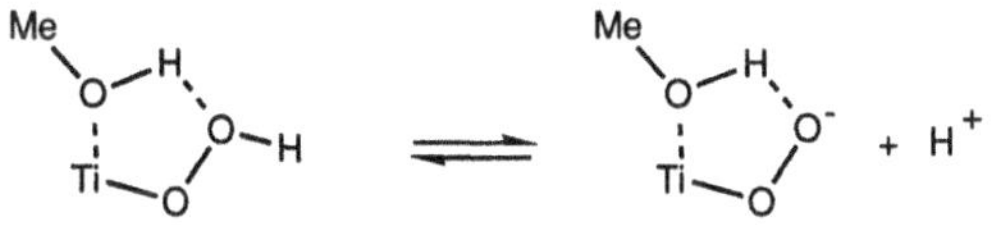

(19)

A possible transition state for the oxygen transfer is shown below.

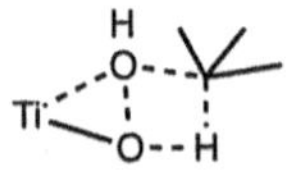

This view accomodates both the effect of the σ-inductive donor properties of alkyl substituents ($CH_3 < C_2H_5 < ... < C_nH_{2n+1}$) and the primary H/D kinetic isotope effect (4.1). However, as shown above for Vanadium, the oxidation of aromatics and alkanes is likely to be radicalic in nature.

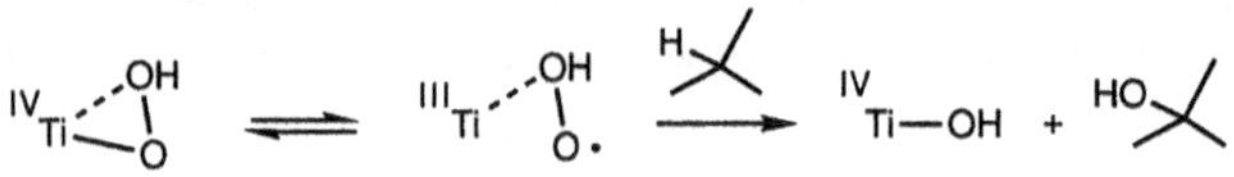

(20)

Due to the easy reducibility of Ti(IV), the intermediate peroxo species could be viewed also as a Ti(III) hydroperoxo radical. This would easily account for hydrogen abstraction from the alkane and subsequent formation of the alcohol (eq. 20).

3.4. Osmium Complexes

As reported above in section 3.1., osmium tetroxide based systems are the oxidant of choice for the direct stoichiometric conversion of olefins into *cis* -glycols. As shown in eq. 12, the peculiarity of this system relies mainly in the ability of the oxide itself to interact with olefins and produce Os(VI) ester intermediates that can be easily hydrolyzed to the glycol product either reductively or oxidatively [103]. The reaction can be carried out also asymmetrically in the presence of dihydroquinine or dihydroquinidine acetate [160]. An excellent review on the *cis* -hydroxylation of unsaturated substrates by OsO_4 has appeared in 1980 [103].

There has been much debate as to whether the initial interaction takes place (Scheme 4) via a concerted 2+2 cycloaddition followed by evolution to the Os(VI) ester (path A) or directly to the ester intermediate via 3+2 concerted cycloaddition (path B). The subject has been recently reviewed [161] and theoretical calculations [162] have indeed supported the direct formation of the osmate ester intermediate, at least in the presence of added ligands like pyridine, although conclusive evidence favoring either pathway has not been given.

Scheme 4:

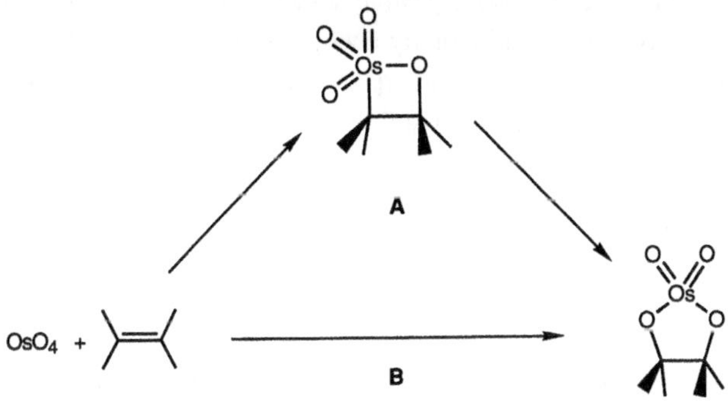

As said above the ester intermediate can be easily hydrolyzed with reducing agents like sodium or potassium sulfite or bisulfite, lithium aluminum hydride, hydrogen sulfide etc. [103] which yield the diol together with lower forms of osmium. More pertinent to the present review is the use of oxidants, particularly hydrogen peroxide. These allow the regeneration of OsO_4 making the process catalytic. However, in this case the formation of the intermediates occurs in a different fashion. The oxidizing mixture, i.e. the so-called "Milas reagent" is made up by mixing together OsO_4, excess H_2O_2 and *t* -BuOH as solvent. This procedure is believed to produce the peroxyosmic acid H_2OsO_6 [163] that is relatively stable in *t* -BuOH. Addition of the olefin results [163] in formation of an Os(VIII) ester through a stepwise process (eq. 21). Hydrolysis of the ester yields the *cis* -diol and OsO_4.

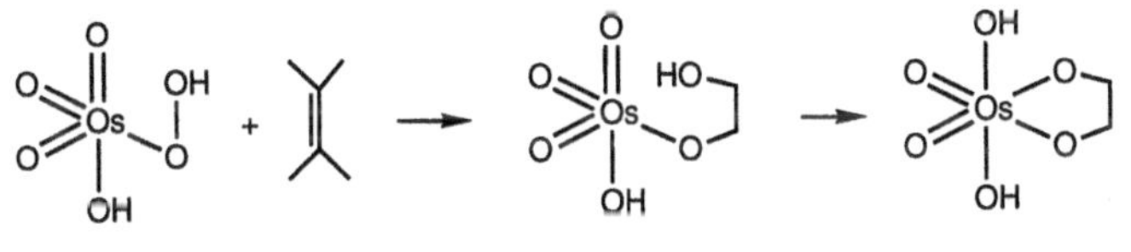

(21)

"peroxyosmic acid"

The initial interaction with H_2O_2 appears to be in line with the behavior of the other d^0 transition metal oxides in producing a peroxo species. It seems reasonable to suggest that the attack of the olefin to the complex will be driven by the same Lewis acid-Lewis base interaction demonstrated for e.g. Mo and W complexes. However, at variance with the latter, epoxides are not intermediates in the reaction (see above) [106]. It has to be pointed out that formulation of the so-called peroxyosmic acid H_2OsO_6 as a normal side-on peroxo complex (eq. 4) would equally account (eq. 22) for the formation of the intermediate Os(VIII) ester as in eq. 21. The latter complex can be observed also by reaction of the glycol with OsO_4, i.e. the reverse of the hydrolysis reaction (eq. 23).

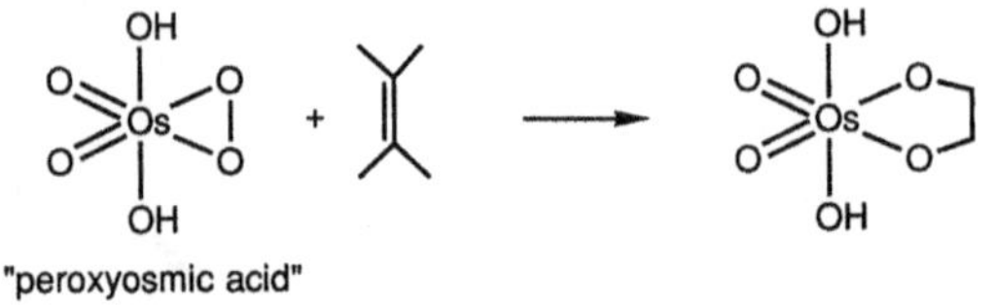

(22)

(23)

The main problem with this catalytic method is the overoxidation to give carbonyl compounds through carbon-carbon cleavage of the *cis* -diol. Milas et al. have proposed [163] that H_2O_2 attacks the ester intermediate through the pathway shown in 24.

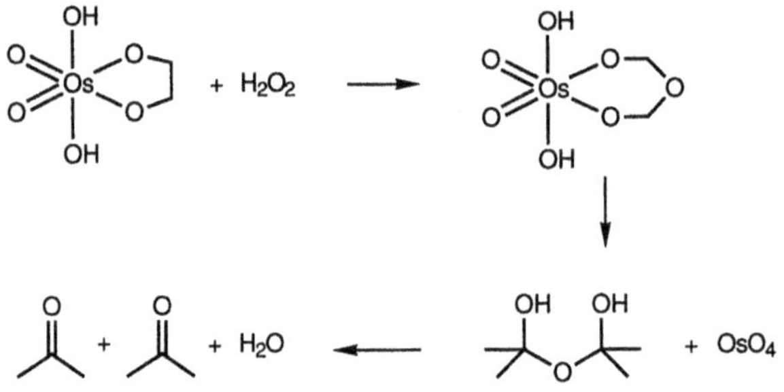

(24)

Other systems based on potassium osmate [164] or osmium trichloride [165] and hydrogen peroxide have been proposed as an alternative to the use of OsO_4 that is highly volatile and toxic.

Better methods for the direct *cis* -dihydroxylation of olefins which avoid overoxidation have been proposed. All of them employ OsO_4 associated with a different oxidant, i.e. *t* -BuOOH under basic conditions [166] or N-methylmorpholine N-oxide [167].

Although very effective as oxidant in a variety of catalytic oxidations[168], the closely related RuO_4 has not been reported to operate catalytically with hydrogen peroxide. Conversely, the parallel use of Re_2O_7 as catalyst in the epoxidation and ketonization of olefins with H_2O_2 has been reported in the patent literature [169].

3.5. Iron and Cobalt Complexes

Although Fe and Co simple salts are the typical reagents for Fenton and hydroxyl radical chemistry (see chapter 4), the use of biomimetic systems based on iron porphyrins to accomplish a wide variety of oxidation reactions has been one of the most intense field of endeavor over the past 10-15 years. These systems are believed to proceed via the formation of highly electrophilic Fe^{V}=O intermediates generated from a variety of monooxygen donors. A review of the biomimetic systems available starting from hydrogen peroxide is reported in chapter 5.

Recently, a series of papers by Sawyer and coworkers [170-173] has attracted considerable interest since a different type of H_2O_2 activation is believed to be involved. The key feature of these systems consists in the use of water free H_2O_2 in dry solvents (mainly MeCN) where the Fe(III)/Fe(II) redox potential was found to be sufficiently high to prevent one electron processes involving H_2O_2 and hence Fenton type chemistry.

Under these conditions, simple iron species like $[Fe(MeCN)_4]\ (ClO_4)_2$ or $FeCl_3$ can exploit their electrophilic properties toward H_2O_2 without promoting electron transfer. In this respect the initial interaction of the complexes with H_2O_2 seems to be in line with the general view reported in Scheme 1. Some analogies with the behavior of the Ti(IV) centers in TS-1 are also possible, although in the case of Fe, being the center involved either a d^6 or d^5 group VIII transition metal, the chemistry involved is more complex. Independently of the oxidation state of iron, the resulting Lewis acid-Lewis base adduct is believed to exist in a dynamic equilibrium between a side-on and an end-on form (eq. 25)

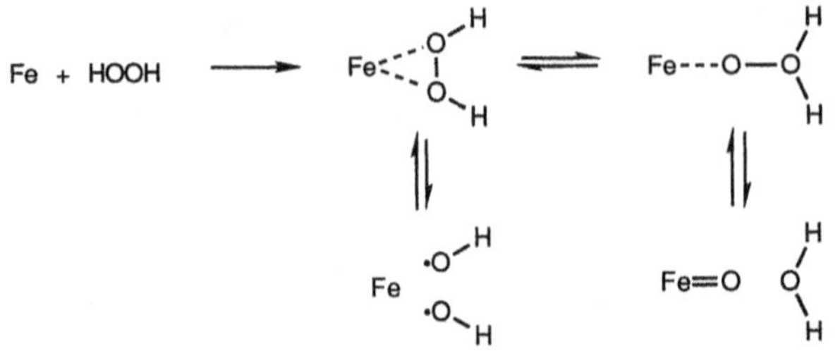

(25)

Fe = [Fe(MeCN)4] (ClO4)2 or FeCl3

inducing to coordinated H_2O_2 either a biradical or an oxene character [171-173]. The reactivity of these species results in a series of monooxygenation or oxidative dehydrogenation reactions like for example the epoxidation of cyclohexene and other simple olefins (with $FeCl_3$ [172]), the oxidation of cyclohexane to cyclohexanol, that of alcohols to aldehydes and aldehydes to carboxylic acids, the oxidation of sulfides to sulfoxides and sulfoxides to sulfones, the oxidative cleavage of ethers and epoxides, the dehydrogenation of cyclohexadiene to benzene and that of thiols to disulfides (all with $Fe(MeCN)_4^{2+}$ [170, 171]). All these reactions depend critically on the amount of free H_2O_2 present in the system, which therefore must be added carefully, since in the presence of an excess of H_2O_2 the active species is further modified (eq. 26) [173] and

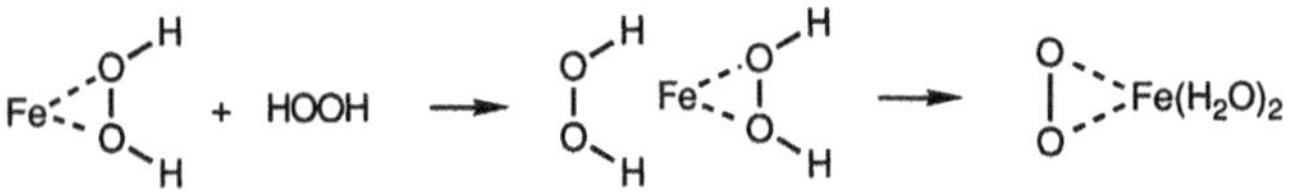

(26)

Fe = [Fe(MeCN)4] (ClO4)2

results in a series of dioxygenation reactions typical of singlet oxygen. These include the cleavage of cis and trans stilbenes to give two moles of benzaldehyde or the conversion of acetylenes to α-diketones [170].

More recently, a complete description [174] of the complex sequence of reactions leading to the active species and their respective oxidation properties has been thoroughly analyzed in the case of iron picolinate complexes of the type $Fe(pic)_2$ in pyridine/HOAc (2/1 m/m)/H_2O_2. The starting species is initially converted into a μ -oxo dimer $(pic)_2FeOFe(pic)_2$ from which binuclear adducts similar to those shown in eqs. 25, 26 are formed. The dioxygen intermediate, which is now seen as a bridging peroxo species, has proved particularly useful for the direct ketonization of methylenic carbon in cyclohexane and for the dioxygenation of acetylenes and arylolefins (Scheme 4). Essentially similar results are obtained using $Co(bipy)_2^{2+}$ and H_2O_2 in a MeCN/py (4/1 m/m) solvent mixture [175].

Scheme 5:

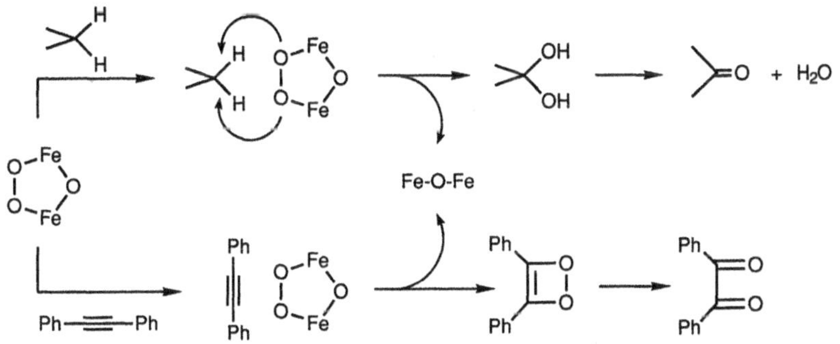

3.6. Palladium and Platinum Complexes

The first example of an alkylhydroperoxide activation with a d^8 transition metal complex was reported by Mimoun over ten years ago in the reaction between some Pd(II) acetates and *t* -BuOOH (eq. 27) to give stable tetrameric alkylperoxo complexes [176].

$$4\,Pd(RCOO)_2 \; + \; 4\,t\text{-BuOOH} \; \rightarrow \; [(RCOO)Pd\text{-}OOBu^t]_4 \qquad (27)$$

$$R = CF_3,\ CCl_3,\ CH_3$$

Some of these species were found to be effective stoichiometric oxidants for the conversion of a variety of simple olefins into the corresponding ketones (eq. 28).

(28)

olefin = 1-hexene, 1-octene, allyl acetate, etc.

The reaction was found moderately catalytic when hydrogen peroxide was used as the primary oxidant [36].

More recently, it has been reported [47] that the reaction between $Pd(OAc)_2$ and 30% H_2O_2 in chloroform produces a mixture containing 1% of a new superoxo species formulated as $(AcO)PdO_2\cdot$ that was found to oxidize stoichiometrically simple olefins to a mixture of ketones and epoxides.

In the same period in which Pd alkylperoxides were reported, the synthesis of stable hydroperoxo and *t*-butylperoxo complexes of Pt(II) was accomplished [37-39] through the simple acid-base exchange shown in eq. 29 starting from Pt-OH complexes and exploiting the moderate acidity of the hydroperoxides.

$$P_2Pt(Rx)(OH) + ROOH \rightarrow P_2Pt(Rx)(OOR) + H_2O \qquad (29)$$

P = PPh_3, PPh_2Me, $PPhMe_2$

P_2 = diphoe, dppe, dppp, dppb, o-dppb, dppx, diop, prophos, chiraphos

Rx = CF_3, CH_2CN, CH_2CF_3, $CH{=}CCl_2$, $CCl{=}CCl_2$, Ph, Me, C_6F_5

R = H, *t*-Bu

Through this synthetic route a wide class of mononuclear, square planar, homologous complexes can be prepared.

Some of these Pt compounds are useful catalysts for the selective (> 99%) epoxidation of terminal olefins using commercial grade hydrogen peroxide as oxidant [39, 79, 177]. The reaction can be performed at room temperature either in monophasic (THF, EtOH, dimethoxyethane/H_2O) or biphasic (CH_2Cl_2 or $C_2H_4Cl_2/H_2O$) reaction media. At the end of the catalytic reaction the catalyst can be easily recovered and after minimal work-up it can be almost quantitatively recycled. In Table 7 some data on the catalytic epoxidation of olefins with these complexes is reported. Significantly, both with Pt and with the above reported Pd catalysts, best results are obtained with electron poor olefins, internal or substituted olefins being not oxidized [177]. The opposite occurs with d^0 transition metal based systems (in agreement with Scheme 1) and in this respect the two classes of catalysts (d^0 and d^8) seem to complement each other very well.

Table 7: Epoxidation of olefins with H_2O_2 catalyzed by $[(P\text{-}P)Pt(CF_3)(CH_2Cl_2)]^+$.

P-P	olefin	time	turnover	ref.
diphoe	propylene	35 min	325	177
	1-hexene	1 h	115	177
	1-octene	1 h	84	177
	styrene	48 h	9	177
dppe	1-octene	2 h	21	39
dppp	1-octene	2 h	6	39
dppb	1-octene	2 h	0.6	39

The requirements necessary for an effective catalyst with the above Pt system are the presence of the CF_3 ligand that is very important in reducing the electron density at the metal center thereby promoting an efficient activation of the olefin and the presence of a diphosphine, which, because of the chelating effect, prevents the degradation of the catalysts avoiding the dissociation of one P ligand and the subsequent oxidation by excess H_2O_2 (eq. 30) [79].

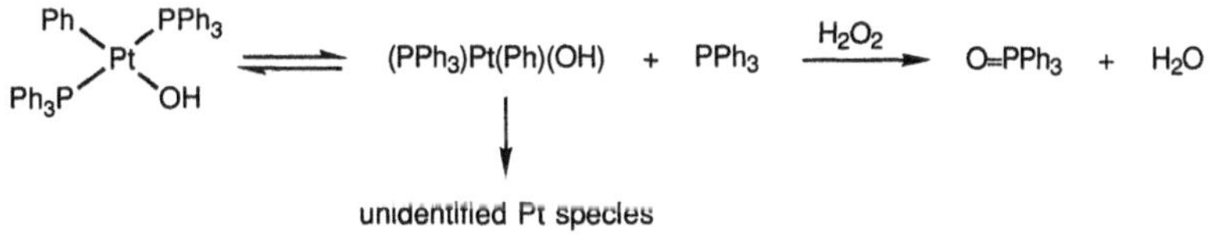

(30)

As shown in eq. 29, the above Pt(II) catalysts can be easily prepared also with chiral diphosphines. These chiral catalysts have proved useful for the enantioselective epoxidation of simple olefins with H_2O_2 [178]. Table 8 indicates that optical purities up to 41% can be obtained and the absolute configuration of the chiral epoxide can be changed by changing the diphosphine. Working with simple olefins the favorable "secondary interaction" [179] arising from other functional groups in the molecule in directing the enantiotopic discrimination cannot be exploited. This observation has led to the design of catalysts with appropriate structural and conformational rigidity in order to minimize the number of diastereoisomers during the asymmetric transformation [180]. The use of chiral diphosphines with the above Pt(II) catalysts seems to fit this scope since the above e.e.'s, although not useful for synthetic purposes, are still unsurpassed for such simple aliphatic olefins.

Table 8: Enantioselective epoxidation of olefins with H_2O_2 catalyzed by $[(P\text{-}P^*)Pt(CF_3)(CH_2Cl_2)]^+$.

P-P*	Olefin	% e.e.
S,S-chiraphos	propylene	41 (S)
	1-octene	36 (S)
R,R-pyrphos	propylene	40 (R)
	1-octene	37 (R)
R-prophos	propylene	35 (R)
	1-octene	28 (R)

These epoxidation reactions have been studied kinetically [181] and have shown typical second order dependence on Pt concentration. This is consistent with the presence of two different Pt species in the rate determining step and the mechanism that has been suggested is reported in Scheme 6. As shown, the balance between (P-P)Pt(CF_3)(OH)

Scheme 6 :

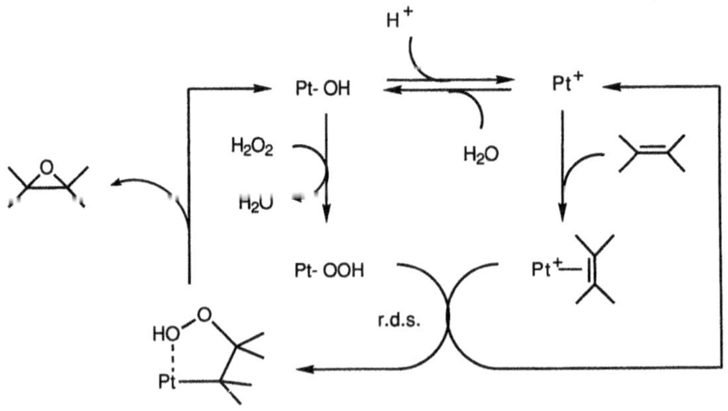

and $[(P-P)Pt(CF_3)(solv)]^+$ is determined essentially by the acidity of the medium, a factor that is very important in maximizing the catalytic activity [181]. The two Pt complexes perform two different roles: the former activates H_2O_2 by increasing its nucleophilicity, the latter activates the olefin toward nucleophilic attack through coordination. Interaction of the active species in the rate determining step produces the epoxide and restores the starting complexes. The system bears close similarities with the well known insertion of cyano olefins into $(PPh_3)_2Pt(O_2)$ reported by Sheldon and Van Doorn [182, 183] (eq. 31). In Scheme 6 the activation performed by the electron-withdrawing cyano substituents is carried out by the $-CF_3$ containing, cationic Pt center thereby allowing the oxidation of ordinary olefins.

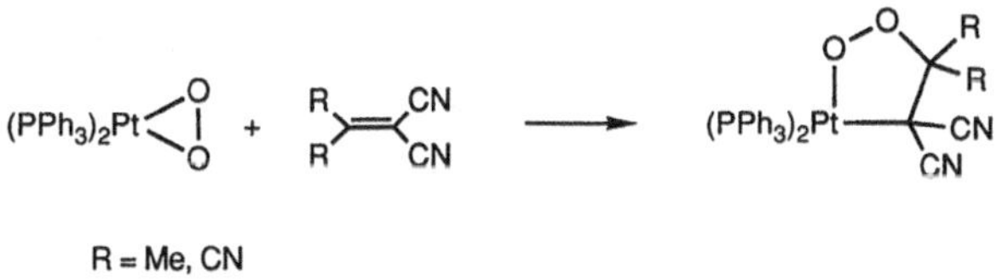

(31)

The intermediate quasi-peroxymetallacycle of Scheme 6 has been suggested by analogy with the peroxymetallacycle of eq. 31. The facile migration of the axial -OH group on Pt provides the driving force necessary to oxirane ring closure and product release.

The chemistry described in Scheme 6 may be viewed also as an example of bifunctional catalysis. This term is used here to indicate a situation where the catalyst behaves independently toward each reactant resulting in separate activation on different metal centers. This is typical of most heterogeneous catalysts which are known to carry out reactions through "ensembles" of surface atoms [184, 185] on which the reactants are individually activated through chemisorption. This characteristic of the above Pt catalyzed epoxidations offers the possibility to exploit the effect of a different metal as co-catalyst. This has been achieved by using a 1/1 mixture of $[(dppe)Pd(CF_3)(solv)]^+$ and $(diphoe)Pt(CF_3)(OH)$ as catalysts for the oxidation of a variety of olefins with H_2O_2 [186]. The use of Pd as olefin activator results in a more active catalyst mixture and the preferential formation of ketones instead of epoxides (eq. 32). Since after each oxygen transfer the two functions are inverted (eq. 32), the original roles are reestablished via eq. 33 as was shown by NMR [186].

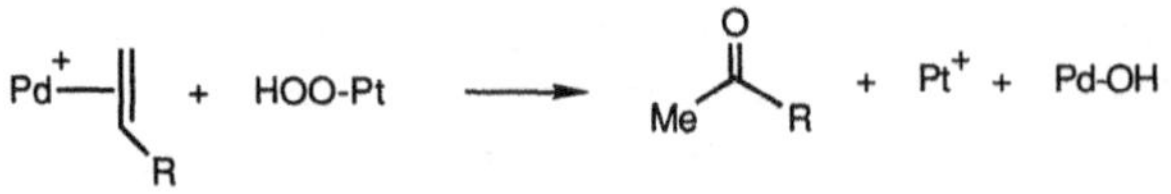

(32)

$$Pt^+ + Pd\text{-}OH \longrightarrow Pt\text{-}OH + Pd^+ \qquad (33)$$

A similar promotion effect on the reaction rate (but not on the selectivity) is observed in the epoxidation of 1-octene with hydrogen peroxide when (diphoe)$Pt(CF_3)X$ ($X = CH_2Cl_2$, OH) catalysts are dispersed on a variety of inorganic oxides like B_2O_3, TiO_2, MoO_3, WO_3 or even Al_2O_3 or SiO_2 [187]. It has to be pointed out that under the conditions used none of the oxides alone displays any activity at all. It is however significant that all the oxides used are able to activate H_2O_2 to some extent.

The same inversion of selectivity (ketones vs. epoxides) observed using the Pd^+/PtOH mixture can be achieved with (diphoe)$Pt(CF_3)X$ complexes as catalysts when the oxidant is changed from H_2O_2 or $KHSO_5$ to the more sterically demanding *t*-BuOOH [188]. These and the above reported results obtained by Mimoun with Pd [36, 176] raise the fundamental question concerning the origin of selectivity in d^8 transition metal catalyzed oxidation of olefins. To account for the experimental data, the mechanism shown in Scheme 7 is suggested. In all cases the involvement of the critical quasi-peroxymetallacyclic intermediate has been suggested [36, 176, 178, 181, 186, 188]. In principle, this can exist also in equilibrium with an open-chain configuration, the position of the equilibrium being determined essentially by steric hindrance of the X group.

Scheme 7:

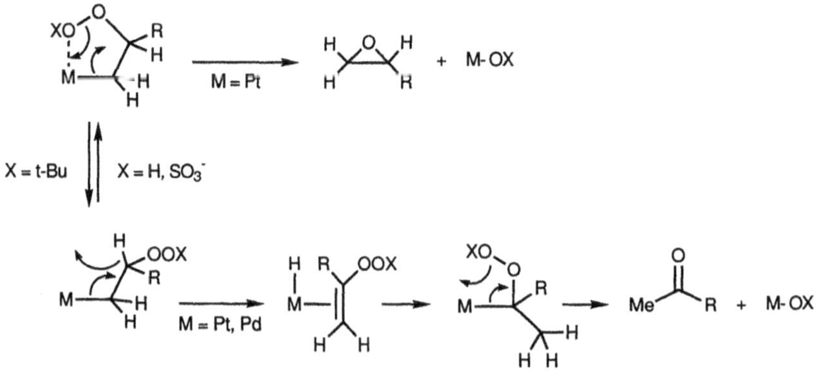

Product formation can be easily accounted for via either oxirane ring closure or β-hydride elimination addition. The choice between the two possible pathways applies only to Pt depending on the experimental conditions (oxidant, temperature etc.), while for Pd the β-hydride elimination addition appears to be overriding in any case, and the ketone is always the preferred product independently of the oxidant used.

The fine tuning of steric requirements related to the coordination of the olefin and the rearrangement of the quasi-peroxymetallacycle in epoxidation with a series of different (P-P)Pt(CF_3)X complexes (see below) has been examined in the case of 1-octene [39]. By varying systematically the shape and size of the diphosphine-metal ring through the use of a variety of tetraphenyldiphosphines, a strong influence on the catalytic activity has been evidenced. So, for example a two orders of magnitude difference in catalytic activity was observed when going from dppe to dppb. Flat, five-membered rings appear to be best suited to maximize the activity of the catalyst. It is suggested that these leave the maximum space available for olefin coordination on the metal and for the subsequent binding of the peroxy-oxygen from the axial position, which seem to be the most critical factors in determining the catalytic activity.

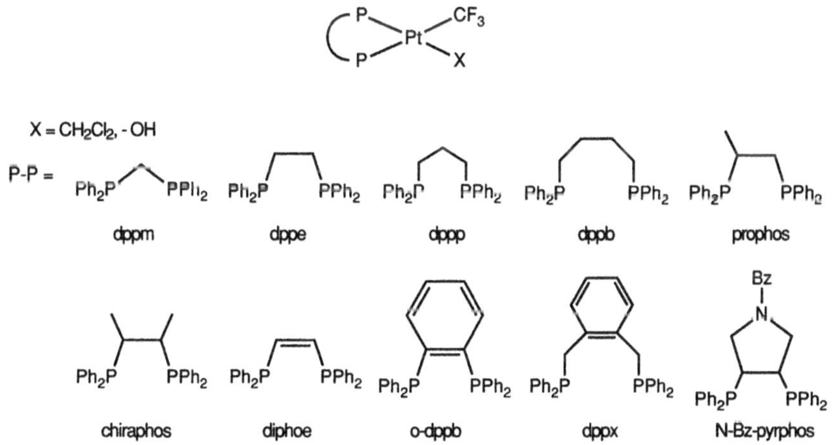

The ability of the (P-P)Pt(CF_3)X complexes to increase the nucleophilicity of H_2O_2 and the additional property to make olefins susceptible to nucleophilic attack can be extended to other substrates. An example is the oxidation of carbon monoxide. This is known to interact

(34)

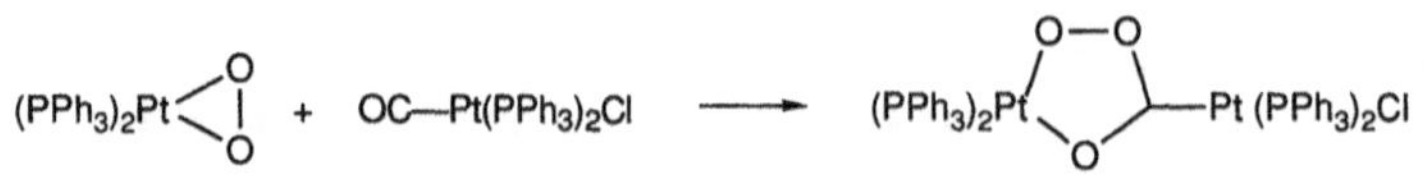

(35)

Scheme 8:

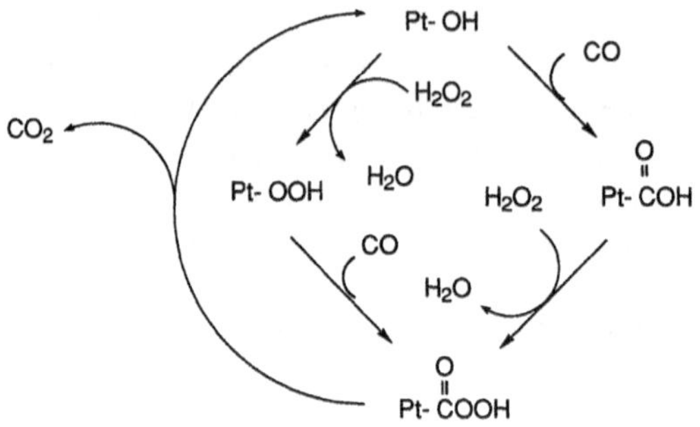

with $(PPh_3)_2Pt(O_2)$ to give either a carbonato complex [189] or, via precoordination on a second Pt center, a bridging peroxycarbonyl species [190] (eq. 34, 35). A similar reaction with CO carried out with the PtOOH complexes leads to the formation of CO_2 and the corresponding PtOH species [37]. Identical results are obtained when the homologous (P-P)Pt(CF_3)(COOH) are reacted with hydrogen peroxide [191]. The reactions are believed to proceed through a common hydroperoxycarbonyl intermediate and the process can be made catalytic under CO atmosphere (Scheme 8).

Other substrates susceptible to nucleophilic attack are ketones. It has been already said (see above) that the Baeyer-Villiger oxidation of ketones with peroxo complexes of Molybdenum can be accomplished with only modest results due to the difficulty to bring together two electrophiles like the carbonyl carbon and the peroxygens. Conversely, a variety of cyclic ketones can be successfully converted into the corresponding lactones under very mild conditions using (dppe)Pt(CF_3)X complexes (X = CH_2Cl_2, OH) and diluted 35% H_2O_2 (Table 9) [192]. The carbon atom migratory aptitude follow the same order found for the classical Baeyer-Villiger reaction with peracids and due to the high turnovers that are observed in certain cases the reaction has some potential in organic synthesis.

Table 9: Baeyer-Villiger oxidation of cyclic ketones to lactones with H_2O_2 catalyzed by $[(dppe)Pt(CF_3)(CH_2Cl_2)]\ BF_4$.

Ketone	Turnover	Time (min)
cyclohexanone	43	35
2-methyl-cyclohexanone	60	49
cyclopentanone	68	108
2-methylcyclopentanone	17	51
cyclobutanone	333	8

As stated above in the case of olefins and carbon monoxide, even in this case the reaction closely resembles the behavior of $(PPh_3)_2Pt(O_2)$ (eq. 36), that is known to react with ketones to form peroxymetallacyclic insertion products.

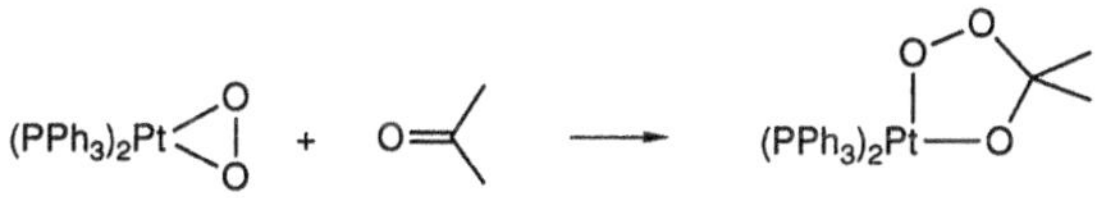

(36)

In a kinetic study of eq. 36 [193] Ugo and coworkers have demonstrated that the reaction proceeds via prior coordination of the ketone to the complex followed by nucleophilic attack of one of the peroxygens to the carbonyl carbon. Similarly, it is suggested [192] that the Baeyer-Villiger oxidation of cyclic ketones involves a quasi-peroxymetallacyclic intermediate, analogous to that involved in the epoxidation of olefins, where the easy migration of the -OH group on Pt provides the driving force necessary to break the carbon-carbon bond and lead to lactone formation.

The possible oxygen transfer pathways have been examined (eq. 37 - 39) and only eq. 37, 38 have been found feasible under the experimental conditions employed [194].

$$Pt(ketone)^+ + Pt\text{-}OOH \rightarrow lactone + Pt^+ + Pt\text{-}OH \quad (37)$$

$$Pt(ketone)^+ + H_2O_2 \rightarrow lactone + Pt^+ + H_2O \quad (38)$$

$$ketone + Pt\text{-}OOH \rightarrow lactone + Pt\text{-}OH \quad (39)$$

However, a mechanistic study of the catalytic reaction in the case of cyclobutanone [194] has restricted the choice to eq. 38 and the mechanism suggested is outlined in Scheme 9.

Scheme 9:

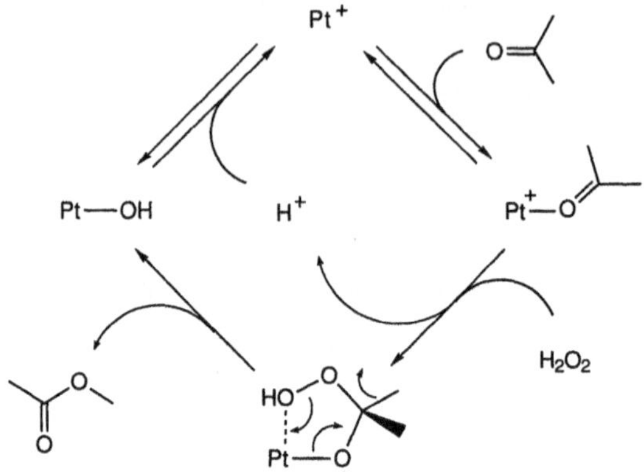

Pt = (dppe)Pt(CF_3)

As shown, at least for cyclobutanone the role of Pt is not to increase the nucleophilicity of H_2O_2, but mainly to activate the substrate and assist the leaving of the -OH group from the reactive intermediate. Nevertheless, this is an important feature since in the classical, non-catalyzed Baeyer-Villiger oxidation of ketones the availability of a good leaving group appears to be a crucial factor (see chapter 3) and is the main reason why hydrogen peroxide is a very poor oxidant while, on the other hand, trifluoroperacetic acid is very often the reagent of choice. The reaction proceeds [9a] through addition of the peracid to the ketone to form the so-called Criegee adduct with subsequent migration of an electron-rich group to an electrophilic oxygen and leaving of the acid anion (eq. 40). If the two reaction intermediates are compared (i.e. the Criegee adduct and

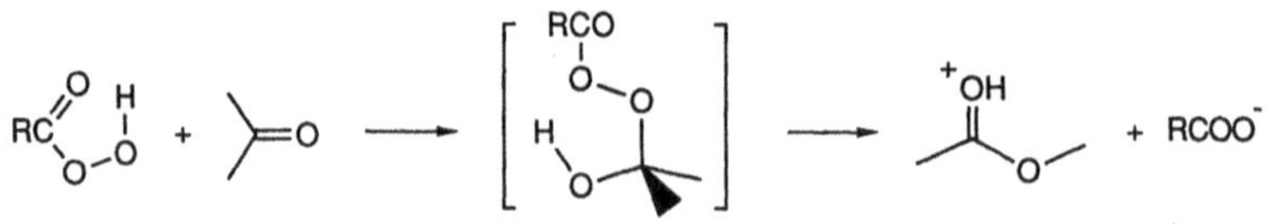 **(40)**

the quasi-peroxymetalla-cycle) it is evident that the additional ability of Pt to assist the leaving of the -OH group is crucial in making H_2O_2 a viable oxidant for the reaction.

Other substrates that have been successfully oxidized with H_2O_2 in Pt(II) catalyzed reactions are susbstituted aromatic compounds [195]. The reaction can bc accomplishcd at moderatly high temperatures using 70% H_2O_2 and (P-P)Pt(CF_3)X (P-P = various diphosphincs; X = CH_2Cl_2, OH) as catalyst. Under appropriate experimental conditions an interesting selectivity toward ortho products is observed.

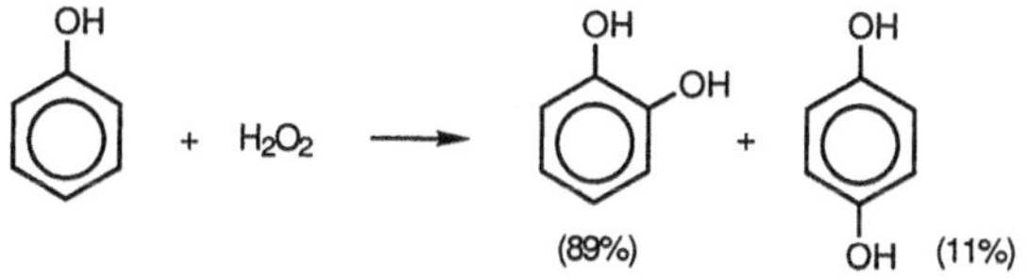

(41)

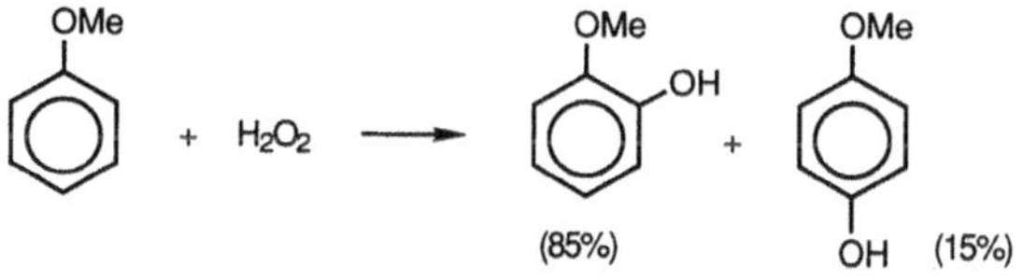

(42)

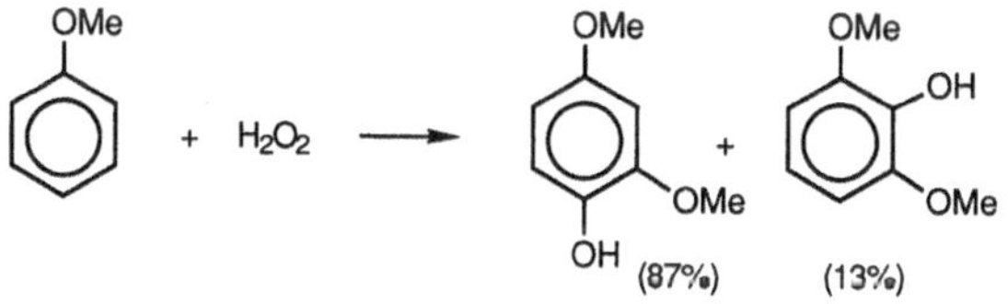

(43)

A mechanistic study [195] has again proved the dual function of the catalyst (Scheme 10) and the involvement of PtOOH species as the actual oxidant. Interestingly, the substrate activation seems to consist in an electrophilic metallation of the aromatic ring leading to C-H breaking. The subsequent oxygen transfer from PtOOH to the Pt-aryl opens the way to product formation and the overall catalytic process represents one of the rare examples of direct functionalyzation of an aromatic C-H.

Scheme 10:

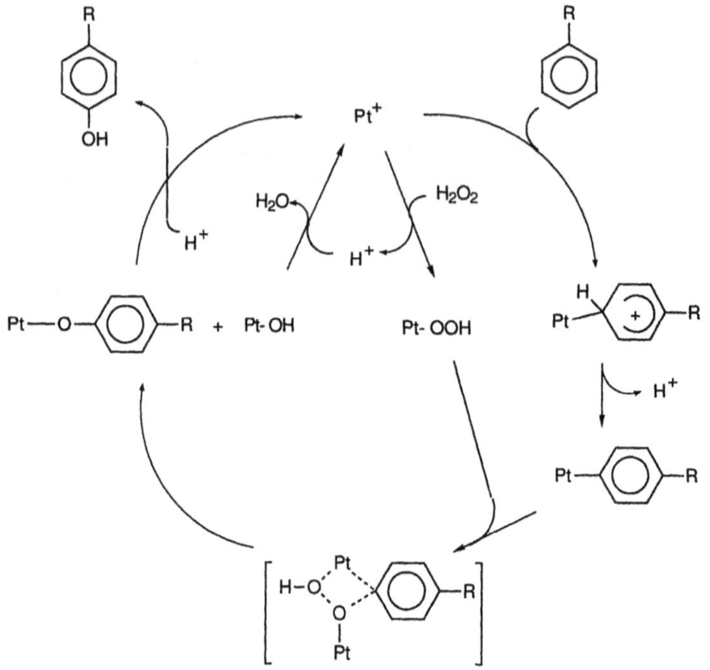

4. Conclusions

The above review has proposed a general guideline, summarized in Scheme 1, in directing the search of a catalyst that is likely to be the most appropriate for the oxidation of a given substrate with H_2O_2. In many examples with nucleophilic substrates like amines, sulfides, internal or substituted olefins, d^0 metal centers have proved more effective catalysts, while on the other hand with electrophilic substrates like coordinated simple terminal olefins or ketones, d^8 metal centers are, in general, to be preferred. Significant exceptions to the above are the synthetically very interesting oxidations of aliphatics and aromatics that are in most cases radicalic in nature.

In trying to set forth some perspectives for future work in this field, only some general considerations are suggested.

Early transition metals have been more widely studied and have already proved their utility in synthetic organic chemistry as is testified by chapter 7. However, many of these reactions are still stoichiometric in nature, the steps necessary to the regeneration of the catalyst after oxygen transfer being still to be worked out under the conditions employed. In this respect the major guideline so far explored has been the use of phase-transfer catalysis which may allow also the separation of the catalyst from the reaction mixture, but certainly more work is called for to comply with this task.

The major advancement in this respect is certainly constituted by Titanium silicalites, that are solid materials that can be used in the liquid phase. To date, these are the only catalysts for the synthesis of organic chemicals with hydrogen peroxide that have gone into commercial production. Indeed, the idea of anchoring a transition metal ion to a solid matrix (possibly inorganic) is always of considerable interest for industry and Ti silicalites suggest some basic requirements (site isolation, oxidatively resistant ligands) that will help future research.

Another problem to be tackled concerns selectivity. Although considerable results have been obtained as far as chemo-, regio- and stereo-selectivity are concerned, the problem of enantio-selectivity with these d^0 species has not been adequately addressed.

As to late transition metals, the present state of knowledge is at much lower level, since their reactivity has just started to be explored and to date only one class of Pt(II) complexes has been found to be adequately reactive toward different substrates. To date, the major virtue of these catalysts has been to render catalytic some of the stoichiometric insertion reactions with electrophiles known for the $P_2Pt(O_2)$ complexes. Besides all the general considerations already suggested for d^0 complexes, future work in this field will have first of all to increase the number of possible catalysts for oxygen transfer reactions and broaden their scope in order to be of some synthetic utility.

Acknowledgements

The author wishes to thank the National Research Council of Italy (CNR, Progetto Finalizzato "Chimica Fine II"), the Italian Ministry for University, Scientific and Technological Research (MURST 40%), Degussa AG and the European Community (BRITE Program) for financial support to the work cited in the references. Thanks are also expressed to Dr. B. Notari of ENI for providing preprints of his work prior to pubblication.

References

1. W. R. Busing and H. A. Levy, *J. Chem. Phys.*, **42**, 3054 (1965).
2. S. C. Abrahams, R. L. Collin and W. N. Lipscomb, *Acta Cryst.*, **4**, 15 (1951).
3. O. Exner in '*The chemistry of Peroxides*' Ed. S. Patai, Wiley-Interscience, New York 1983, chapter 2, p. 85 and references therein.
4. B. F. Pedersen and B. Pedersen, *Acta Chem. Scand.*, **18**, 1454 (1964).
5. V. A. Sarin, V. Ya. Dudarev, T. A. Dobrynina and V. E. Zavodnik, *Kristallografiya* , **24**, 472 (1979).
6. M. A. A. F. de C. T. Carrondo, W. P. Griffith, P. D. Jones and A. C. Skapski, *J. Chem. Soc. Dalton Trans.*, 2323 (1977).
7. D. Cremer in '*The Chemistry of Peroxides*' Ed. S. Patai, Wiley-Interscience, New York 1983, chapter 1, p. 43 and references therein.
8. A. J. Everett and G. O. Minkoff, *Trans. Faraday Soc.*, **49**, 410 (1953).
9. See for example: (a) B. Plesnicar in '*Oxidation in Organic Chemistry*', part C, Ed. W. Trahanovsky, Academic Press, New York 1978, chapter III, p. 214; (b) G. Bouillon, C. Lick and K. Shank in '*The Chemistry of Peroxides*' Ed. S. Patai, Wiley-Interscience, New York 1983, chapter 10, p. 287; (c) D. Swern in '*Organic Peroxides*', vol I, Ed. D. Swern, Wiley-Interscience, New York 1970, chapter 6, p. 313; and references therein.
10. L. M. Hjelmeland and G. Loew, *Tetrahedron* , **33**, 1029 (1977).
11. B. Plesnicar, M. Tasevski and A. Azman, *J. Am. Chem. Soc.*, **100**, 743 (1978).
12. B. Plesnicar in '*The Chemistry of Peroxides*' Ed. S. Patai, Wiley-Interscience, New York 1983, chapter 17, p. 521 and references therein.
13. W. M. Schubert and R.R. Kintner in '*The Chemistry of Carbonyl Group*' Ed. S. Patai, Wiley-Interscience, New York 1966, chapter 14, p. 749 and references therein.
14. I. O. Sutherland in '*Comprehensive Organic Chemistry*' Eds. D.H.R. Barton and W. D. Ollis, Pergamon, Oxford 1979, vol. 2, chapter 9.8, p. 895 and references therein.
15. K. B. Wiberg, *J. Am. Chem. Soc.*, **77**, 2519 (1953); *ibid.*, **77**, 3961 (1953).
16. J. E. McIsaac jr., R. E. Ball and E. J. Behrman, *J. Org. Chem.*, **36**, 3048 (1971).
17. Y. Sawaki and Y Ogata, *Bull. Chem. Soc. Japan* , **54**, 793 (1981).
18. H. O. House and R. S. Ro, *J. Am. Chem. Soc.*, **80**, 2428 (1958).
19. C. A. Bunton and G. O. Minkoff, *J. Chem. Soc.*, 665 (1949).
20. H. Mimoun, I. Seree de Roch and L. Sajus, *Bull. Soc. Chim. France*, 1481 (1969).
21. H. Mimoun, I. Seree de Roch and L. Sajus, *Tetrahedron* , **26**, 37 (1970).
22. J. M. Le Carpentier, A. Mitschler and R. Weiss, *Acta Cryst. Sect. B*, **28**, 1288 (1972).
23. See for example: H. Mimoun in '*Comprehensive Coordination Chemistry*' Eds. G. Wilkinson, R.D. Gillard and J. McCleverty, Pergamon, Oxford 1987, chapter 61.3 p 317 and references therein.
24. F. Di Furia and G. Modena, *Rev. Chem. Intermediates*, **6**, 51 (1985) and references therein.
25. (a) R. Stomberg, *Ark. Kemi*, **22**, 29 (1964); (b) G. W. J. Fleet and W. Little, *Tetrahedron Lett.*, **42**, 3749 (1977).
26. H. Mimoun, M. Postel, F. Casabianca, A. Mitschler and J. Fischer, *Inorg. Chem.*, **21**, 1303 (1982).

27. H. Mimoun, L. Saussine, E. Daire, M. Postel, J. Fischer and R. Weiss, *J. Am. Chem. Soc.*, **105**, 3101 (1983).
28. O. Bortolini, S. Campestrini, F. Di Furia, G. Modena and G. Valle, *J. Org. Chem.*, **52**, 5467 (1987).
29. (a) J. E. Guerchais and B. Spinner, *Bull. Soc. Chim. France*, **4**, 1123 (1965); (b) G. Mathern and R. Weiss, *Acta Cryst. Sect. B*, **27**, 1572 (1971).
30. P. Chaumette, H. Mimoun, L. Saussine, J. Fischer and A. Mitschler, *J. Organomet. Chem.*, **250**, 291 (1983).
31. A. D. Westland and M. T. H. Tarafder, *Inorg. Chem.*, **21**, 3228 (1982).
32. I. Bkouche Waksman, C. Bois, J. Sala-Pala, and J. E. Guerchais, *J. Organomet. Chem.*, **195**, 307 (1980).
33. J. Sala-Pala, J. Roue and J. E. Guerchais, *J. Mol. Catal.*, **7**, 141 (1980).
34. R. Guilard, J. M. Latour, C. Lecomte, J. C. Marchon, J. Protas and D. Ripoll, *Inorg. Chem.*, **17**, 1228 (1978).
35. R. Chevrier, T. Diebald and R. Weiss, *Inorg. Chim. Acta* , **19**, L57 (1976). H. Ledon and M. Bonnet, *J. Mol. Catal.*, **7**, 309 (1980).
36. M. Roussel and H. Mimoun, *J. Org. Chem.*, **45**, 5381 (1980).
37. G. Strukul, R. Ros and R.A. Michelin, *Inorg. Chem.*, **21**, 495 (1982).
38. R. A. Michelin, R. Ros and G. Strukul, *Inorg. Chim. Acta* , **37**, L491 (1979).
39. A. Zanardo, R. A. Michelin, F. Pinna and G. Strukul, *Inorg. Chem.*, **28**, 1648 (1989).
40. E. I. Ochiai, *Inorg. Nucl. Chem. Lett.*, **9**, 987 (1973).
41. W. L. Johnson and J. F. Geldard, *Inorg. Chem.*, **17**, 1675 (1978).
42. B. Bosnich, W. G. Jackson, S. T. D. Lo and J. W. McLaren, *Inorg. Chem.*, **13**, 2605 (1974).
43. C. A. Reed and W. R. Roper, *J. Chem Soc. Dalton Trans.*, 1370 (1973).
44. M. Green, T. A. Kuc and S. H. Taylor, *J. Chem. Soc. Dalton Trans.*, 2334 (1971).
45. G. Gregorio, G. F. Pregaglia and R. Ugo, *Inorg. Chim. Acta* , **3**, 89 (1969).
46. G. Strukul, unpublished results.
47. E. P. Talsi, V. P. Babenko, V. A. Likholobov, V. M. Nekipelov and V. D. Chinakov, *J. Chem. Soc. Chem. Commun.*, 1768 (1985).
48. L. E. Johnston and J. A. Page, *Can. J. Chem.*, **47**, 4241 (1969).
49. H. L. Roberts and W. L. Symes, *J. Chem. Soc. A* , 1450 (1968).
50. J. H. Bayston and M. E. Winfield, *J. Catal.*, **3**, 123 (1964).
51. M. T. Atlay, M. Preece, G. Strukul and B. R. James, *J. Chem. Soc. Chem. Commun.*, 406 (1982).
52. M. T. Atlay, M. Preece, G. Strukul and B. R. James, *Can. J. Chem.*, **61**, 1332 (1983).
53. (a) C.-Y. Sue and B. R. James, *Proc. 6th Int. Symp. Homog. Catal.*, Vancouver 1988, Abstract P-30; (b) C.-Y. Sue, *Ph.D. Dissertation* , University of British Columbia, Vancouver 1988.
54. T. T. Wenzel, *Stud. Surf. Sci & Catal.* , vol. 66, Ed. L. Simandi, Elsevier, Amsterdam 1991 p. 545.
55. F. Igersheim and H. Mimoun, *Nouv. J. Chim.*, **4**, 711 (1980).
56. S. Muto, H. Ogata and Y. Kamiya, *Chem. Lett.*, 809 (1975).
57. H. Mimoun, *Pure Appl. Chem.*, **53**, 2389 (1981).
58. H. Mimoun in '*The Chemistry of Peroxides* ' Ed. S. Patai, Wiley-Interscience, New York 1983, chapter 15, p. 463 and references therein.
59. R. Stomberg, *Acta Chem. Scand.*, **17**, 1563 (1963).
60. N. W. Terry, E. L. Amma and L. Vaska, *J. Am. Chem. Soc.*, **94**, 653 (1972).
61. C. Giannotti, C. Fontaine, A. Chiaroni and C. Riche, *J. Organomet. Chem.*, **113**, 57 (1976).

62. M. Matsumoto and K. Nakatsu, *Acta Cryst. Sect. B*, **31**, 2711 (1975).
63. W. Winter, C. Mark and V. Schurig, *Inorg. Chem.*, **19**, 2045 (1980).
64. M. Laing, M. J. Nolte and E. Singleton, *J. Chem. Soc. Chem. Commun.*, 660 (1975).
65. H. Mimoun, R. Charpentier, A. Mitschler, J. Fischer and R. Weiss, *J. Am. Chem. Soc.*, **102**, 1047 (1980).
66. F. W. B. Einstein and B. R. Penfold, *Acta Cryst.*, **17**, 1127 (1964).
67. J. A. McGinnety, R. J. Doedens and J. A. Ibers, *Inorg. Chem.*, **6**, 2243 (1967).
68. T. Kashiwagi, N. Yasuoka, N. Kasai, M. Kakudo, S. Takahashi and N. Hagihara, *J. Chem. Soc. Chem. Commun.*, 743 (1969).
69. G. Ferguson, M. Parvez, P. K. Monaghan and R. J. Puddephatt, *J. Chem. Soc. Chem. Commun.*, 267 (1983).
70. H. Mimoun, P. Chaumette, M. Mignard, L. Saussine, J. Fischer and R. Weiss, *Nouv. J. Chim.*, **7**, 467 (1983).
71. R. D. Bach, G. J. Wolber and B. A. Coddens, *J. Am. Chem. Soc.*, **106**, 6098 (1984).
72. K. A. Jørgensen and R. Hoffmann, *Acta Chem. Scand.*, **B40**, 411 (1986).
73. K. A. Jørgensen, R. A. Wheeler and R. Hoffmann, *J. Am. Chem. Soc.*, **109**, 3240 (1987).
74. K. A. Jørgensen, *Chem. Rev.*, **89**, 431 (1989).
75. Comprehensive reviews on group VIII metal-dioxygen complexes have appeared in the past; see for example: (a) L. Vaska, *Chem. Rev.*, **9**, 175 (1976); (b) V. J. Choy and C. J. O'Connor, *Coord. Chem. Rev.*, **9**, 145 (1972/73); and references therein.
76. L. Vaska, *Science*, **140**, 809 (1963).
77. T. G. Appleton and M. A. Bennett, *Inorg. Chem.*, **17**, 738 (1978).
78. G. Strukul, R. A. Michelin, J. D. Orbell and L. Randaccio, *Inorg. Chem.*, **22**, 3706 (1983).
79. G. Strukul and R. A. Michelin, *J. Am. Chem. Soc.*, **107**, 7563 (1985).
80. T. Yoshida, T. Matsuda, T. Okano, T. Kitani and S. Otsuka, *J. Am. Chem. Soc.*, **101**, 2027 (1979).
81. D. P. Arnold and M. A. Bennett, *J. Organomet. Chem.*, **199**, 119 (1980).
82. G. Villain, P. Kalck and A. Gaset, *Tetrahedron Lett.*, **21**, 2901 (1980).
83. A. Gaset, G. Constant, P. Kalck and G. Villain, *J. Mol. Catal.*, **7**, 355 (1980).
84. D. P. Arnold, M. A. Bennett, G. T. Crisp and J. C. Jeffery, *Adv. Chem. Ser.*, **196**, 195 (1982).
85. D. P. Arnold and M. A. Bennett, *J. Organomet. Chem.*, **202**, 107 (1980).
86. M. A. Bennett and T. Yoshida, *J. Am. Chem. Soc.*, **95**, 3030 (1973).
87. S. E. Diamond, B. Grant, G. M. Tom and H. Taube, *Tetrahedron Lett.*, 4025 (1974).
88. C. M. Jensen and W. C. Trogler, *J. Am. Chem. Soc.*, **108**, 723 (1986).
89. C. M. Jensen and W. C. Trogler, *Science*, **233**, 1069 (1986).
90. H. E. Bryndza and W. Tam, *Chem. Rev.*, **88**, 1163 (1988).
91. R. A. Sheldon and J. K. Kochi, '*Metal-Catalyzed Oxidations of Organic Compounds*', Academic Press, New York 1981, chapter 4, p. 81.
92. S. Campestrini, V. Conte, F. Di Furia, G. Modena and O. Bortolini, *J. Org. Chem.*, **53**, 5721 (1988).
93. R. Curci and J. O. Edwards in '*Organic Peroxides*' Ed. D. Swern, Wiley, New York 1970, vol I, p. 199 and references therein cited.
94. O. Bortolini, F. Di Furia, G. Modena, C. Scardellato and P Scrimin, *J. Mol. Catal.*, **11**, 107 (1981).
95. O. Bortolini, S. Campestrini, F. Di Furia and G. Modena, *J. Org. Chem.*, **52**, 5093 (1987).
96. O. Bortolini, F. Di Furia, G. Modena and R. Seraglia, *J. Org. Chem.*, **50**, 2688 (1985).

97. P. Burckard, J. P. Fleury and F. Weiss, *Bull. Soc. Chim. France* , 2730 (1965).
98. K. Kahr, *Angew. Chem.*, **72**, 135 (1960).
99. O. L. Lebedev and S. N. Kazarnovskii, *Zh. Obsch. Khim.*, **30**, 1631 (1960).
100. N. A. Milas and S. Sussman, *J. Am. Chem. Soc.*, **58**, 1302 (1936).
101. N. A. Milas and S. Sussman, *J. Am. Chem. Soc.*, **59**, 5345 (1937).
102. N. A. Milas and L. S. Maloney, *J. Am. Chem. Soc.*, **62**, 1841 (1940).
103. M. Schröder, *Chem. Rev.*, **80**, 187 (1980).
104. M. Hudlicky, *'Oxidations in Organic Chemistry'* ACS Monograph 186, American Chemical Society, Washington 1990, chapter 1, p. 38, 39.
105 J. A. Connor and E. A. V. Ebsworth, *Adv. Inorg. Chem. Radiochem.*, **6**, 279 (1964).
106. M. Mugdan and D. P. Young, *J. Chem. Soc.*, 2988 (1949).
107. G. B. Payne and P. H. Williams, *J. Org. Chem.*, **24**, 54 (1959).
108. H. C. Stevens and A. J. Kaman, *J. Am. Chem. Soc.*, **87**, 734 (1965).
109. Z. Raciszewski, *J. Am. Chem. Soc.*, **82**, 1267 (1960).
110. J. Itakura, H. Tanaka and H. Ito, *Bull. Chem. Soc. Japan* , **42**, 1604 (1969).
111. R. A. Sheldon, *J. Mol. Catal.*, **7**, 107 (1980).
112. K. A. Jørgensen, *Chem. Rev.*, **89**, 431 (1989).
113. J. Prandi, H. B. Kagan and H. Mimoun, *Tetrahedron Lett.*, **23**, 2617 (1986).
114. C. Venturello, E. Alneri and M. Ricci, *J. Org. Chem.*, **48**, 3831 (1983).
115. D. Prat and L. Lett, *Tetrahedron Lett.*, **27**, 707 (1986).
116. D. Prat, B. Delpech and R. Lett, *Tetrahedron Lett.*, **27**, 711 (1986).
117. K. S. Kirschenbaum and K. B. Sharpless, *J. Org. Chem.*, **50**, 1979 (1985).
118. M. A. Beg and I. Ahmad, *J. Catal.*, **39**, 260 (1975).
119. Y. Ogata and K. Tanaka, *Can. J. Chem.*, **59**, 718 (1981).
120. H. B. Kagan, H. Mimoun, C. Mark and V. Schurig, *Angew. Chem. Int. Ed. Engl.*, **18**, 485 (1979).
121. O. Bortolini, F. Di Furia, G. Modena and A.Schionato, *J. Mol. Catal.*, **35**, 47 (1986)..
122. G. Amato, A. Arcoria, F. P. Ballistreri, G. A. Tomaselli, O. Bortolini, V. Conte, F. Di Furia, G. Modena, and G. Valle, *J. Mol. Catal.*, **37**, 165 (1986).
123. K. B. Sharpless, J. M. Townsend and D. R. Williams, *J. Am. Chem. Soc.*, **94**, 295 (1972).
124. H. Mimoun, *J. Mol. Catal.*, **7**, 1 (1980).
125. A. Arcoria, F. P. Ballistreri, G. A. Tomaselli, F. Di Furia and G. Modena, *J. Mol. Catal.*, **18**, 177 (1983).
126. A. Arcoria, F. P. Ballistreri, G. A. Tomaselli, F. Di Furia and G. Modena, *J. Mol. Catal.*, **24**, 189 (1984).
127. H. Mimoun, *Angew. Chem. Int. Ed. Engl.*, **21**, 734 (1982).
128. S. L. Regen and G. M. Whitesides, *J. Organomet. Chem.*, **59**, 293 (1973).
129. N. J. Lewis, S. Y. Gabhe and M. L. DeLaMater, *J. Org. Chem.*, **42**, 1479 (1977).
130. E. Vedejs, *J. Am. Chem. Soc.*, **96**, 5944 (1974).
131. E. Vedejs, D. A. Engler and J. E. Telschow, *J. Org. Chem.*, **43**, 188 (1978).
132. S. E. Jacobson, R. Tang and F. Mares, *J. Chem. Soc. Chem. Commun.*, 888 (1978).
133. O. Bortolini, F. Di Furia, G. Modena and P. Scrimin, *J. Mol. Catal.*, **7**, 59 (1980).
134. F. Di Furia and G. Modena, *Rec. Trav. Chim. Pays-Bas* , **98**, 181 (1979).
135. O. Bortolini, F. Di Furia, G. Modena and P. Scrimin, *J. Mol. Catal.*, **9**, 323 (1980).
136. V. Conte, F. Di Furia, G. Modena and O. Bortolini, *J. Org. Chem.*, **53**, 4581 (1988).
137. M. Postel, C. Brevard, H. Arzoumanian and J. G. Riess, *J. Am. Chem. Soc.*, **105**, 4922 (1983).
138. R. Curci, C. Fusco, O. Sciacovelli and L. Troisi, *J. Mol. Catal.*, **32**, 251 (1985).
139. I. P. Gerothanassis and M. Momenteau, *J. Am. Chem. Soc.*, **109**, 6944 (1987).

140. O. Bortolini, V. Conte, F. Di Furia and G. Modena, *J. Org. Chem.*, **51**, 2661 (1986).
141. S. Campestrini, F. Di Furia, G. Modena and O. Bortolini, *J. Org. Chem.*, **55**, 3658 (1990).
142. K. B. Sharpless and T. R. Verhoeven, *Alrichimica Acta* , **12**, 63 (1979).
143. M. Bonchio, V. Conte, F. Di Furia and G. Modena, *J. Org. Chem.*, **54**, 4368 (1989).
144. E. Daire, H. Mimoun and L. Saussine, *Nouv. J. Chim.*, **8**, 271 (1984).
145. H. P. Wulff, *U.S. Patent* , 3 923 843 (1975), *Br. Patent* , 1 249 079 (1975), both to Shell Oil.
146. See for example: M. G. Finn and K. B. Sharpless in '*Asymmetric Synthesis* ', Ed. J. D. Morrison, Academic Press, New York 1986, vol. 5, p. 247.
147. B. E. Rossiter in '*Asymmetric Synthesis* ', Ed. J. D. Morrison, Academic Press, New York 1986, vol. 5, p. 193.
148. B. Notari, *Stud. Surf. Sci. & Catal.*, vol. 37, Eds. P. J. Grobet, W. J. Mortier, E. F. Vansant and G. Schulz-Ekloff, Elsevier, Amsterdam 1989, p. 413.
149. B. Notari in '*Chemistry of Microporous Crystals* ', Ed. T. Inui, Kodansha, Tokyo 1991, p. 343.
150. C. Neri, A. Esposito, B. Anfossi and F. Buonomo, *Eur. Patent* , 100 119 (1984).
151. C. Neri, B. Anfossi and F. Buonomo, *Eur. Patent* , 100 118 (1984).
152. A. Thangaraj, R. Kumar and P. Ratnasamy, *Appl. Catal.*, **57**, L1 (1990).
153. A. Esposito, M. Taramasso, C. Neri and F. Buonomo, *Br. Patent* , 2 116 974 (1985).
154. P. Roffia, G. Leofanti, A. Cesana, M. Mantegazza, M. Padovan, G. Petrini, S. Tonti and P. Gervasutti, *Stud. Surf. Sci. & Catal.*, vol. 55, Eds. G. Centi and F. Trifirò, Elsevier, Amsterdam 1990, p. 43.
155. M. G. Clerici, *Appl. Catal.*, **68**, 249 (1991).
156. U. Romano, A. Esposito, F. Maspero, C. Neri and M. Clerici, *Stud. Surf. Sci. & Catal.*, vol. 55, Eds. G. Centi and F. Trifirò, Elsevier, Amsterdam 1990, p. 33.
157. (a) D. R. C. Huybrechts, L. De Bruycker and P. A. Jacobs, *Nature* , **345**, 240 (1990); (b) T. Tatsumi, M. Nakamura, S. Negishi and H. Tominaga, *J. Chem. Soc. Chem. Commun.*, 476 (1990).
158. M. R. Boccuti, K. M. Rao, A. Zecchina, G. Leofanti anf G. Petrini, *Stud. Surf. Sci. & Catal.*, vol. 48, Eds. C. Morterra, A. Zecchina and G. Costa, Elsevier, Amsterdam 1989, p. 133.
159. H. Ledon and F. Varescon, *Inorg. Chem.*, **23**, 2735 (1984).
160. S. G. Hentges and K. B. Sharpless, *J. Am. Chem. Soc.*, **102**, 4263 (1980).
161. K. A. Jørgensen and B. Schiøtt, *Chem. Rev.*, **90**, 1483 (1990).
162. K. A. Jørgensen and R. Hoffmann, *J. Am. Chem. Soc.*, **108**, 1867 (1986).
163. N. A. Milas, J. H. Trepagnier, J. T. Nolan and M. I. Iliopulos, *J. Am. Chem. Soc.*, **81**, 4730 (1959).
164. W. D. Lloyd, B. J. Navarette and M. F. Shaw, *Synthesis*, 610 (1972).
165. C. J. Norton, *US Patent* , 3 335 174 (1967).
166. (a) K. B. Sharpless and K. Akashi, *J. Am. Chem. Soc.*, **98**, 1986 (1976); (b) K. Akashi, R. E. Palermo and K. B. Sharpless, *J. Org. Chem.*, **43**, 2063 (1978).
167. (a) V. Van Rheenan, R. C. Kelly and D. Y. Cha, *Tetrahedron Lett.*, 1973 (1976); (b) E. N. Jacobsen, I. Markò, W. S. Mungall, G. Schröder and K. B. Sharpless, *J. Am. Chem. Soc.*, **110**, 1968 (1988).
168. For a recent review see: W. P. Griffith, *Platinum Metals Rev.*, **33**, 181 (1989).
169. (a) D. M. Fenton, *US Patent*, 3 316 279 (1967); (b) D. M. Fenton and L. G. Wolgemuth, *US Patent*, 3 518 285 (1970); both to Union Oil.
170. H. Sugimoto and D. T. Sawyer, *J. Am. Chem. Soc.*, **106**, 4283 (1984).
171. H. Sugimoto and D. T. Sawyer, *J. Am. Chem. Soc.*, **107**, 5712 (1985).

172. H. Sugimoto and D. T. Sawyer, *J. Org. Chem.*, **50**, 1784 (1985).
173. D. T. Sawyer, *CHEMTECH* , 369 (1988).
174. C. Sheu, S. A. Richert, P. Cofrè, B. Ross jr., A. Sobkowiak, D. T. Sawyer and J. R. Kanofsky, *J. Am. Chem. Soc.*, **112**, 1936 (1990).
175. H.-C. Tung and D. T. Sawyer, *J. Am. Chem. Soc.*, **112**, 8214 (1990).
176. H. Mimoun, R. Charpentier, A. Mitschler, J. Fischer and R. Weiss, *J. Am. Chem. Soc.*, **102**, 1047 (1980).
177. G. Strukul and R. A. Michelin, *It. Patent* , 1 170 243 (1983); *J. Chem. Soc. Chem. Commun.*, 1358 (1984).
178. R. Sinigalia, R. A. Michelin, F. Pinna and G. Strukul, *Organometallics*, **6**, 728 (1987).
179. B. Bosnich and M. D. Fryzuk, *Top. Stereochem.*, **12**, 119 (1981).
180. C. Bolm, *Angew. Chem. Int. Ed. Engl.*, **30**, 403 (1991) and references therein.
181. A. Zanardo, F. Pinna, R. A. Michelin and G. Strukul, *Inorg. Chem.*, **27**, 1966 (1988).
182. R. A. Sheldon and J. A. Van Doorn, *J. Catal.*, **31**, 427 (1973).
183. R. A. Sheldon and J. A. Van Doorn, *J. Catal.*, **31**, 438 (1973).
184. G. C. Bond '*Heterogeneous Catalysis: Principles and Applications* ', 2nd ed., Oxford Science Pubblications, Oxford, 1987.
185. W. M. H. Sachtler, *CHEMTECH* , 434 (1983).
186. G. Strukul, A. Zanardo and F. Pinna, *Stud. Surf. Sci. & Catal.*, vol. 55, Eds. G. Centi and F. Trifirò, Elsevier, Amsterdam 1990, p. 81.
187. G. Strukul, A. Zanardo, F. Pinna, G. Goor and M. Schmidt, *Recl. Trav. Chim. Pays-Bas.*, **109**, 107 (1990).
188. G. Strukul, R. Sinigalia, A. Zanardo, F. Pinna and R. A. Michelin, *Inorg. Chem.* **28**, 554 (1989).
189. J. E. Lyons, *Asp. Hom. Catal.*, vol 3, Ed. R. Ugo, Reidel, Dordrecht 1977, p. 1.
190. M. Kubota, F. S. Rosenberg and M. J. Sailor, *J. Am. Chem. Soc.*, **107**, 4558 (1985).
191. I. Torresan, R. A. Michelin, A. Marsella, A. Zanardo, F. Pinna and G. Strukul, *Organometallics* , **10**, 623 (1991).
192. G. Strukul, M. Del Todesco Frisone, R. Giovanetti and F. Pinna, *Stud. Surf. Sci. & Catal.*, vol. 66, Ed. L. Simandi, Elsevier, Amsterdam 1991, p. 405.
193. R. Ugo, G. M. Zanderighi, A. Fusi and D. Carreri, *J. Am. Chem. Soc.*, **102**, 3745 (1980).
194. M. Del Todesco Frisone, F. Pinna and G. Strukul, in preparation.
195. A. Marsella, F. Pinna and G. Strukul, presented at the 7th International Symposium on Homogeneous Catalysis, Lyon 1990, paper C-4; A. Marsella, F. Pinna and G. Strukul, *Organometallics*, submitted.

CHAPTER 7

Peroxometal Complexes Derived from Hydrogen Peroxide. Some Applications in Organic Synthesis

Valeria CONTE and Fulvio DI FURIA

Centro di Studio sui Meccanismi di Reazioni Organiche del CNR, Dipartimento di Chimica Organica, Via Marzolo 1, I-35131 Padova, Italy.

1. Introduction

1.1. Historical Background.

The rate accelerating effect of even minute amounts of some metal ion derivatives such as Ti(IV), V(V), Mo(VI) and W(VI) on the oxidation of organic and inorganic substrates by hydrogen peroxide[1,2] has been known and synthetically exploited for a long time since the pioneering work of Milas.[3-5] However, only in the last twenty years such a catalytic effect has been satisfactorily rationalized as arising from the *in situ* formation of peroxometal species which are, by far, more effective oxidants than hydrogen peroxide itself.[1,2,6] Indeed, the rate of reaction **(2)**, carried out by **1**, Scheme 1, may be in some cases several orders of magnitude larger than that of reaction **(1)**. To account for the catalytic behavior it is, of course, necessary that, after the oxidation of the substrate, **1** be reformed upon addition of H_2O_2 to $L_zM{=}O$.

Scheme 1

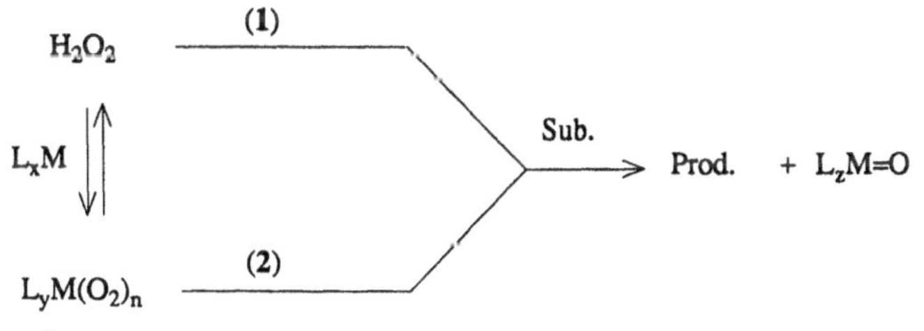

1

M = Ti(IV), V(V), Mo(VI), W(VI); n = 1,2

A key step forward to the understanding of the catalysis has been the finding that peroxo species such as **1** belong to a family of compounds which in several instances may be isolated and fully characterized.[2,7] There has been some debate as to the real structure of

G. Strukul (ed.), Catalytic Oxidations with Hydrogen Peroxide as Oxidant, 223–252.

peroxometal species in solution, particularly in protic, XOH, solvents. The earlier formulation of such compounds as hydroperoxide-like species or "inorganic peracids"[2] formed *via* the equilibrium process shown in Eq. (**3**) is now severely questioned.[6,8]

(3)

In fact, based on spectroscopic and kinetic measurements, it has been reported that hydrogen peroxide adds to the metal ions mentioned above as a bidentate ligand in an equilibrium reaction characterized by an association constant much larger than that of alkyl hydroperoxides of comparable nucleophilicity.[8] Moreover the ^{17}O-NMR spectra of peroxometal derivatives of V(V), Mo(VI) and W(VI) show the equivalence of the peroxidic oxygens bound to the metal.[9-11] Therefore, it appears that the cyclic arrangement of the peroxo groups observed in the solid state is preserved also in solution.

As anticipated above, the relevance of peroxometal species in the context of organic synthesis is related to their ability as oxidizing agents of a large variety of substrates.[1,2,12] This is self-evident for catalytic processes such as those outlined in Scheme 1 but it is also true for stoichiometric oxidations where the isolated peroxospecies may substitute for other oxidants.[12-14] Beside the obvious advantage over species such as Cr(VI) derivatives, which have a much larger enviromental impact, peroxometal complexes derived from hydrogen peroxide, by means of an accurate tuning of their coordination sphere, may be rendered very selective oxidants, capable of carrying out regio-, stereo- and enantioselective reactions.[15-19]

1.2. Limitations and Scope.

This chapter, which by no means intends to be exhaustive, will deal with some synthetically significant examples of oxidation of organic substrates by peroxocomplexes. The choice of such examples has been made with a view toward emphasizing the selectivities obtained, and at the same time, pointing out potentialities which have not yet been exploited. The mechanistic aspects as well as the industrial applications will be dealt with in passing only, because these topics are the subject of separate chapters in this book.

Moreover, for details on already classical oxidations, such as thioether oxidations,[20] olefin[21] and allylic alcohol epoxidations[12] the reader is referred to recently published reviews.

Following these lines we will discuss in some detail the oxidation of alkynes, alcohols and ketones as well as the oxidation of enolates and carbanions. The first section of the chapter will be dedicated to the oxidation of some model substrates under phase transfer conditions[22] in order to show the applicability of such a technique to the field of oxidation. The hydroxylation of aliphatic and aromatic substrates will also be briefly reviewed even though the results so far reported in the literature cannot be considered as satisfactory as for other substrates. They are, however, worthy of mention because of the possible future developments and the synthetic relevance of these processes.

2. Oxidation under Phase-transfer Conditions.

The most attractive feature of hydrogen peroxide, as an oxidant, compared with other peroxidic species, resides in the obvious fact that its reduced form, water, can be easily disposed off. By contrast, the main disadvantage of dilute aqueous solutions of H_2O_2, i.e. those which are commercially available and safely handled, is their scarce solubility in low polarity organic solvents. This is unfortunate because, as a general rule, the reactivity of peroxospecies is greatly depressed in polar organic solvents, probabily because of an efficient ground state stabilization, through solvation, of the oxidant.

Accordingly, an experimental procedure has been developed which allows to use dilute solutions of H_2O_2 in the presence of a metal catalyst.[22] The idea behind the development of such a procedure is a very simple one even though its realization may not be straightforward.

In fact, see Scheme 2, the technique involves the formation, in the aqueous solution, of the peroxospecies starting from H_2O_2 and a suitable metal precursor. The peroxometal complex is then extracted into an organic solvent immiscible with water where the substrate can be oxidized at a convenient rate. Eventually, the oxo-metal species returns into the aqueous phase thus closing the catalytic cicle.

Scheme 2

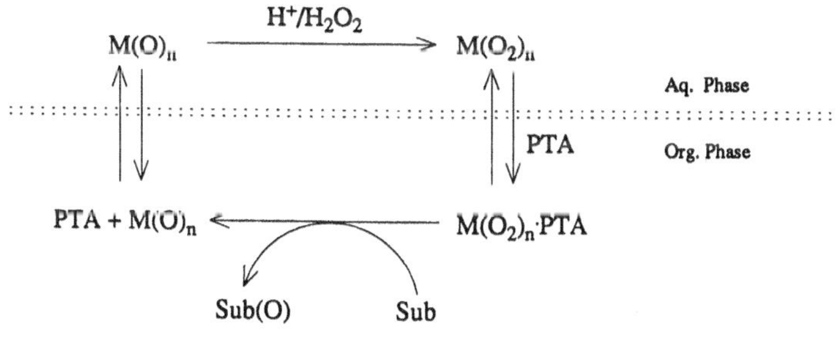

PTA=phase-transfer agent
n = 1,2

For the success of such a procedure, the choice of an appropriate phase-transfer agent is the crucial point.[23] Such a choice is directly related to the nature, either neutral or anionic, of the peroxospecies formed in water.

In fact it has been found that peroxometal derivatives *i.e.* $M(O_2)_n(H_2O)_m$ are fairly strong acids which undergo, in protic solvents, acid-base equilibria[23,24] such as those shown in Eq. 4 for aqueous solutions:

$$M(O_2)_n(H_2O)_m \overset{H_2O}{\rightleftharpoons} [M(O_2)_n(OH)(H_2O)_{m-1}]^- + H_3O^+ \qquad \textbf{(4)}$$

M = Mo(VI) or W(VI)

Therefore, depending on the pH of the solution, both neutral and anionic form of the peroxospecies are present in variable amounts. Such a behavior turned out to be rather useful in synthetic applications because the oxidation chemistry of neutral[25] and anionic[26] Mo(VI) and W(VI)-peroxocomplexes is vastly different. In particular neutral peroxospecies are very effective oxidants of nucleophilic substrates such as thioethers and alkenes, whereas the corresponding anions are able to oxidize alcohols to carbonyl compounds.

As far as the extraction of the anions in the organic phase, generally a chlorinated solvent, is concerned, it has been confirmed that lipophilic quaternary ammonium cations and similar classical cationic phase-transfer agents are suitable reagents.[27] By contrast, the transfer of neutral peroxo compounds into the organic phase can be achieved by using neutral lipophilic phosphoric amides or pyridine oxides. Table 1 collects some data[22,23] on the efficiency as extracting agents of some neutral PTAs.

Table 1 : Efficiency of neutral PTA in the extraction of Mo(VI)-peroxo complexes in DCE.

PTA	(Abbr.)	% extraction[a]
$[(CH_3(CH_2)_{10}CH_2)NH]_3P{=}O$	(TDPT)	98
$Ph\text{-}CH_2CH_2CH_2\text{-}Py(O)$	(RPyNO)	87
$[(CH_3CH_2CH_2CH_2)_2N]_3P{=}O$	(HBPT)	63
$[(CH_3CH_2)N]_3P{=}O$	(HEPT)	63

a) $[Mo]_{st}$ = 0.02 M; $[H_2O_2]$ = 0.8 M; [PTA] = 0.05 M; pH = 1.2 (H_2SO_4); DCE/H_2O = 2.5/5 mL; T = 25°C.

2.1. Oxidation of Thioethers

The oxidation of nucleophiles by peroxometal complexes has been extensively studied. There are several pieces of evidence which indicate that such a process involves in most of the cases the nucleophilic attack of the substrate to the peroxide oxygen in a simple, bimolecular reaction.[28] As a consequence, the order of reactivity found for typical substrates parallel their nucleophilicity as shown below:[28]

$$k^{rel}_{thioethers} > k^{rel}_{sulfoxides} > k^{rel}_{alkenes}$$

This is rather fortunate as far as the oxidation of thioethers to sulfoxides by peroxometal complexes is concerned. In fact, the larger reactivity of the former over the latter substrates is sufficient to achieve a high degree of selectivity, being sulfones present only in traces even when small excesses of thioether over the oxidant are used.

In Table 2 some data are collected referring to the oxidation of model thioethers in a two-phase system where Mo(VI) or W(VI) derivatives are the catalysts.[29]

Table 2 : Oxidation of thioethers with H_2O_2 catalyzed by Mo(VI) or W(VI) derivatives under phase-transfer conditions, at 40°C.[a]

#	Thioether (mmol)		catalyst (mmol)	time (min)	H_2O_2 %cons.	product(s) yield % [b]
1	p-ClPh-S-CH_3	(20.5)	Mo(VI) (0.1)	10	77	77 (94)
2	p-ClPh-S-CH_3	(20.5)	W(VI) (0.1)	10	17	16 (100)
3	p-ClPh-S-CH_3	(20.5)	Mo(VI) (0.5)	10	100	100 (88)
4	p-ClPh-S-CH_3	(20.5)	W(VI) (0.5)	10	75	75 (92)
5	p-ClPh-S-CH_3	(40.5)	Mo(VI) (0.5)	10	100	100 (97)
6	DBTP[d]	(20.5)	Mo(VI) (0.5)	60	78	75 (69)

a) DCE/H_2O = 25 / 2 mL; H_2O_2 20.5 mmol; PTA = HBPT 1 mmol; H_2SO_4 / Na_2MO_4 (M = Mo(VI)) or W(VI)) = 1.2 / 1. b) numbers in parentheses are % of sulfoxides based on the total yields of products obtained (sulfoxide + sulfone). c) DBTP = dibenzothiophene

As anticipated, the selectivities observed are generally very high. Indeed, in the case of Mo(VI) they approach almost 100% even when only a 2-fold excess of *p*-$ClC_6H_4SCH_3$ over H_2O_2 is used. Under otherwise identical conditions W(VI) is somehow a less selective catalyst.

It is worthy of mention that addition of H_2O_2 at the end of the reaction fully restores the catalytic cycle thus demonstrating that the metal complex and the ligand are not significantly degraded during the oxidative process.[29]

2.2. Oxidation of Alkenes

The synthesis of epoxides by direct epoxidation of a simple olefin by H_2O_2 is an important procedure both from a synthetic and an industrial point of view. The reaction, however, proceeds at a convenient rate only when suitable catalysts are present.[2,21] These are usually metal oxides of 5A,B and 6 A,B groups.

In the last few years procedures have been developed by different authors for the catalytic oxidation of olefinic substrates in two-phase systems.

Venturello et al. found[30] that a two-component association of tungstate and phosphate (or arsenate) ions, under acidic conditions, is a good catalyst for the epoxidation of alkenes by dilute hydrogen peroxide.

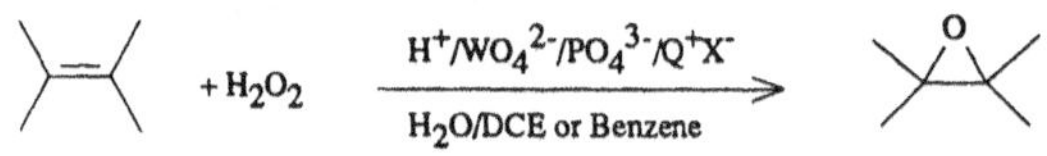

(5)

Q^+X^- = onium salts

It is important to mention that only when tungstate and phosphate (or arsenate) ions are used together the resulting peroxometalspecies is an effective oxidant capable of epoxidizing even poorly reactive 1-alkenes.

Representative results[30] obtained by using such a procedure are collected in Table 3.

It can be seen that high selectivities (80-90%) on both H_2O_2 and substrate at almost complete consumption of the oxidant are obtained. Such a high reactivity of anionic peroxospecies toward nucleophiles is rather surprising. In fact, anions deriving from simple peroxotungsten species are known[31] to be fairly good oxidants of the alcoholic function but they are almost unreactive toward double bonds. A possible rationale offered by the authors is that the anion in the system described above, has a highly delocalized charge (see discussion below) so that the electrophilic reactivity of the peroxo oxygens is almost unaffected by the anionic character of the species.

Table 3 : Epoxidation of model alkenes with H_2O_2 in a two-phase system catalyzed by the association of WO_4^{2-}/PO_4^{3-}

#[a]	Alkene (mmol)	$(Q^+X^-)^b$	pH	T°C	time (min)	H_2O_2 % conv.	epoxide %yield
1	1-octene (200)	**A** (2)	1.6	70	45	98	82
2	1-octene (200)	**B** (2)	1.6	70	60	97	81
3[c]	1-octene (200)	**A** (2)	1.6	70	45	96	79
4	1-octene (200)	**C** (2)	1.6	70	60	95	71
5[d]	allyl chloride (550)	**A** (2)	2	60	150	96	80
6[e,f]	styrene (400)	**B** (3)	3	40	180	93	77
7[f]	α-methylstyrene (750)	**B** (3)	4.5	40	240	93	79
8[g]	cyclohexene (400)	**B** (0.8)	3	70	25	98	88

a) DCE (15mL) / 8% aqueous H_2O_2 (120 mmol) with a molar ratio $PO_4^{3-}:WO_4^{2-}$: Q^+ of 5:2.5:1 unless otherwise indicated. b) A = Aliquat 336; B = Arquad 2HT; C=hexadecyltributylphosphonium chloride. c) AsO_4^{3-}, $Na_2HSO_4 \cdot 7\,H_2O$. d) C_6H_6 (30 mL) as solvent. e) a mixture of 40% w/v $H_3PO_4 \cdot H_2O$ and $NaH_2PO_4 \cdot H_2O$ in a 1:2 molar ratio was used. f)15% H_2O_2 and C_6H_6 (15 mL). g) same as in e but with a molar ratio of 1:1.

The isolation and characterization of the anionic peroxocomplexes have been subsequently carried out[32] and it has been shown that their formula is:

$$Q^+_3 \{PO_4[W(O)(O_2)_2]_4\}^{3-}$$

2

a; $Q^+ = [(C_6H_{13})_4N]^+$

b; $Q^+ = [(C_8H_{17})_3NCH_3]^+$

c; $Q^+ = \{[C_{18}H_{37}(76\%) + C_{16}H_{33}\,(24\%)]_2N(CH_3)_2\}^+$

Such peroxocomplexes have been utilized for the catalytic epoxidation of alkenes with hydrogen peroxide in a two-phase system.[32] Examples are given in Table 4, which confirm the efficiency of the procedure.

Table 4 : Epoxidation of alkenes with H_2O_2 catalyzed by tetrakis(diperoxotungsto)phosphates 1a and 1c, in a two-phase system.

#[a]	Olefin (mmol)	catalyst (mmol)	solvent (mL)	time (min)	T°C	product(s)	yield % [d]
1[b]	1-dodecene (100)	**2b** (0.31)	DCE (10)	60	70	1,2-epoxydodecane	94
2[c]	2,4,4-trimethyl-2-pentene (100)	**2b** (0.31)	DCE (10)	120	70	3-*t*-butyl-2,2-dimethyloxirane	91
3[b]	allyl chloride (150)	**2b** (0.38)	C_6H_6 (15)	150	refl.	epichlorohydrin	85
4[c]	styrene (200)	**2c** (0.51)	C_6H_6 (10)	120	50	Styrene oxide	81
5[c]	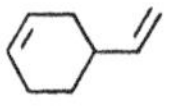 (100)	**2c** (0.20)	C_6H_6 (20)	60	60	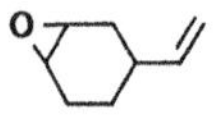	83
6[b]	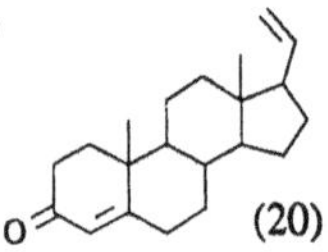 (20)	**2c** (0.04)	C_6H_6 (10)	45	50	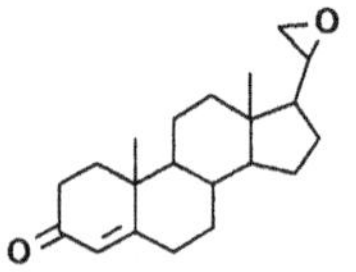	92

a)The reactions were run in an aqueous/organic two-phase system by using 60 mmol of H_2O_2, 12.5 mL of aqueous solution. b) only H_2O_2 present in the aqueous phase. c) $Na_2WO_4 \cdot 2H_2O$ and H_3PO_4 (3 and 15 mol\mol of catalyst respectively) also present in the aqueous phase, pH adjusted to 4 with NaOH. d) the yields are based on the H_2O_2 initially added.

It is interesting to find that even simple terminal alkenes or weak nucleophiles, such as allyl chloride, are reactive under the experimental conditions adopted. Moreover, an adjustment of the pH of the aqueous phase to higher values (*ca.*3) renders this method suitable also for the synthesis of acid-sensitive epoxides.

In the field of olefin epoxidations an alternative two-phase procedure, still based on Mo(VI) and W(VI) catalysts has also been reported. This method closely resembles the one already presented for thioether oxidation being based on the very same components:[29]

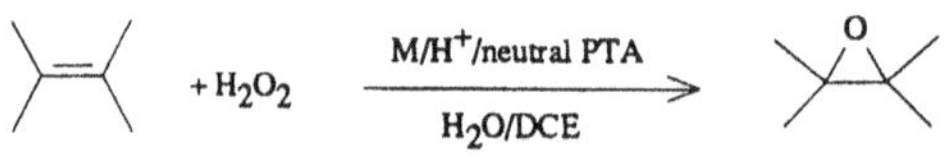

(6)

M = Mo(VI) or W(VI)

Data referring to the oxidation of representative olefins are collected in Table 5

Table 5 : Oxidation of model alkenes with H_2O_2 catalyzed by Mo(VI) in the presence of neutral PTAs under phase-transfer conditions at 50°C.

#[a]	Alkene (mmol)	PTA[b] (mmol)	H^+ (mmol)	time (h)	H_2O_2 %consumed	product(s) yield % [c]
1	*c*-hexene (100)	HMPT (1.0)	2.5	5	82	35 (14)
2	*c*-hexene (100)	HBPT (1.0)	0.6	5	35	21 (91)
3	*c*-hexene (100)	TCyPT[d] (1.0)	0.8	5	78	33 (100)
4	*c*-hexene (100)	RPyNO (1.0)	0.8	5	100	64 (98)
5	*t*-2-octene (100)	HBPT (1.0)	0.6	14	92	29 (100)
6	*t*-2-octene (100)	HBPT (1.0)	2.5	14	100	98 (80)
7	*t*-2-octene (100)	TDPT (1.0)	2.5	14	100	70 (63)
8	*t*-2-octene (100)	RPyNO (1.0)	2.5	14	100	76 (85)

a) the oxidations were run in a two-phase system DCE (25 mL):aqueous H_2O_2 (2 mL, 20.5 mmol); Mo(VI) 0.5 mmol. b) see abbreviation in Table 1. c) numbers in brackets are % of epoxide based on the total yield of products obtained (epoxides + glycols). d) TCyPT = tricyclohexyl phosphorous triamide.

It can be observed that the yields and the selectivities obtained in alkene epoxidations are rather good, even if they do not reach the values obtained in thioether oxidations.

An important drawback of this reaction is connected with the acid catalyzed hydrolysis of epoxides to glycols. In fact it has been shown that the mineral acid used to adjust the pH of the aqueous solution can be extracted by the PTA agent into the organic phase. Direct experiments have shown that by using the same PTA, under otherwise identical experimental conditions, $HClO_4$ is extracted much more than H_2SO_4.[23] Therefore the use of the latter is recommended.

As far as the epoxidation of olefinic double bond with hydrogen peroxide in a two-phase system is concerned there is one other procedure reported in the literature where tungsten is used as a catalyst.[33] Such a catalyst, formulated as a pertungstate salt

$W_2O_{11}^{2-}.2Ph_3PCH_2Ph^+$ where the cation is lipophilic enough to act also as PTA agent, can be either isolated or prepared *in situ.*

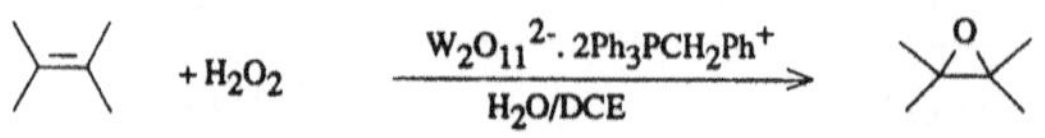

(7)

This system possesses the advantage that the reaction may be carried out under almost neutral conditions thus avoiding the problems mentioned above connected with the hydrolysis of the epoxides.

The pertinent results are summarized in Table 6.

Table 6 : Epoxidation of alkenes with H_2O_2 catalyzed by $W_2O_{11}^{2-}\cdot 2Ph_3PCH_2Ph^+$ in a two-phase system.

#	Olefin	catal [a] mol %	time h	T°C	product(s) yield % [c]	
1	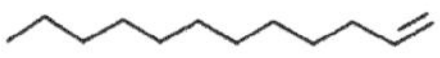	10	22	40	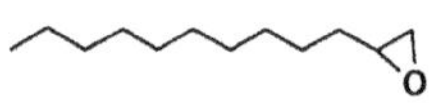	65
2		10[b]	24	40	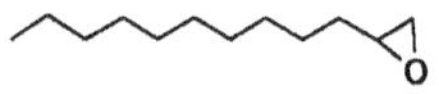	70
3	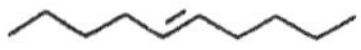	10[b]	22	40	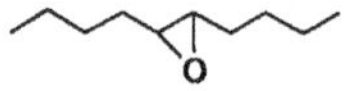	72
4	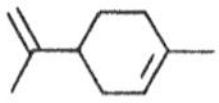	10	20	40	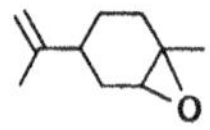 cis/trans = 1/1	60
5	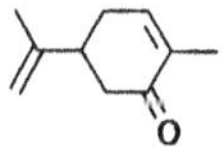	10	24	50	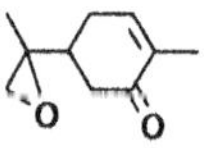 erythro/threo = 1/1	63
6	geraniol	0.2	24	0	geraniol-2,3-epoxide	98[d,e]
7	nerol	1	15	20	nerol-2,3-epoxide	70

a) isolated catalyst. b) catalyst prepared in situ. c) isolated yields, reactions performed with 2 mmol of substrate. d) yield by ^{1}H-NMR on the crude product. e) 100 mmol of substrate.

2.3. Oxidation of Alcohols.

Mechanistic studies on the behavior of Mo(VI) peroxocomplexes revealed that anionic peroxospecies of general formula $[MO(O_2)_n(OH)(H_2O)_{m-1}]^-$ were able to oxidize secondary alcohols to the corresponding carbonyl compounds.[26] In the light of such indications a two-phase system was developed, similar to the one outlined in equation 6 except that the acidity of the aqueous phase and the nature of PTA was modified. In particular cationic PTAs and a less acidic pH were used so that anionic Mo(VI) or W(VI) peroxo species were extracted and employed as oxidants of alcohols.[27]

Table 7 collects data referring to the oxidation of secondary alcohols to the corresponding ketones.

The data reported indicate that also in the oxidation of the alcoholic function the two-phase oxidative procedure has synthetic significance. In fact the selectivities observed are rather high thus allowing the complete conversion of the substrate. There is, however, a further oxidation to carboxylic acids when benzyl alcohol or aliphatic primary alcohols are employed.

Table 7 : Oxidation of model alcohols to the corresponding carbonyl compounds with H_2O_2 catalyzed by Mo(VI) or W(VI) complexes under phase-transfer conditions, at 75°C.

#[a]	Alcohol	catalyst	H_2O_2 (mmol)	time min	yield %[b]
1	*c*-hexanol	Mo(VI)	40	150	88
2	*c*-hexanol	W(VI)	40	50	97
3	2-octanol	W(VI)	40	75	97
4	menthol	Mo(VI)	60	210	97
5	menthol	W(VI)	50	120	89
6	borneol	Mo(VI)	40	120	100
7	borneol	W(VI)	20	45	96
8	benzyl alcohol	Mo(VI)	20	30	85

a) Alcohol 10mmol; PTA=Aliquat 336, 2 mmol; catalyst 1 mmol; pH= 3.0 for Mo(VI); pH=1.4 for W(VI); added acid H_2SO_4. b) yields based on the alcohol initially added.

3. Oxidation of Alkynes

The oxidation of acetylenic derivatives by peroxides has received so far little attention in spite of its potential synthetic utility.[34] This is mainly due to the fact that the triple bond in such compounds is usually much less reactive than the double bond of structurally similar alkenes, as observed in the reactions with peroxyacids.[34] Moreover, mixtures of products are obtained upon oxidation. As an example, the oxidation of di-*tert*-butyl-acetylene by *m*-chloro-perbenzoic acid (MCPBA) leads to the product distribution shown below:

Scheme 3

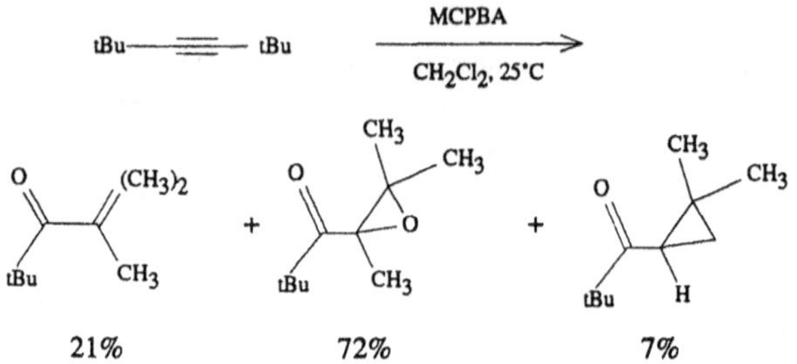

The reaction is believed to proceed through the intermediacy of an oxirene derivative, which could also be in equilibrium with an oxocarbene intermediate:

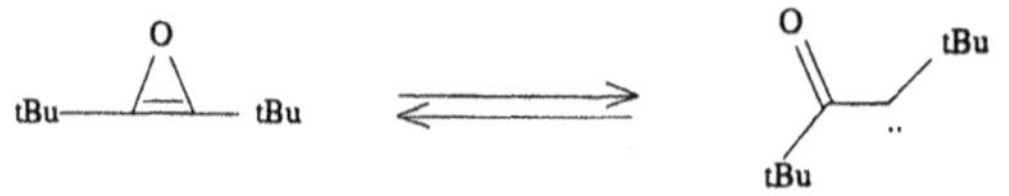

(8)

Recently Tomaselli *et al.*[35] reported that peroxocomplexes of molybdenum and tungsten, either used as stoichiometric oxidant or formed *in situ* from hydrogen peroxide and suitable metal precursors in a catalytic fashion, may carry out a fairly selective oxidation of the triple bond leading mainly to α-dicarbonyl compounds.

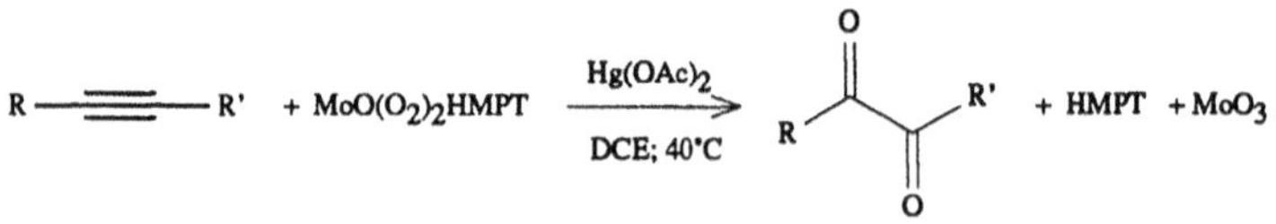

(9)

However, in such a reaction, the presence of a mercury derivative, either as catalyst or a co-catalyst is a necessary requisite. The role played by such a metal has not been unequivocally unveiled. The observation that other Lewis acids, though less efficiently, enhance the triple bond oxidation rate by peroxometal complexes may suggest the occurrence of an acid catalyzed reaction.

At any rate, the selectivity attained in the oxidation, particularly in the stoichiometric process,[35] is fairly good as demonstrated by the examples reported in Table 8.

Table 8 : Oxidation of model alkynes to the corresponding α-dicarbonyl compounds with $MoO(O_2)_2HMPT$ catalyzed by $Hg(OAc)_2$ in DCE; at 40°C.

#[a]	Alkyne[b] R	R'	Sub/ oxidant	peroxide/ $Hg(OAc)_2$	time h	conversion[c]	yield[d] %
1	Ph	Ph	1.0	1.0	20	—	80
2	Ph	Ph	0.4	2.5	20	94	77
3	Ph	H	1.1	1.1	0.25	—	90
4	$CH_3(CH_2)_5$	$CH_3(CH_2)_5$	1.0	1.0	3.5	76	64
5	$CH_3(CH_2)_3$	H	1.1	1.0	0.25	—	86
6	$CH_3(CH_2)_2$	Ph	1.1	1.1	2.0	65	70

a) alkyne initial concentration ranged 0.01-0.1 M. b) see equation 8. c) based on the substrate. d) based on the reagent present in smaller concentration.

Oxidation of alkynes, both terminal and internal, mainly to products resulting from the cleavage of the C-C triple bond can be also carried out in a two-phase system by using hydrogen peroxide, catalytic amounts of $(cetylpyridium)_3PM_{12}O_{40}$ where M = Mo(VI) **CMP** or W(VI) **CWP** and $Hg(OAc)_2$ as co-catalyst.[36] Some examples related to these oxidations are collected in Table 9.

A comparison has also been made between the catalytic activity of the polyoxometalates $PM_{12}O_{40}^{3-}$ (CMP and CWP) and that of H_2MO_4 (M = Mo(VI) or W(VI)) in the oxidation of the same substrates with H_2O_2. Such a comparison demonstrates that the former species are by far more reactive than the simple acids.[36] It has recently reported[37,38] that by treating with 35% H_2O_2 12-tungstophosphoric and 12-molybdophosphoric acid and cetylpyridium chloride in water, peroxoderivatives (PCWP and PCMP) are formed whose general structure is

Table 9 : Oxidation of alkynes with H_2O_2 catalyzed by CMP or CWP in a two-phase system at 40°C.

#[a]	Alkyne(mmol) R	R'		cat.	conv. %	time h	products (yields%)[b]
1	Ph	Ph	(2.8)	CMP	88	22	PhCOCOPh (6) PhCOOH (196)
2	Ph	H	(3.0)	CMP	100	0.15	$PhCOCH_3$ (18) PhCOOH (60) HCOOH (60)
3	n-C_3H_7	n-C_3H_7	(2.6)	CMP	100	13	$C_3H_7COCOC_3H_7$ (1) C_3H_7COOH (106) $C_3H_7COCH(CH_3)_2$ (22) $C_2H_5CH{=}CHCOC_3H_7$ (7)
4	n-C_4H_9	H	(2.2)	CMP	100	0.3	$C_4H_9COCH_3$ (11) C_4H_9COOH (69) HCOOH (72)
5	Ph	Ph	(3.1)	CWP	93	20	PhCOCOPh (7) PhCOOH (175)
6	Ph	H	(3.1)	CWP	100	0.15	$PhCOCH_3$ (3) PhCOOH (90) HCOOH (91)
7	n-C_3H_7	n-C_3H_7	(3.2)	CWP	96	19	$C_3H_7COC_3H_7$ (15) C_3H_7COOH (94) $C_2H_5CH{=}CHCOC_3H_7$ (14) $C_3H_7COCOC_3H_7$ (3)
8	n-C_4H_9	H	(2.2)	CWP	100	0.3	$C_4H_9COCH_3$ (21) C_4H_9COOH (66) HCOOH (62)

a) all the experiments were run in the presence of $Hg(OAc)_2$ 0.8mmol under nitrogen atmosphere, 20.5 mmol of H_2O_2 in the aqueous phase 2 mL, DCE 15 mL. b) yields are based on the stoichiometry of the reaction which produces, in case of C-C cleavage, 2 moles of products per mole of substrate.

identical, except for the counter cation, to the one prepared by Venturello.[32] Ishii et al. used PCWP complex to catalyze the oxidation of internal alkynes mainly to the corresponding α-ß-epoxyketones by using hydrogen peroxide as oxidant in a two-phase system.[39] Table 10 reports the pertinent results obtained.

Table 10 : Oxidation of model alkynes with H_2O_2 catalyzed by PCWP in refluxing $CHCl_3$.

#[a]	Alkyne R	R'	conv. %	products[b] (yields%)
1	Ph	Ph	45	PhCOCOPh (42)
2[c]	Ph	Ph	66	PhCOOH (66)
3	n-C_3H_7	n-C_3H_7	98	n-C_3H_7COOH (5) n-$C_3H_7COCH{=}CHC_2H_5$ (15) 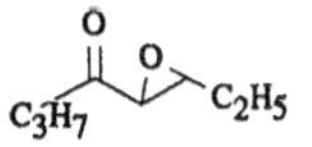 (62)
4	C_2H_5	n-C_4H_9	98	$C_2H_5COCH{=}CHC_3H_7$ (2) C_2H_5COOH (5) C_4H_9COOH (8) $CH_3CH{=}CHCOC_4H_9$ (9) 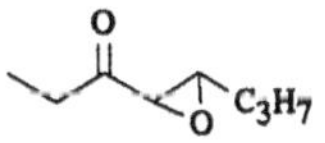 (57) 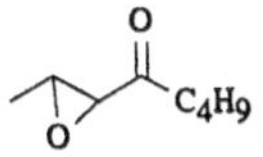 (11)
5	CH_3	n-C_5H_{11}	98	$CH_3COCH{=}CHC_4H_9$ (14) $C_5H_{11}COOH$ (13) $CH{=}CHCOC_5H_{11}$ (7) 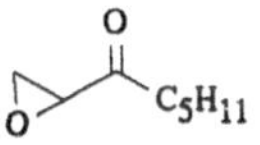 (6) 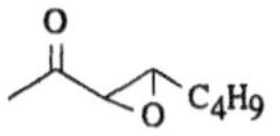 (40)
6	t-Bu	n-Bu	34	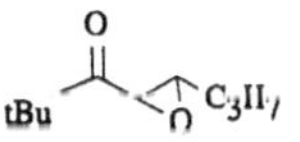 (13) 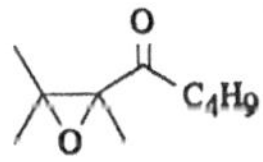 (6)

a) alkyne 3mmol, catalyst 30 wt %; H_2O_2 18 mmol; $CHCl_3$ 7.5 mL, 24 hrs. b) based on alkyne used. Remainders were products not identified. c) t-BuOH as solvent.

It is interesting to notice that such a procedure does not need the presence of any co-catalyst. On the other hand the products distribution observed is rather different from the one observed in the presence of Hg(II) where α-dicarbonyl compounds are the major products.[36]

The scope of polyoxometalates in oxidation reactions with hydrogen peroxide as far as other classes of substrates are concerned is reviewed in ch.8

4. Oxidation of Alcoholic Functions

Although several procedures are already available for the oxidation of the alcoholic function,[14] there is a continuous effort to develop novel reagents characterized by larger selectivities and also by smaller enviromental impact than the classical oxidation reagents such as those employing Cr(VI) derivatives. From this point of view, peroxometal complexes, formed from hydrogen peroxide and suitable metal precursors, may have a synthetic significance. In particular the anionic peroxomolybdenum complex $[MoO(O_2)_2(pico)]^-[(n\text{-}Bu)_4N]^+$ (MoO_5pico) (pico = picolinate N-oxido anion) appears to be the most effective.[40-42]

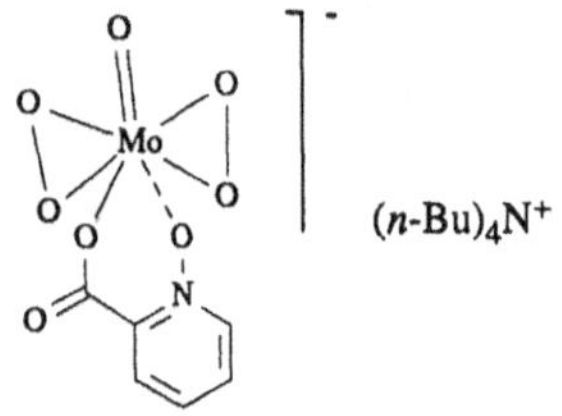

MoO_5pico

The presence of the lipophilic cation renders such a complex fairly soluble in organic solvents where it is able to oxidize both primary and secondary alcohols to the corresponding carbonyl compounds.[40-42] Worthy of notice is the selectivity of the oxidation of primary alcohols which are oxidized to aldehydes without any overoxidation to carboxylic acids. The oxidation of the alcoholic function by MoO_5pico may be carried out also for molecules where other oxidizable (double and triple bonds, pyridine nitrogen) or labile (oxirane ring) groups are present. Such features are illustrated by the examples[40-42] collected in Table 11 where a comparison is also made, in some instances, with two other widely used oxidation procedures, employing activated DMSO[14] or the perruthenate anion in combination with a tertiary amine oxide[43] respectively.

Such a comparison indicates that the oxidations carried out by MoO_5pico though slower than those obtained by the other two reagents, are generally somehow more selective.

The major drawback of the procedure based on MoO_5pico is, of course, the fact that it is a stoichiometric reaction. No data are so far available in the literature concerning the development of catalytic system by employing an excess of hydrogen peroxide.

There are other reagents based on molybdenum that may carry out the catalytic oxidation of alcohols. As an example the procedure developed by Trost *et al.*, based on the utilization of ammonium molybdate and hydrogen peroxide in THF or DMF in the presence of a quaternary

Table 11 : Oxidation of model alcohols with $MoO(O_2)_2PICO^-(n\text{-}Bu)_4N^+$ in DCE at 50°C[a].
[Substrate]/[oxidant] = 1

#	Substrate	Product	%Yields (Time, hrs) Act.DMSO	$(n\text{-}Pr_4N)(RuO_4)/R_3NO$	PICO
1	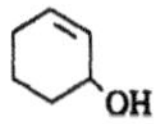	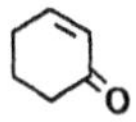	87(0.25)	/	99(3)
2	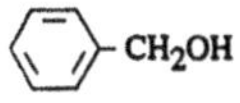	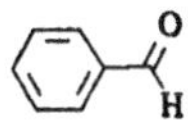	100(0.25)	80(2)	96(5)
3		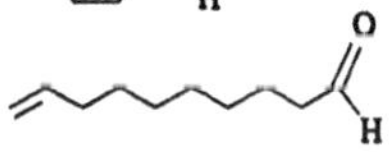	/	70(3)	99(9)
4	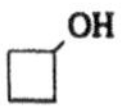	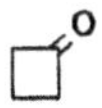	96(0.25)	95(11)	98(5)
5	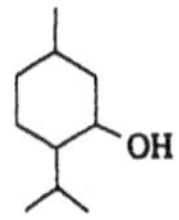	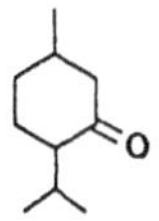	/	85(1)	97(9)
6	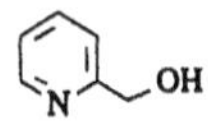	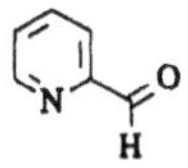	70(0.25)	/	97(48)
7	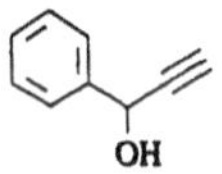	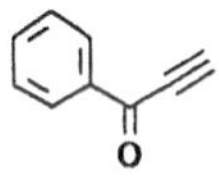	/	/	80(0.5)
8	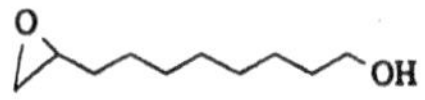	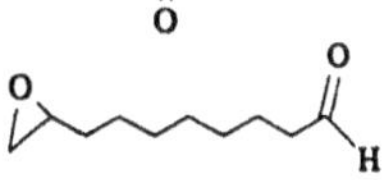	/	/	96(10)
9	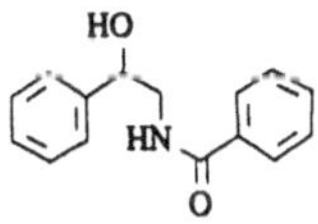	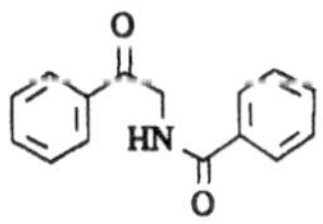	/	/	94(7)
10	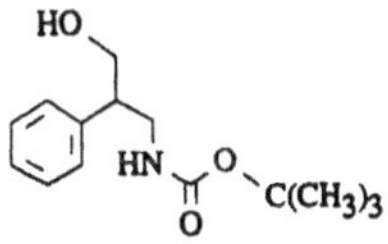	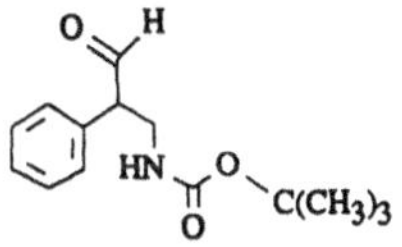	/	/	81(15)[b]
11	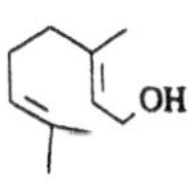	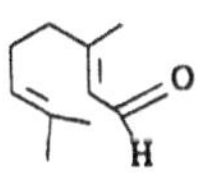	/	/	95(4)[c]

a) for comparison purposes, data referring to the oxidation of selected alcohols with other two oxidants are reported, see ref. 41 and 43. b) at 60 C. c) in DMF ammonium salt and potassium carbonate,[44,45] appears a rather effective one. In such a system it is likely that an anionic peroxocomplex is formed which is the active species in alcohol oxidation (see previous discussion). Some data illustrating the scope of the procedure are collected in Table 12

Table 12 : Oxidation of model alcohols with H_2O_2 catalyzed by Mo(VI)[a] in THF at r.t.

#	Substrate	Product	%Yields (Time, hrs)
1	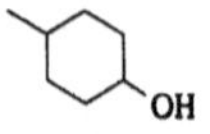	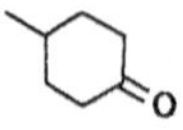	88 (96)
2	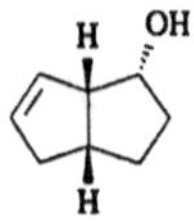	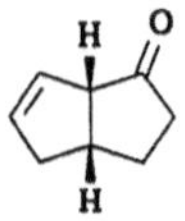	49 (120)
3	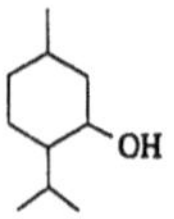	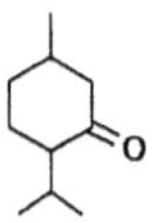	94 (120)
4	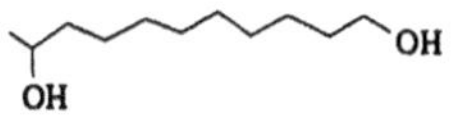	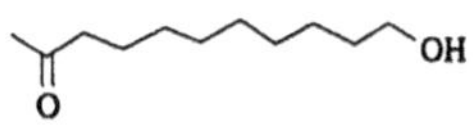	89 (168)
5	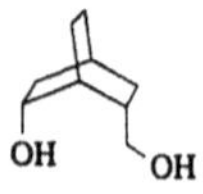	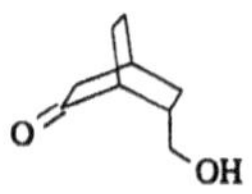	88 (24)
6	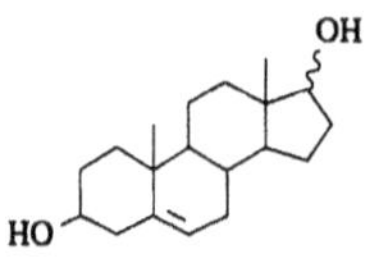	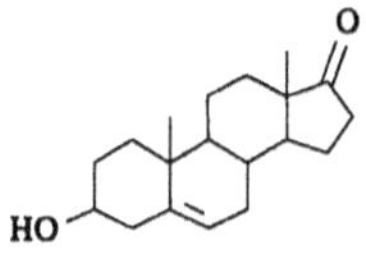	90 (144)

a) Mo(VI) as $(NH_4)_6Mo_7O_{24}$ (0.2-1 mmol); 10 mmol of H_2O_2 (1mL 30% solution in water).

An example referring to the stoichiometric oxidation of alcohols by a neutral molybdeum monoperoxocomplexes, containing two molecules of hydroxamic acid as ligands, is also reported in the literature.[46]

5. Oxidation of Enolates and Carbanions

The α-hydroxylation of enolates of ketones and esters is a synthetically useful reaction which may be carried out by several oxidants:[47]

Scheme 4

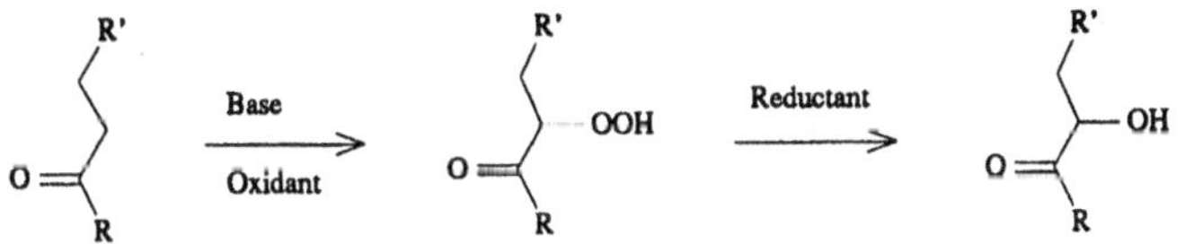

Among these, the $MoO(O_2)_2(HMPT)(py)$ complex (MoOPH) containing both the hexamethylphosphorous amide and pyridine ligands, has been found to be a particularly effective one.[48]

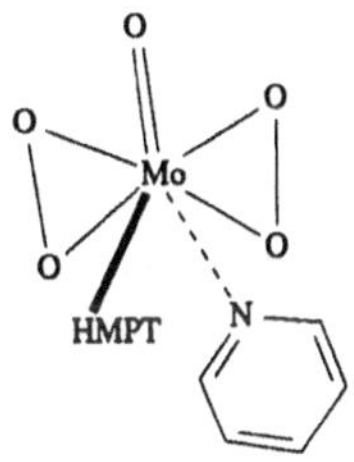

MoOPH

Although it is not clear why MoOPH should be superior to other neutral peroxomolybdenum complexes in this kind of transformation, it may be suggested that its stability in strongly alkaline solutions plays a crucial role. A major advantage of such a reaction is the easiness of the experimental procedure which usually involves the addition of an excess of solid MoOPH to a solution in THF of the carbonyl compound and the base maintained at low temperature.[47] The recovery of the hydroxylated product is then obtained simply by quenching the reaction mixture with water and by subsequent extraction with an organic solvent.

Table 13 collects some typical results obtained by the described procedure.[47]

Table 13 : Oxidation of carbonyl compounds with MoOPH

#[a]	Carbonyl compound	product %yield
1	Methyl heptanoate	Methyl 2-hydroxyheptanoate (74)
2[c]	Ethyl bicyclo [2,2,2]oct-2-ene 5-carboxylate	Ethyl 5-hydroxybicyclo [2,2,2] oct-2-ene-5-carboxylate (85)
3	α-butylbutyrolactone	α-hydroxy-αbutylbutyrolactone (73)
4	γ-phenyl-γ-methylbutyrolactone	α-hydroxy-γ-phenyl-γ-methyl-butyrolactone (56)
5	Valerophenone	α-hydroxy ketone (70)[b]
6	Camphor	α-hydroxy ketone (70 endo OH)[b]
7	2-phenylcyclohexanone	α-hydroxy ketone (70, 4:1 trans:cis)[b]

a) ester oxidations were run by adding the MoOPH reagent to the enolate whereas ketone oxidations were performed by adding the enolate to the oxidant. b) main by product being α-diketone.

Such a procedure is now currently used in organic synthesis also because of the high degree of selectivity attainable when chiral substrates are subjected to the MoOPH hydroxylation. As an example, such a reaction is used in the synthesis of the product **3**, which is a key intermediate in the synthesis of the natural product Azadirachitin, as reported below:[49]

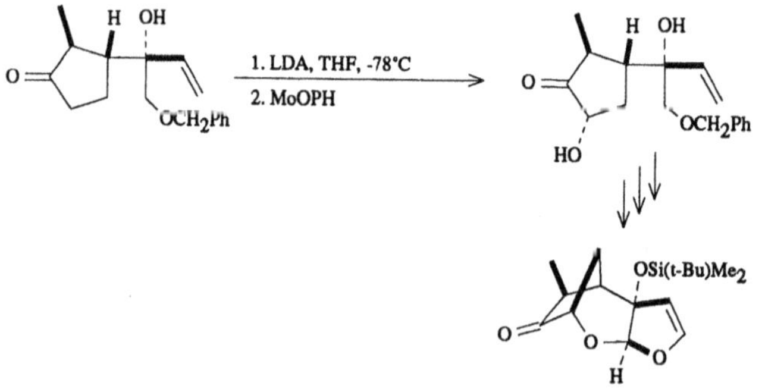

(10)

3

Similar selective hydroxylations have been carried out on β-amino carbanions.[50] The reaction is a key step in a synthesis for (1S,2S)-N-Methylpseudoephedrine recently reported.[50]

The mechanism of the MoOPH hydroxylation has not been yet examined in detail. However, the hypothesis has been formulated of a nucleophilic attack of the carbanionic center to the peroxide oxygen leading to the two alternative intermediates or transition states:[47]

Scheme 5

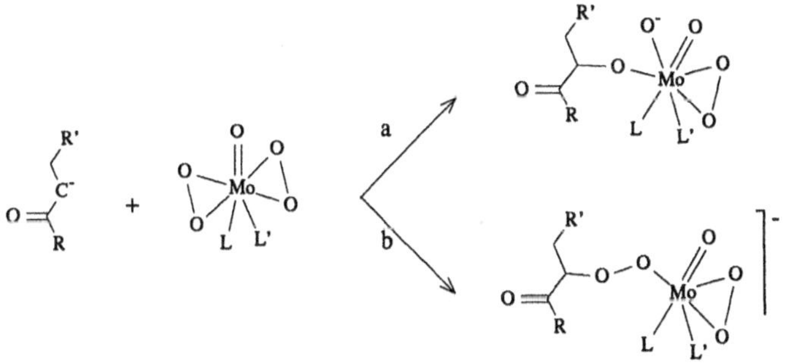

Along this line, a parallelism may be drawn between the hydroxylation of enolates discussed above and the conversion of arylbromides to phenols with MoOPH through the intermediacy of the Grignard reagent:[51]

$$\mathrm{ArX} \xrightarrow{\mathrm{Mg/THF}} \mathrm{ArMgX} \xrightarrow[\text{2. } H_2O]{\text{1. MoOPH, THF, -78}^{\circ}\text{C}} \mathrm{ArOH} \tag{11}$$

In fact, also in this case, a nucleophilic attack at the peroxide oxygen may be envisaged. Such a reaction has also synthetic relevance as demonstrated by the examples collected in Table 14:[51]

Table 14 : Synthesis of phenols from arylbromides with MoOPH

#	Arylbromide	Phenol (%yield)[a]
1	Bromobenzene	phenol (89)
2	1-bromo-4-methoxybenzene	4-methoxyphenol (67)
3	1-bromo-4-ethylbenzene	4-ethylphenol (70)
4	1-bromonaphtalene	1-hydroxynaphtalene (85)

a) based both on substrate and oxidant.

More in general, it might be predicted that MoOPH could react with other carbanions such as those formed from nitro and sulfonyl compounds. There are, in fact, reports on the use of the MoOPH reagent as oxidant in the transformation of a nitro alkane into a ketone under basic conditions, see Table 15.[52]

Table 15 : Synthesis of carbonyl compounds from nitro derivatives with MoOPH

#	Substrate	Product (%yield)[a,b]	
1	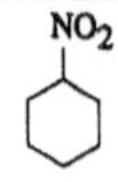	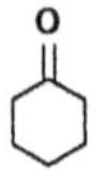	(86)
2	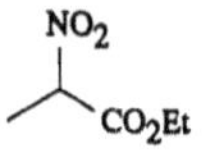	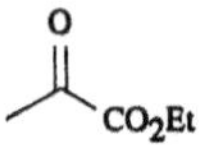	(73)
3		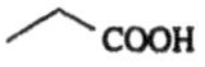	(69)
4	$C_6H_5CH_2NO_2$	C_6H_5COOH	(74)
5	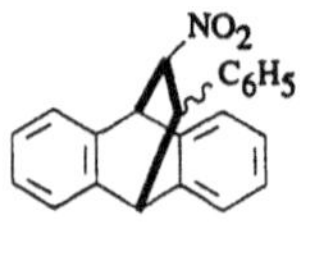	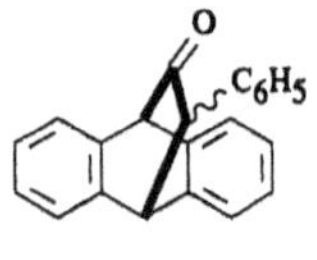	(96)

a) yields refer to isolated product. b) triethyl-amine was used for nitronate anion formation except for example 5 where LDA was used.

It should be mentioned that there are data referring to the oxidation of nitro derivatives to carbonyl compounds by hydroperoxides under basic conditions in the presence of $VO(acac)_2$ as catalyst.[53]

6. Oxidations of Ketones

The oxidative process leading from ketones to esters has been discovered at the end of the last century, by Baeyer and Villiger, who employed as oxidizing agent the inorganic Caro's acid, H_2SO_5.[54]

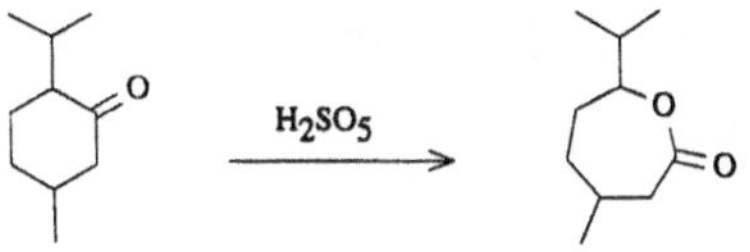

(12)

Nowadays the B.V. reaction is routinely carried out by using organic peracids.[55] A procedure employing supported perarsonic acids[56] has been also reported. Owing to the low reactivity of some ketones, particularly the open chains ones, the reaction requires rather drastic conditions and often a powerful oxidant, such as trifluoroperacetic acid, is needed. For cyclic ketones, such as cyclopentanone and cyclohexanone, which are generally more easily oxidized than acyclic derivatives, peracetic acid can be employed.[55] Among non metallic peroxidic oxidants, dimethyldioxirane and methyl-(trifluoro-methyl)-dioxirane, generated by reaction of potassium caroate ($KHSO_5$) and the corresponding ketones, has been recently reported to be rather effective reagents.[57]

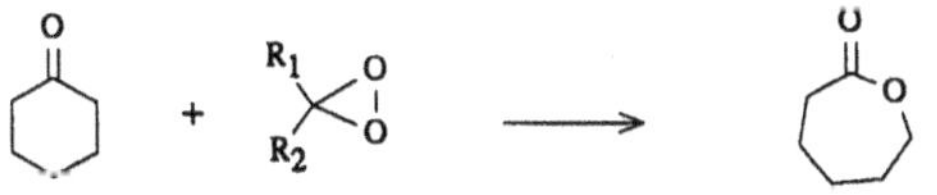

(13)

As far as the peroxometalcomplexes are concerned, it has been shown that the molybdenum oxo-peroxo-dipicolinato complex $MoO(O_2)(dipic)(H_2O)$ (dipic = dipicolinate dianion):

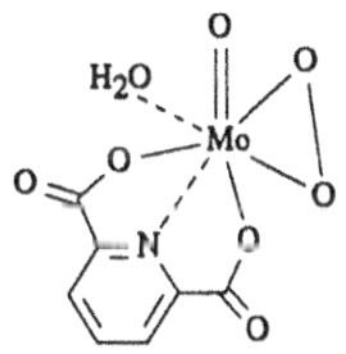

Modipic

is able to carry out the catalytic B.V. oxidation of ketones by H_2O_2 as exemplified by the results reported in Table 16.[58]

Rather curiously no data are available concerning the reactivity of Modipic used as stoichiometric oxidant in non polar solvents.

Table 16 : Oxidation of Cyclic Ketones with H_2O_2 Catalyzed by MoDipic[a]

#	Substrate	Product (%yield)
1	Cyclopentanone	δ-Valerolactone (40) + 5-hydroxypentanoic acid (20)
2	Cyclohexanone	ϵ-caprolactone (10)
3	2-methylcyclopentanone	5-methylvalerolactone (82)
4	2-methylcyclohexanone	6-methylcaprolactone (10)

a) all reactions were carried out in CH_3CN at 60^oC, MoDipic : Substrate : H_2O_2 = 1 : 30 : 30.

7. Hydroxylation of Alkanes and Arenes

Aside from the classical methods of synthetic organic chemistry, which usually require a sequence of reactions, the direct hydroxylation of alkanes may be obtained by strongly electrophilic reagents, such as $H_3O_2^+$, under superacidic conditions.[59] This reaction is believed to proceed *via* electrophilic OH insertion by protonated H_2O_2 in the C-H bond of the substrate:

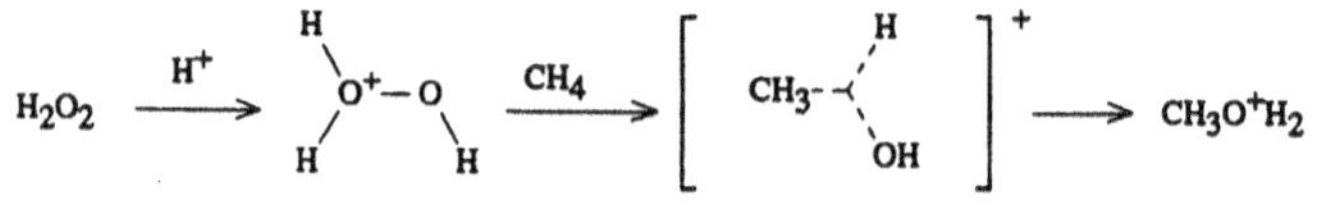

(14)

Reactions of oxygen with anions derived from alkanes possessing acidic protons, to form hydroxylated products, have been also reported.[60]

The metal catalyzed hydroxylation of alkanes can be achieved with a variety of oxygen donors (PhIO, NaOCl etc.) by using a metalloporphyrin as catalyst. In such systems the oxidant is believed to be an oxo-derivative.[61]

It has been known for a long time that H_2O_2 and Fe(II), the Fenton reagent, is able to oxidize aliphatic hydrocarbons but the yields and the selectivities are not of synthetic interest.[2]

As far as peroxocomplexes are concerned it has been recently shown that the hydroxylation reaction can be performed by using, in a stoichiometric process, species such as

$VO(O_2)(pic)(H_2O)_2$ derived from V(V) and H_2O_2:[62]

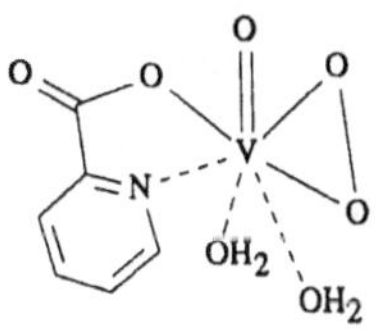

$VO(O_2)pic$

Although these complexes are almost unique in their ability to oxidize aliphatic C-H bonds, the data of Table 17, indicate that the yields and the selectivities attainable are not yet satisfactory.

Table 17 : Oxidation of aliphatic and aromatic hydrocarbons with complex $VO(O_2)Pic$ in CH_3CN at 20°C.

#[a]	Substrate	product(s) yield % [b]		
1	*cis*-butene	*cis*-epoxide (19)	*trans*-epoxide (8)	acetaldehyde (20)
2	*trans*-butene	*cis*-epoxide (2)	*trans*-epoxide (12)	acetaldehyde (3)
3	norbonene	norbonen-epoxide (22)		
4	benzene	phenol (56)		
5	cyclohexane	cyclohexanol (17)	cyclohexanone (8)	

a) complex 0.04 M; substrate 2 M. b) based on $[VO(O_2)Pic]_o$

There has been an effort toward the utilization of $VO(O_2)pic$ as a catalyst in the hydrocarbon hydroxylations by hydrogen peroxide.[63]

In the case of aromatic substrates, which are generally more reactive than alkanes, the catalytic hydroxylation with H_2O_2 in the presence of $VO(O_2)pic$ does proceed, characterized by a fairly good selectivity, as shown by the data reported in Tables 18.

Table 18 : Hydroxylation of benzene with H_2O_2 catalyzed by $VO(OiPr)_3$ in the presence of picolinic acid, in CH_3CN at 20°C.

#[a]	H_2O_2 (mmol)	Phenol mmol	yield %[b]
1	6	0.034	70
2	6	0.067	140
3	6	0.099	206
4	6	0.124	258
5	6	0.137	285

a) each reaction corresponds to a subsequent addition of 1.25 equivalent of H_2O_2 with respect to $[VO(OiPr)_3]_o$; picolinic acid : V(v) = 1.2 : 2.
b) calculated on $[VO(OiPr)_3]_o$.

It may be also noticed that this reaction may be carried out under phase-transfer conditions by using as lipophilic extracting agent the 4-(3-heptyl)pyridine-2-carboxylic acid (HPCA).[63]

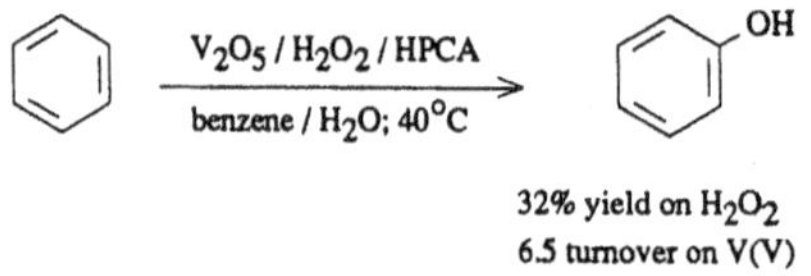

(15)

Clearly, both procedures still need to be optimized. At any rate the turnover numbers and the yields obtained so far can be considered of interest also in the light of the very mild conditions used.

8. Conclusions

The examples presented in this chapter demonstrate the synthetic relevance of some peroxometal complexes as oxidizing agents both in stoichiometric and catalytic reactions. At the same time, they also suggest that the scope of such reagents is far from being fully exploited. There is little doubt that many of the reactions examined here may be significantly improved, from a general point of view, as a result of a better understanding of the role played by the various parameters such as the solvent, the nature of the coordination sphere of the metal catalyst and the presence of a co-catalyst and additives.

More in particular, a major improvement would be the development of procedures capable of enantioselective oxidations. Such a goal, though an ambitious one, may be considered not too far away as demonstrated by the spectacular results obtained in the asymmetric oxidation of allylic alcohols[64] and thioethers[64-70] by hydroperoxides-based reagents.

Besides the reagents and the reactions already studied, it may be anticipated that several other oxidative transformations will be explored in the near future. Such a prediction is more easily made today than it would have been some years ago, when the emphasis was on the efficiency of the oxidant. Nowadays it is widely recognized that, together with the chemioselectivity, the ideal oxidation reagent should posses a high degree of stereo-, regio- and enantioselectivity, *i.e.* exactly the features displayed by peroxometalcomplexes.

9. References

1. *The Chemistry of Peroxides*, S. Patai Ed., Wiley Interscience, Chichester (1983)
2. R.A. Sheldon and J.K. Kochi "*Metal Catalyzed Oxidations of Organic Compounds*", Academic Press, New York (1981).
3. N.A. Milas and S. Sussman, *J. Am. Chem. Soc.* **58**, 1302 (1936).
4. N.A. Milas and S. Sussman, *J. Am. Chem. Soc.*, **59**, 342 (1937).
5. N.A. Milas and L.S. Maloney, *J. Am. Chem. Soc.*, **62**, 1841 (1940).
6. F. Di Furia and G. Modena, *Rev. Chem. Interm.*, **6**, 51 (1985).
7. cfr. in particular ref.1 ch.15.
8. F. Di Furia, G. Modena, R. Curci, S. J. Bachofer, J. O. Edwards and M. Pomerantz, *J. Mol. Catal.*, **14**, 219 (1982).
9. R. Curci, G. Fusco, O. Sciacovelli and L. Troisi, *J. Mol. Catal.*, **32**, 251 (1985).
10. M. Postel, C. Brevard, H. Arzoumanian and J. G. Riess, *J. Am. Chem. Soc.*, **105**, 4922 (1983).
11. M. Camporeale, L. Cassidei, R. Mello, O. Sciacovelli, L. Troisi and R. Curci in "*The role of Oxygen in Chemistry and Biochemistry*" vol.33 of "*Studies in Organic Chemistry*", W. Ando and Y. Moro-Oka Eds., Elsevier Science, Amsterdam (1988) p.201.
12. S. Campestrini, V. Conte, F. Di Furia, G. Licini and G. Modena, *Chim. Ind.*(Milan), **72**, 408 (1990).
13. A.H. Haines "*Methods for the Oxidation of Organic Compounds, Alkanes, Alkenes, Alkynes and Arenes*", Academic Press, London (1985).
14. A.H. Haines "*Methods for the Oxidation of Organic Compounds, Alcohols, Alcohol Derivatives, Alkyl Halides, Nitroalkanes, Alkyl Azides, Carbonyl Compounds, Hydroxyarenes and Aminoarenes*", Academic Press, London (1988).
15. O. Bortolini, V. Conte, F. Di Furia and G. Modena, *J. Mol. Catal.*, **19**, 331 (1983).
16. A. Arcoria, F.P. Ballistreri, G.A. Tomaselli, F. Di Furia and G. Modena, *J. Org. Chem.*, **51**, 2374 (1986).
17. H.B. Kagan, H. Mimoun, C. Mark and V. Schurig, *Angew. Chem. Int. Ed. Engl.*, **19**, 485 (1979).
18. V. Schurig, K. Hintzer, U. Leyrer, C. Mark, P. Pitchen and H.B. Kagan, *J. Organomet. Chem.*, **370**, 81 (1989).
19. O. Bortolini, F. Di Furia, G. Modena and A. Schionato, *J. Mol. Catal.*, **35**, 47 (1986).
20. M. Madesclaire, *Tetrahedron*, **42**, 5459 (1986).

21. K.A. Jørgensen, *Chem. Rev.* **89**, 431 (1989).

22. O. Bortolini, V. Conte, F. Di Furia and G. Modena in "*The Role of Oxygen in Chemistry and Biochemistry*" vol.33 of "*Studies in Organic Chemistry*", W. Ando and Y. Moro-Oka Eds., Elsevier Science, Amsterdam (1988) p.301.

23. O. Bortolini, L. Bragante, F. Di Furia and G. Modena, *Can. J. Chem.*, **64**, 1189 (1986).

24. A. Arcoria, F.P. Ballistreri, G.A. Tomaselli, F. Di Furia and G. Modena, *J. Mol. Catal.*, **24**, 189 (1984).

25. see as an example H. Mimoun, I. Sèrèe de Roch and L. Sajus, *Tetrahedron*, **26**, 37 (1970).

26. see as an example V. Conte, F. Di Furia and G. Modena, *J. Org. Chem.*, **53**, 1665 (1988).

27. O. Bortolini, V. Conte, F. Di Furia and G. Modena, *J. Org. Chem.*, **51**, 2661 (1986).

28. F. Di Furia and G. Modena, *Pure & Appl. Chem.*, **54**, 1853 (1982).

29. O. Bortolini, F. Di Furia, G. Modena and R. Seraglia, *J. Org. Chem.*, **50**, 2688 (1985).

30. C. Venturello, E. Alneri and M. Ricci, *J. Org. Chem.*, **48**, 3831 (1983).

31. S. E. Jacobson, D. A. Mucigrosso and F. Mares, *J. Org. Chem.*, **44**, 921 (1979).

32. C. Venturello and R. D'Aloisio, *J. Org. Chem.*, **53**, 1553 (1988).

33. J. Prandi, H. B. Kagan and H. Mimoun, *Tetrahedron Lett.*, **27**, 2617 (1986).

34. B. Plesnicar in "*Oxidation in Organic Chemistry*", W.S. Trahanovsky Ed., Academic Press, New York (1978) vol 5-C ch.3.

35. F.P. Ballistreri, S. Failla, G. A. Tomaselli and R. Curci, *Tetrahedron Lett.*, **27**, 5139 (1986).

36. F. P. Ballistreri, S. Failla, E. Spina and G. A. Tomaselli, *J. Org. Chem.*, **54**, 947 (1989).

37. Y. Ishii, K. Yamawaki, T. Ura, H. Yamada, T. Yoshida and M. Ogawa, *J. Org. Chem.*, **53**, 3587 (1988).

38. Y. Ishii, K. Yamawaki, T. Ura, and M. Ogawa, *J. Org. Chem.*, **52**, 1868 (1987).

39. Y. Ishii and Y. Sakata, *J. Org. Chem.*, **55**, 5545 (1990).

40. O. Bortolini, L. Bragante, F. Di Furia, G. Modena and L. Cardellini, *Chimica Oggi*, **4**, 69 (1986).

41. O. Bortolini, S. Campestrini, F. Di Furia, G. Modena and G. Valle, *J. Org. Chem.*, **52**, 5467 (1987)

42. S. Campestrini, F. Di Furia, G. Modena and O. Bortolini, *J. Org. Chem.*, **55**, 3658 (1990).

43. W.P. Griffith and S.V. Ley, *Aldrichim. Acta*, **23**, 13 (1990).

44. M. Trost and Y. Masuyama, *Isr. J. Chem.*, **24**, 134 (1984).

45. B. M. Trost and Y. Masuyama, *Tetrahedron Lett.*, **25**, 173 (1984).

46. H. Tomioka, K. Takai, K. Oshima and H. Nozaki, *Tetrahedron Lett.*, **21**, 4843 (1980).
47. E. Vedejs, D. A. Engler and J. E. Telschow, *J. Org. Chem.*, **43**, 188 (1978) and references therein.
48. E. Vedejs, *J. Am. Chem. Soc.*, **96**, 5944 (1974).
49. Y. Nishikimi, I. Iimori, M. Sodeoka and M Shibasaki, *J. Org. Chem.*, **54**, 3354 (1989).
50. J. Blagg and S.G. Davies, *Tetrahedron*, **43**, 4463 (1987).
51. N.J. Lewis, S.Y. Gabhe and M.R. De La Mater, *J. Org. Chem.*, **42**, 1479 (1977).
52. M.R. Galobardes and H.W. Pinnick, *Tetrahedron Lett.*, **22**, 5235 (1981).
53. P.A. Bartlett, F.R. Green and T.R. Webb, *Tetrahedron Lett.*, 331 (1977).
54. A. von Baeyer and V. Villiger, *Ber. Dtsch. Chem. Ges.*, **32**, 3625 (1899).
55. H.O. House "*Modern Synthetic Reactions*", The Benjamin Cummings Publishing Co., New York (1972) 2nd ed. pp. 321-329.
56. S.E. Jacobson, F. Mares and P.M. Zambri, *J. Am. Chem. Soc.*, **101**, 6938 (1979).
57. M. Camporeale, T. Fiorani, L. Troisi, W. Adam, R. Curci and J.O. Edwards, *J. Org. Chem.*, **55**, 93 (1990) and references therein.
58. S.E. Jacobson, R. Tang and F. Mares, *J. Chem. Soc. Chem. Commun.*, 888 (1978).
59. G.A. Olah, *Acc. Chem. Res.*, **20**, 422 (1987).
60. R.D. Guthrie in "*Comprehensive Carbanion Chemistry*", vol.5 part A of "*Studies in Organic Chemistry*", E. Buncel and T. Durst Eds., Elsevier Science, Amsterdam (1980), ch.5.
61. B. Meunier, *Gazz. Chim. It.*, **118**, 485 (1988).
62. H. Mimoun, L. Saussine, E. Daire, M. Postel, J. Fischer and R. Weiss, *J. Am. Chem. Soc.*, **105**, 3101 (1983).
63. M. Bonchio, V. Conte, F. Di Furia and G. Modena, *J. Org. Chem.*, **54**, 4368 (1989).
64. M.G. Finn and K.B. Sharpless in "*Asymmetric Synthesis*", vol. 5, J.D. Morrison Ed., Academic Press, New York (1985) ch.7.
65. P. Pitchen, E. Dunach, M.N. Deshmukh and H.B. Kagan, *J. Am. Chem. Soc.*, **106**, 8188 (1984).
66. O. Samuel, B. Ronan, H.B. Kagan, *J. Organomet. Chem.*, **370**, 43 (1989).
67. E. Duñach and H.B. Kagan, *Nouv. J. Chim.*, **89**, 1 (1985).
68. F. Di Furia, G. Modena and R. Seraglia, *Synthesis*, 325 (1984).
69. F. Di Furia, G. Licini, G. Modena and G. Valle, *Bull. Soc. Chim. Fr.*, **127**, 734 (1990).
70. F. Di Furia, G. Licini and G. Modena, *Gazz. Chim. It.*, **120**, 165 (1990).

CHAPTER 8

The Use of Polyoxometalates in Reactions with Hydrogen Peroxide

Craig L. HILL

Department of Chemistry, Emory University, Atlanta, GA 30322, U.S.A.

1. Introduction and General Considerations

The development of selective catalytic methods to oxidize organic substrates is often a nontrivial exercise. Few classes of organic reactions are as mechanistically complex and often as difficult to control and render selective as metal catalyzed oxidations [1, 2]. The inherent kinetic behavior of oxidants and substrates as well as other factors can give rise to formidable obstacles in this area. A fact of more than passing significance in all organic oxidation processes, and one often overlooked in research and development efforts is that both the reactants (substrates) and the products are thermodynamically unstable with respect to further oxidation to CO_2 and H_2O. Within this inherent limitation, it is the systematic control of a number of kinetic processes that dictate the observed chemistry. The title subject is of very recent vintage and as a consequence most of the literature cited in this chapter is quite recent. In order to review this subject properly, the fundamental and pertinent chemical and physical features of both hydrogen peroxide reactions and polyoxometalates must be presented. Before we turn to an elaboration of these subjects in the appropriately designated sections of the chapter below, we first discuss several points relevant to the development of oxidation and/or catalytic processes in general.

The impetus for developing catalysts and catalytic processes is multifold. In the absence of catalysts, many reactions of technical or commercial interest do not proceed at all, some economically significant processes proceed but not at adequate rates, and many processes proceed but do not do so with adequate selectively. With the rate and selectivity benefits that catalysts and catalyzed processes provide, come liabilities. Chief among these are catalyst degradation and its consequent effect on rates and selectivities. We turn to a brief examination of these points, in general, and their role in organic oxidations, in particular.

G. Strukul (ed.), *Catalytic Oxidations with Hydrogen Peroxide as Oxidant*, 253–280.

The consequences of the presence of a catalyst on reaction rates are fairly trivial and well defined both in technical and practical contexts -- reactions proceed faster over a wide range of temperatures. Given this well appreciated point, we turn immediately to the affect of catalysts on selectivities. In synthetic processes, three types of selectivity are defined: chemoselectivity (what types of groups or molecules are attacked/ transformed), regioselectivity (in what region(s) or location(s) of the molecule is the transformation manifested), and stereoselectivity (what stereoisomer is produced). Catalysis directly affects all three types of selectivity. One general type of chemoselectivity is often critical in organic oxidation processes and more important than in other generic families of processes such as hydrolyses, reductions, additions, etc. Namely, the initial products, (often the desired products of the oxidation), are as or more reactive than the reactants themselves. This type of kinetic selectivity precludes the accumulation of the initial product as it is consumed as fast or nearly as fast as it is generated. Consider the oxidation of a representative alkane, cyclohexane, eq 1, as an illustration of this point. The relative rate of oxidation of the substrate alkane (k_1) is substantially less than the rate of oxidation of the initial product, cyclohexanol (k_2) and the rate of the further oxidation of its oxidation product, the ketone (k_3). If the desired product is the alcohol, the reaction will be doomed to failure by this chemoselectivity ($k_{product} > k_{substrate}$) unless the product is removed by some means (chemical or physical) as it is produced. Removal of such intermediate oxidation state species is usually impractical technically or economically.

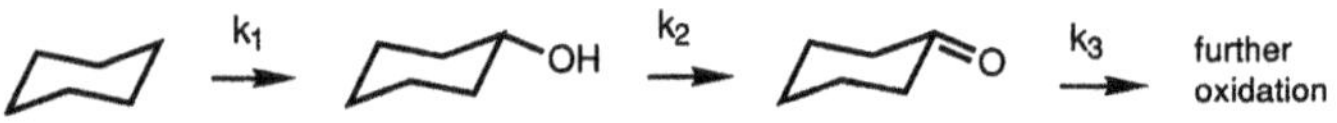

(1)

These points lead to a related issue of general pertinence in formulating effective syntheses of any kind, including catalytic ones, namely the development of processes that display the desired selectivities at *high conversion of substrate*. A host of oxidation processes for a number of types of organic substrates have been reported over the years that proceed at very high selectivities but at modest or low conversion. The motivation for many of these studies has been to elucidate mechanism or to model biological catalytic oxidation processes and not to formulate practical synthetic procedures. In many of these studies the conversions are not reported at all and thus to the uninitiated such reactions at first glance might appear to be synthetically attractive and useful when quite the opposite may be the case.

The conversions in many organic oxidation studies are not reported for three related reasons: *i.* the inherent kinetic selectivities (e.g. $k_{product} > k_{substrate}$) are such that the desired product never builds up to a satisfactory concentration at any conversion, *ii.* the reactions are run under pseudo-first-order conditions, that is, with a large excess of substrate such that the reaction can only be taken experimentally to a low conversion, and *iii.*, product yields are reported as based, not on substrate, which is the most conventional and synthetically significant way to do so, but on oxygen donor or even catalyst. In some reactions, yields as such are not reported at all but selectivities are; selectivity = moles of the product of interest / the moles of all detectable products. One family of oxidation processes that illustrates these points perhaps most definitively is the oxo transfer oxidation (oxygenation) of hydrocarbons by oxygen donors catalyzed by transition metal complexes [3-16]. A survey of the 30 or so papers on alkane oxygenation (hydroxylation) catalyzed by metalloporphyrins, indicates that all but 3 or so of these studies report yields based on oxygen donor, or even on catalyst, and not on substrate. The conversions, which are not even discussed as such in most of these studies, average between 1 and 2% [3-17]!

The regiochemical and stereochemical characteristics exhibited by an organic oxidation reaction depend on both the reaction mechanism and the reaction conditions. In some but not all cases, the process(es) in the mechanism involving attack of oxidizing species or intermediates on the organic substrate itself dictate the regiochemistry and stereochemistry. Given the importance of mechanism in these selectivities, further discussion in this area will be cited in context with mechanistic points made later.

In the early stages of development of a catalytic organic oxidation process, the mechanism and factors that determine rates and selectivities are probed; the factors that determine catalyst stability are often ignored. As much of the basic research and most of that pursued in academic laboratories is aimed at the former issues, stability is often not addressed. The importance of catalyst stability, or in more general terms, stability of the catalytic process, depends on the type, scale, and cost of the reaction as well as other factors. Whereas stability of a catalyst system is not critical in some processes, including production of some fine chemicals, it is critical in most large scale processes or processes where economic factors dictate the necessity of high turnover numbers and long catalyst lifetimes.

A related point, and one of importance that is often not fully considered, is the affect of catalyst degradation not just on rates but on selectivities. It is often the case that a catalyst under the reaction conditions can be transformed to other catalytically competent species. The latter, although they still exhibit catalytic activity, do not catalyze the same manifold of processes catalyzed by the parent catalyst. Some of the secondary catalytic materials generated in situ can catalyze nonselective processes. In the overall synthetic reaction, a drop in selectivity rather than rate may dictate when the reaction must be terminated.

In addition to the general thermodynamic and kinetic points above, there are other important considerations that drive research and development in catalysis and that impact the H_2O_2 oxidations catalyzed by polyoxometalates addressed below. These include: *i.* minimizing waste and using minimally toxic solvents, *ii.* recovering catalyst readily and in high yield after reaction, and *iii.* other issues that enter in the process development stage.

To summarize, one seeks to develop catalytic organic systems that have the following attributes: *i.* the desired selectivities (chemo, regio, and stereo) at high conversion of substrate, *ii.* high rates, and *iii.* high stabilities. In addition, a sophisticated understanding of the energetic and mechanistic features of the reaction are needed so that selectivities, rates, and stabilities can be controlled as systematically and as much as possible. Very few catalytic processes, including some of the most commercially significant ones, are sufficiently well understood at a detailed mechanistic level that the above features have been rationally optimized.

2. Hydrogen Peroxide: General Chemistry and Transition Metal-Catalyzed Oxidations

The focus of much of this book, and on much chemistry involving H_2O_2 at present, is to devise selective organic oxidations using H_2O_2 and catalysts for such oxidations. Many factors make H_2O_2 an attractive oxidant at present. Two of the most significant are the current availability and low cost of H_2O_2 and the fact that many processes involving H_2O_2 generate H_2O as the only H_2O_2-derived product. In other words, toxicity is often minimal and the chemistry is deemed environmentally innocuous.

Hydrogen peroxide is almost always associated with water. Not only is this oxidant available as an aqueous solution of varying concentration, but its principal product in a number of processes is also water. This has a number of consequences for catalysis. Often the water must be removed from H_2O_2-based oxidations. Water can can affect water-sensitive functions in substrates and/or products, and more commonly, it can interfere with the catalyst and its molecular mode(s) of operation. Water can inactivate or render many transition metal catalysts less selective by the simple expedient of competing effectively with the substrate or H_2O_2 itself for binding at the catalyst active site.

Hydrogen peroxide is a good oxidizing agent in both acidic and basic media, a weak reducing agent, and it is thermodynamically unstable with respect to disporportionation, eq 2-5 [18, 19]. In addition to these thermodynamic indicators, the dominant kinetic predisposition of H_2O_2 is as an oxidant. For many processes catalysis by transition metal species is observed and sometimes the catalytic enhancement can be orders of magnitude [20] The are a number of organic oxidations by H_2O_2 catalyzed by

$$H_2O_2 + 2H^+ + 2e^- \rightarrow 2H_2O \qquad E^o = 1.77\ V \qquad (2)$$

$$O_2 + 2H^+ + 2e^- \rightarrow H_2O_2 \qquad E^o = 0.68\ V \qquad (3)$$

$$HO_2^- + H_2O + 2e^- \rightarrow 3OH^- \qquad E^o = 0.87\ V \qquad (4)$$

$$2H_2O_2 \rightarrow 2H_2O + O_2 \qquad \Delta H = -99\ kJ\ mol^{-1} \qquad (5)$$

$$S + H_2O_2 \xrightarrow{\text{TM catalyst}} SO + H_2O \qquad (6)$$

where S = alkene or other organic substrate, SO = epoxide or oxidized organic product, and TM catalyst = some transition metal containing species.

a range of transition metal species, in which the net transformation is transfer of oxygen from the H_2O_2 to the substrate, eq 6. As such H_2O_2 is a source of "active oxygen" and a formal "oxygen donor". These net processes (eq 6) may or may not take place by an oxo transfer mechanism, however. Of the other oxygen donors which have been investigated in context with transition metal-catalyzed oxo transfer to organic substrates, H_2O_2 probably functions in a mechanistic sense as a formal oxo donor less often than any of them. This list of oxygen
donors includes iodosylarenes, amine N-oxides, oxaziridines, persulfate, peracids, organoperoxides, and some main group and transition metal oxo compounds, as well as H_2O_2. Thorough reviews of oxo transfer between inorganic centers [21] and between inorganic centers and organic C-H bonds

[22] have been published recently. Regardless of mechanism, optimization of the desired process, often some form of eq 6, and minimization of side reactions involving the oxidant, often but not always eq 5, is sought in the development of H_2O_2-based catalytic organic oxidation processes. If eq 5 or other side reactions are too fast then the overall reaction, again, kinetically controlled, will be of little value as H_2O_2 will simply be consumed before it is used in productive chemistry, e.g. eq 6.

There are several mechanisms for the general processes given in eq 5 and 6 [1, 12b, 23]. Substrate oxidation, eq 6, can proceed by homolytic (or one-electron) mechanisms, heterolytic (or two-electron) mechanisms or both, depending on the transition metal complex that is used as the catalyst and to a lesser extent on the other reaction conditions. The principal homolytic mechanisms involving metal-catalyzed organic substrate oxidation by hydrogen peroxide are analogous to those in Fenton chemistry (eq 7-11) and have been thoroughly reviewed in chapter 4. That is, they involve formal one-electron changes in the oxidation state of the metal complex. The principal metals that facilitate these type of reactions, and fairly analogous ones with alkyl

$$Fe^{II} + H_2O_2 \rightarrow Fe^{III}OH + HO\cdot \qquad (7)$$

$$Fe^{III} + H_2O_2 \rightarrow Fe^{II} + HO_2\cdot + H^+ \qquad (8)$$

$$Fe^{II} + HO\cdot \rightarrow Fe^{III}OH \qquad (9)$$

$$Fe^{III} + HO_2\cdot \rightarrow Fe^{II} + O_2 + H^+ \qquad (10)$$

$$HO\cdot + H_2O_2 \rightarrow H_2O + HO_2\cdot \qquad (11)$$

hydroperoxides, are Co(II)/Co(III), Mn(II)/Mn(III), Fe(II)/Fe(III), Cu(I)/Cu(II), and a few other metal species which have one-electron redox couples with potentials in thermodynamically appropriate ranges. Several of these intermediates in eq 7-11, and analogous reactions with other metals, are kinetically competent to directly oxidize various organic substrates [24] as are oxometal complexes which can arise from deprotonation of the hydroxy complexes of the high valent forms of the metal complexes with or without further metal centered oxidation. The fact that several substrate oxidation processes are also present, and a host of less prominent but potentially significant side reactions are present, renders metal-catalyzed homolytic oxidation processes quite difficult to control and often as nonselective as they are mechanistically complex [12b]. One of the potentially significant side reactions and one that occurs in catalyzed organoperoxide based oxidations involves redox initiated radical chain autoxidation using the O_2 generated in eq 10 and other processes.

The principal heterolytic mechanisms involving metal-catalyzed organic substrate oxidation by hydrogen peroxide typically proceed with fewer elementary processes than their homolytic counterparts [1,23]. Some of these processes, including ones that directly impact on selectivity, are more amenable to experimental investigation and preparative control than for homolytic mechanisms. In the heterolytic mechanisms, the metal center, usually a transition metal center, facilitates heterolytic cleavage of the O-O peroxide bond in one of several ways, and the substrate is often (but not always) oxygenated in the coordination sphere of the metal complex. Formation of electrophilically activated peroxo functions is common. Unlike homolytic peroxide activation mechanisms, heterolytic ones do not necessarily involve a change in the oxidation state of the metal and do not involve *freely diffusing* organic radical intermediates. A tacit assumption in the literature is that metal-catalyzed organic oxidations by peroxides of all kinds that proceed by heterolytic mechanisms do not involve radicals. Many of these processes may in fact involve radical intermediates but these intermediates are not kinetically distinguishable, that is, they do not diffuse away from the metal center at any point along the reaction coordinate.

Eq 12 depicts the principal electrophilic activation process in generic terms. Metal oxo groups that are converted to metal hydroxy peroxy groups by this simple addition process

$$\mathrm{M{=}O} + \mathrm{ROOH} \rightarrow \underset{\text{inorganic perester}}{\mathrm{M(OH)(OOR)}} \qquad (12)$$

(which may or may not be catalyzed), can occur in discrete monomeric species as in the case of cytochrome P-450 [25,26], other heme-dependent oxygenase enzymes and most synthetic transition metal catalysts, or on metal oxide surfaces. Polyoxometalates reside in between these two extremes in size and experimental tractability. The principal metals that catalyze the oxidation of organic substrates by H_2O_2 via heterolytic mechanisms are Cr, Mo, V, Ti, and W, usually in their d^0 electronic configurations.

As for the organic substrate oxidation process, eq 6, the peroxide disproportionation process, eq 5, can also take place by homolytic or heterolytic mechanisms or by both depending on the catalyst and reaction conditions [1]. Under some conditions, d^0 transition metal species catalyze eq 5 generating singlet oxygen [27]. No other chemical process associated with H_2O_2 is more subject to catalysis than dismutation, eq 5. This fact impacts on the preparation and storage of H_2O_2 as well as its use in organic oxidation processes. Although many processes exist where the catalysis of substrate oxygenation, eq 6, is significantly more effective than catalysis of dismutation, eq 5, it is often hard to avoid catalyzing the former without also catalyzing the latter. Experimental and theoretical work should continue to focus on which structural and electronic features of molecular catalysts optimize the ratio of substrate oxygenation to dismutation in H_2O_2 reactions.

3. Polyoxometalates and Polyperoxometalates

The basic properties of polyoxometalates and polyperoxome-talates will be given in this section in preparation for discussion of the catalytic work executed to date that is presented in the following section. The early transition metal polyoxometalates, or polyoxometalates for short, are a substantial family of anionic inorganic cluster-like compounds. The structural skeletons of these complexes are networks of bonds between transition metal and oxide ions [28]. The transition metal ions that form polyoxometalates are usually in their d^0 electronic configurations. These metal ions, in order of increasing number of structural families associated with each, are Ta(V), Nb(V), V(V), W(VI) and Mo(VI). The latter two ions both form a large array of structures and this number is currently growing at a significant rate. There are both isopoly compounds (also known as isopolyanions, isopoly complexes, or isopolyoxometalates) which contain only metal and oxide ions, and heteropoly compounds (also known as heteropolyanions, heteropoly complexes or heteropolyoxometalates) which contain so-called "heteroatoms" in addition to the metal and oxygen atoms. Most of the elements in the periodic table have now been reported to function as heteroatoms in one type of polyoxometalate or other. Figure 1 illustrates in polyhedral notation, representative isopoly and heteropoly compounds.

(A)

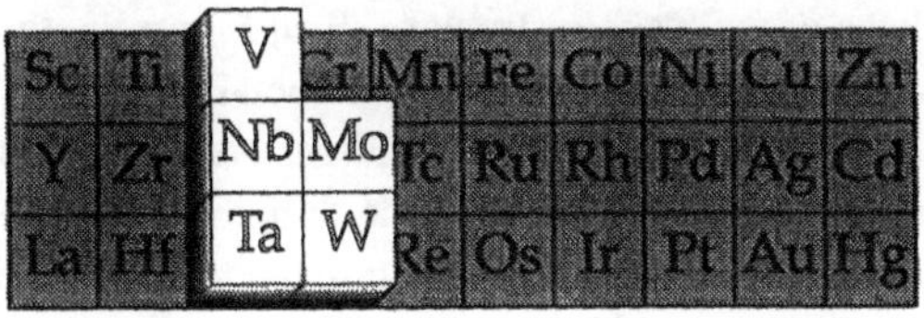

(B)

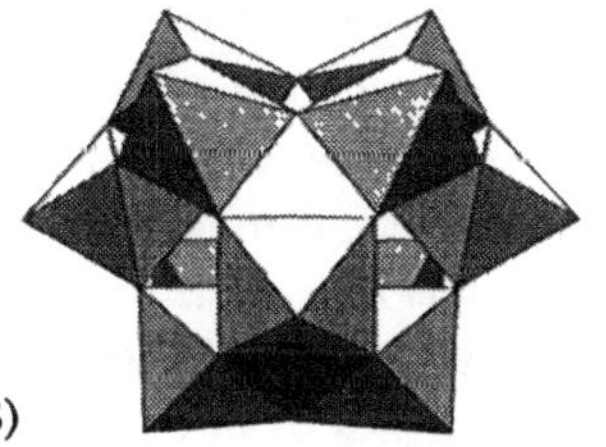

$X^{n+}M_{12}O_{40}^{(8-n)-}$ ("Keggin")
a heteropolytungstate (T_d);
heteroatom (X^{n+}) not shown.

(C)

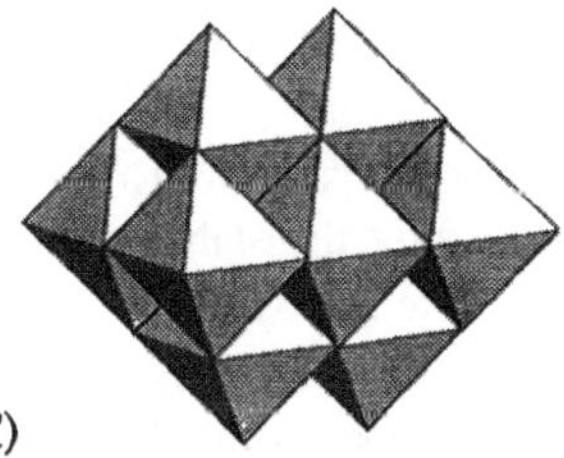

$V_{10}O_{28}^{6-}$ (decavanadate),
an isopolyvanadate (D_{2h})

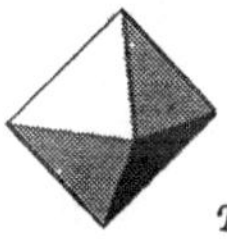

Each octahedron represents one metal surrounded by six oxygen atoms, located at each vertex.

Figure 1. A: the elements that form polyoxometalates; B and C: structures of representative families of polyoxometalates in polyhedral notation; B is the Keggin heteropolytungstate, α-$X^{n+}M_{12}O_{40}^{(8-n)-}$, M = Mo^{VI} or W^{VI} (T_d point group symmetry). The X^{n+} "heteroatom" that resides in the central tetrahedral hole of the Keggin structure has been omitted here for clarity; C is the isopolyvanadate, decavanadate, $V_{10}O_{28}^{6-}$, (D_{2h} point group symmetry). In polyhedral notation each octahedron represents one metal atom surrounded by 6 oxygen atoms (or a tetrahedron in the case of 4-coordinate metal ions). The metal atom in each octahedron is displaced toward its terminal oxo group(s). The metal atom is not seen directly in polyhedral notation.

Although the conventional wisdom is that the thermal stability of polyoxometalates is generally quite high, it actually varies considerably with the type of structure, the type of transition metal ions in the structure, the nature of the cationic counterions, and the extent of hydration. While polyoxometalates of the Keggin structure (B in Figure 1), the most common, accessible, and investigated class of polyoxometalates, are relatively robust to thermal decomposition, other structural classes are significantly less so. Just within the Keggin polyoxomolybdates, the temperature at which thermal decomposition takes place varies over hundreds of degrees depending on the identity of the heteroatom, X^{n+}, occupying the tetrahedral central hole in the molecule, and on the cations. Nearly all polyoxometalates decompose at temperatures below 600°C to their corresponding metal oxides [28]. Polyoxotungstates decompose to WO_3, polyoxomolybdates to MoO_3, polyoxovanadates to V_2O_5, and so on. As such, polyoxometalates are not thermodynamically stable although the high kinetic thermal stability of some of them make it easy in some instances to forget this point. The potential surface that quantitatively defines the stability of various isopolyoxometalates and their corresponding d^0 metal oxides is complex in all cases. Not only are there many isopolyoxometalate structures known for most of the polyoxometalate-forming d^0 transition metal ions, but there are also multiple forms of the metal oxides themselves, only one of which being the truly thermodynamically stable form under a given set of physical conditions. Several distinct solid state structures for both WO_3 and MoO_3 metal oxides have been characterized thus far and the list may not yet be complete. More experimental data and theoretical work will be required before such polyoxometalate-metal oxide surfaces for any d^0 transition metal are well defined, or at least defined to the point where they have some predictive value. It is not clear how soon this potentially useful information will be available even on the most investigated and utilized systems, those of W(VI) and Mo(VI). One point that is clear, however, is that the various isopolyoxometalates each represent localized kinetic minima on such surfaces with some structural families residing in deeper localized minima than others.

Often more pertinent than the thermal stability in conjunction with H_2O_2 and oxidation chemistry, is oxidative stability. By virtue of being composed of d^0 metal and oxide ions, polyoxometalates are highly resistant to oxidation. This fact, coupled with the diversity of synthetically accessible polyoxometalate molecular geometries, indicates that these compounds should have substantial application in oxidation processes. Whereas nearly all other homogeneous catalysts for organic oxidation reactions are composed in part of organic structural components (the ligands are part or entirely organic), polyoxometalates have no organic structural components. Nearly all organic groups are thermodynamically unstable with respect to oxidation to CO_2 and

H_2O as indicated earlier, and kinetic oxidative degradation of such catalysts is observed sooner or later in all processes that effect oxidation of the less reactive classes of organic substrates. Polyoxometalates clearly have an inherent advantage. Whereas conventional homogeneous oxidation catalysts with organic ligands are just kinetically stable (to varying degrees) with respect to irreversible oxidative degradation, polyoxometalates may be thermodynamically stable with respect to oxidative degradation.

In addition to oxidative and thermal stability, other properties of polyoxometalates make them attractive for applications in catalysis and other areas. First, many of these complexes are relatively accessible from commercially available and inexpensive starting materials. Second, many, but not all, polyoxometalates exhibit a rich acid/base and redox chemistry. Furthermore there are often rational and defensible syntheses for groups of polyoxometalates that exhibit a range of these properties. Third, the cationic counterions of polyoxometalates can be used to solubilize the complexes in liquids ranging greatly in polarity from water to hydrocarbons. Fourth, polyoxometalates, and largely heteropolyanions of tungsten or molybdenum, often exhibit useful levels of activity (physical, chemical, or catalytic) either in solution (homogeneous conditions) or in the solid state (heterogeneous conditions) and can be used in either mode. Fifth and finally, most polyoxometalates are relatively low in toxicity. Only polyoxovanadates and vanadium-containing complexes have consistent toxicity in cell culture [29].

Polynuclear complexes of the same polyoxometalate-forming d^0 metals that contain one or more peroxo groups (O_2^{2-} in some protonation state) are of direct pertinence to the H_2O_2 oxidation of organic substrates catalyzed by these metals. The chemistry of a few peroxide/d^0 transition metal ion (polyperoxometalate) systems have been examined both in solution and in the solid state. Recently peroxide/M systems, where M = V(V) [30], W(VI) [31], and Mo(VI) [31] have been studied in aqueous solution by Raman, infrared and heteronuclear NMR techniques to complement a host of earlier studies [28]. In the solid state, a few polyperoxotungstate species and more than 15 polyperoxomolybdate species have been structurally characterized by X-ray crystallography [32-36]. Inasmuch as the polytungstates are currently of greater significance in the title subject of this chapter, emphasis will be placed on these throughout this chapter. The only polyperoxotungstate complex that has been isolated *and* investigated in conjunction with catalysis and oxidation is that from the Venturello system which is discussed below. This complex,

$\{PO_4[W(O)(O_2)_2]_4\}^{3-}$, was isolated from organic media and structurally characterized as its tris(tetra-*n* -hexylammonium) salt

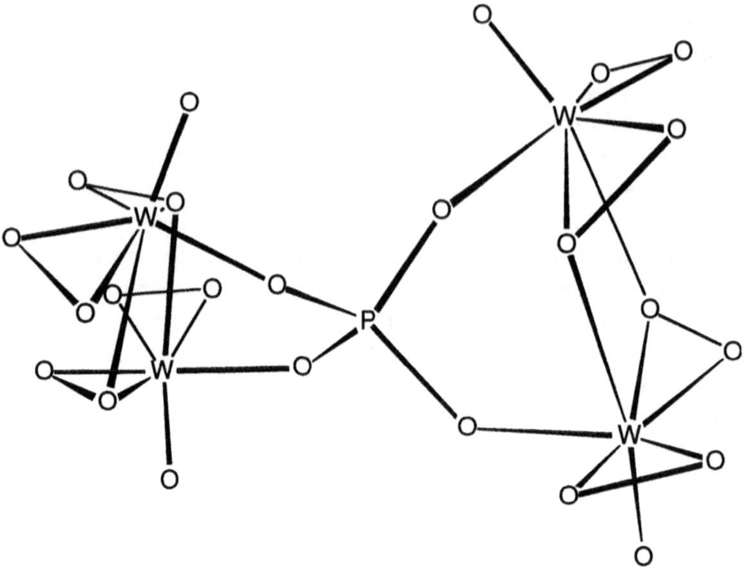

Figure 2. Molecular structure of anionic complex isolated from the Venturello oxidation system, $\{PO_4[W(O)(O_2)_2]_4\}^{3-}$ [32].

(Figure 2) [32]. The structure is comprised of 4 distorted pentagonal bipyramids in edge shared pairs that are bridged by a central phosphate moiety. Each of the W atoms contains a symmetrical and terminal peroxo group (η^2-O_2^{2-}) and an unsymmetrical and bridging peroxo group. Other structurally characterized peroxo polytungstates include $[W_4O_8(O_2)_6(CO_3)]^{6-}$ [33] and $[W_2(O)_3(H_2O)_2(O_2)_4]^{2-}$ [34]. In none of these complexes are any of the peroxo groups unsymmetrically ligated. The degree of asymmetry of peroxo ligation in electrophilic early transition metal ions could well be a major factor in reactivity.

Polyperoxometalates do not approach most families of polyoxometalates in thermal stability. The principal mechanisms for thermal degradation of peroxo moieties of all kinds are homolytic in nature and usually involve cleavage of the weak peroxo O-O bond. Although ligation to metal ions can perturb the O-O bond energy by perhaps up to 20% or so, this bond remains the weakest in peroxo transition metal complexes. Although the oxidative stability of polyperoxometalates has not been seriously addressed, the ready ability of nearly all types of peroxo groups to function as reducing agents (cf. reverse of eq 3), suggests that oxidative degradation of the polyperoxometalates should be facile under many conditions used in organic oxidation reactions.

4. Polyoxometalates as Catalysts - An Overview

The diverse redox and other chemical properties of polyoxometalates have led to the development of organic oxidation processes catalyzed by polyoxometalates where the molecular role(s) of the complex varies from system to system. Based on chemical role, five families of processes can be defined.

I. Polyoxometalates as heterogeneous oxidation catalysts [37]. Some of these processes, and in particular, the catalytic oxidation of methacrolein to methacrylic acid, have achieved commercialization recently in both Japan and the US [38,39]. As with many other heterogeneous catalytic oxidation systems, all the detailed mechanistic role(s) of the polyoxometalate catalyst are not clear. These heterogeneous oxidation catalysts can contain numerous additives but the exact roles of these additives on the molecular origins of the enhancements in selectivity, reactivity, and stability that they provide to these systems is not generally clear. Since these systems do not involve catalysis of H_2O_2 reactions, we will not discuss them further here.

II. Polyoxometalates as co-catalysts. There are a large number of studies, largely by Matveev, Kozhevnikov, and co-workers, in which polyoxometalates (usually vanadium-substituted heteropolymolybdic acids) function as co-catalysts or co-oxidants in processes related to the Wacker oxidation. The role of the heteropoly compound in nearly all of these studies is to reoxidize Pd or other catalytic species and couple the use of O_2 as the terminal oxidant with catalysis by Pd [40]. As H_2O_2 has been minimally deployed in these systems experimentally, we will not discuss them further in this review.

III. Polyoxometalates as catalysts for photochemical processes. The ability of polyoxo-metalates to photooxidize alcohols and other organic molecules has been known for some time. Recently such processes have been investigated in depth [41] and new catalytic systems with unusual selectivities, properties and uses have been devised [42]. Although H_2O_2 might prove to be of real value in processes such as these, it has not been used significantly to date. As a consequence, this chemistry will not be discussed further here.

IV. Polyoxometalates as better d^0 catalysts for peroxide-based organic oxidations. Several papers have appeared recently in this area [43-49] including those by the research groups of Venturello [32,47] and Ishii [48,49] which have attracted considerable attention. These will be discussed in the next section.

V. Polyoxometalate derivatives as inorganic oxidatively resistant analogues of tractable homogeneous catalysis for organic oxidations. These systems, first reported by our group [50] and now under investigation in several groups [51-54], are also discussed below.

5. Specific Types of Systems

5.1. d^0 transition metal systems

The organic oxidation processes catalyzed by d^0 transition metal ions (including most polyoxometalates), that are of the greatest interest are those that take place by heterolytic mechanisms (*vide supra*). The first epoxidation reactions by H_2O_2 catalyzed by tungstate, WO_4^{2-}, were reported in the 1940s [55]. Several papers and reviews on oxidation of aliphatic alkenes, unsaturated alcohols, carboxylic acids or other substrates by H_2O_2 catalyzed by tungstate and related low nuclearity d^0 metal ions appeared subsequently [56-59]. Most of these papers report moderate to excellent yields. Few give high selectivities for the desired product at high conversions of substrate and H_2O_2, however. Although chlorinated solvents have been the most extensively used, ethers, alcohols or other solvents have proved satisfactory in some of these systems [57,58].

Several papers on organic oxidations by H_2O_2 catalyzed by d^0 polyoxometalates have appeared recently. Trost reported selective oxidation of secondary alcohols to ketones and aldehydes to acids in a range of organic substrates bearing other groups [43]. More recently, Misono and co-workers reported selective oxidation of cyclopentene to glutaraldehyde by $H_3PMo_{12-x}W_xO_{40}$ (x = 3-9). Yields (based on feed H_2O_2) of epoxide and diol products, (the latter doubtless derived in these acidic systems from opening of epoxide in situ), were significantly smaller than that of the dialdehyde. No mention was made of over oxidation to glutaric acid although this product would probably have not been seen by the gas chromatographic methods used in the work. A dried solution of H_2O_2 in tributylphosphate (3% residual H_2O) was used in these experiments [44].

Most of the reported research addressing organic oxidations by H_2O_2 catalyzed by d^0 polyoxometalates, has involved the use of two-phase conditions and phase-transfer catalysis. Phase transfer is accomplished by making the hydrophobic quaternary ammonium salts of the polyoxometalates before hand or by generating them in situ. Schwegler and co-workers [45] reported alkene oxidation by both the defect (so-called "lacunary") complexes, $X^{n+}W_{11}O_{39}{}^{(8-n)-}$, $X^{n+} = P^{5+}$ or Si^{4+}, and some d^n, $n \neq 0$ transition metal-substituted derivatives thereof. These investigators did not see any reactivity with $[(n\text{-}C_6H_{13})_4N]_3[PW_{12}O_{40}]$ as the catalyst. The latter result is puzzling and seems to be at odds with the observations from our group and that of Ishii [48,49]. Tomaselli and co-workers [46] as well as Ishii [49] reported the selective oxidation/oxygenation of alkynes by H_2O_2 catalyzed by the Ishii compound, $(\text{cetylpyridinium})_3(PW_{12}O_{40})$.

5.2 Venturello/Ishii systems

Venturello first reported an effective epoxidation system using monomeric tungstate and phosphate that proceeded under 2-phase conditions (eq 13) [47a]. After more work, this group reported not only the isolation from this system of the complex discussed previously, $\{PO_4[W(O)(O_2)_2]_4\}^{3-}$, but also the fact that this complex catalyzed reasonably selective and rapid epoxidation in nonprotic solvents [32,47b]. They further reasonably surmised from the data that this complex was probably the active oxidant under phase transfer conditions although hard mechanistic evidence (time resolved kinetics of product distributions, etc.) was then and is now lacking. Venturello and co-workers went on to document that this chemistry was effective for the oxidative cleavage of 1,2 diols to carboxylic acids and other reactions [47c,d].

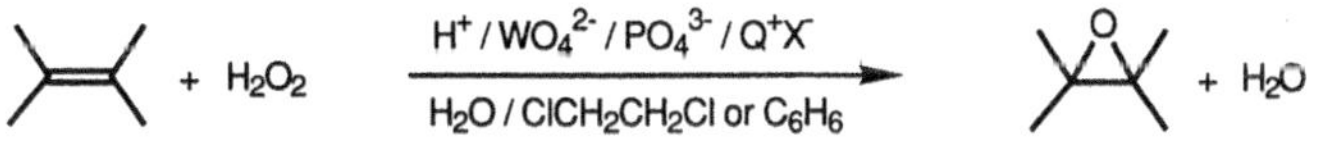

(13)

where Q^+X^- is an onium salt phase transfer catalyst

Table 1: Epoxidation of olefins in CHCl3 with 35% aq H2O2 in the presence of different catalysts (0.4 mol%) under Various Conditions.

Run	Substrate	Catal. (eq.)	Additive[a] (h)	Time (%)	Conv.[b] (%)	Product
1	1-octene	$H_3PW_{12}O_{40}$	CPC (3)	5	82	1,2-epoxyoctane (98)
2[c]	1-octene	$H_3PW_{12}O_{40}$		24	trace	
3	1-octene	$H_3PW_{12}O_{40}$	CPC (3)	24	58	1,2-epoxyoctane (8)
4	1-octene	$H_3PMo_{12}O_{40}$	CPC (3)	24	11	1,2-epoxyoctane (88)
5	1-octene	$H_4SiW_{12}O_{40}$	CPC (3)	24	trace	1,2-epoxyoctane (98)
6	1-octene	$H_3PMo_6W_6O_{40}$	CPC (3)	24	24	1,2-epoxyoctane (98)
7	1-octene	H_2MoO_4	CPC (3)	24	trace	
8	1-octene	H_2MoO_4	CPC (3)	24	trace	
9	1-octene	$H_3PW_{12}O_{40}$	CTMAB (3)[d]	24	80	1,2-epoxyoctane (95)
10	1-octene	$H_3PW_{12}O_{40}$	TOMAC (3)[e]	24	77	1,2-epoxyoctane (98)
11	1-octene	$H_3PW_{12}O_{40}$	THAB (3)[f]	24	11	1,2-epoxyoctane (98)
12	1-octene	CWP[g]		12	79	1,2-epoxyoctane (98)
13	1-octene	CMP[h]		24	41	1,2-epoxyoctane (98)
14	2-octene	$H_3PW_{12}O_{40}$	CPC (3)	5	98	2,3-epoxyoctane (96)
15		$H_3PW_{12}O_{40}$	CPC (3)	2	96	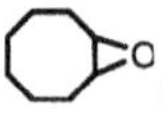 (98)
16	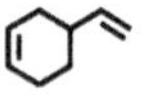	$H_3PW_{12}O_{40}$	CPC (3)	2	96	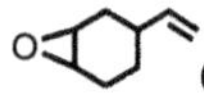 (98)[i]
17[j]		$H_3PW_{12}O_{40}$	CPC (3)	6	91	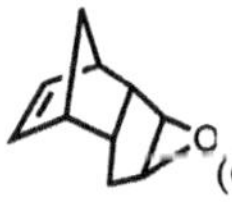 (60)
						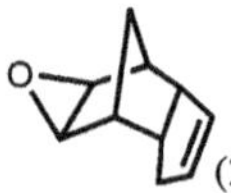 (22)  (18)

[a] CPC = cetylpyridinium chloride. [b] Conversion, determined by gas chromatography. [c] *t*-BuOH was used as the solvent. [d] CTMAB: $C_{16}H_{33}N^+(CH_3)_3Br^-$. [e] TOMAC: $CH_3N^+(C_8H_{17})_3Cl^-$. [f] THAB: $(C_4H_9)_4N^+Br^-$. [g] CWP: tris(cetylpyridinium)12-tungstophosphate.
[h] CMP: tris(cetylpyridinium)12-molybdophosphate. [i] Ratio of cis / trans = *ca.*1. [j] Room temp.

Ishii reported the epoxidation of allylic alcohols under 2-phase conditions using (cetylpyridinium)$_3$($PMo_{12}O_{40}$) [48a]. Subsequent work largely product distribution analyses, indicated that the analogous tungsten system was effective for the high selectivity and high conversion epoxidation of olefins, (ranging from terminal and unactivated to conjugated and functionalized), and also effective for oxidation of alcohols and alkynes, oxidative cleavage of vic-diols to carboxylic acids, and oxidative cyclization (lactonization) of diols [48,49]. Representative reactions are given in Tables 1-3 [49]. The selectivities/conversions in many of these processes are as high as those in any other homogeneous H_2O_2-based catalytic epoxidation system.

Table 2: Epoxidation of α,β-unsaturated acids with 35% aq H_2O_2 by polyoxometalates.

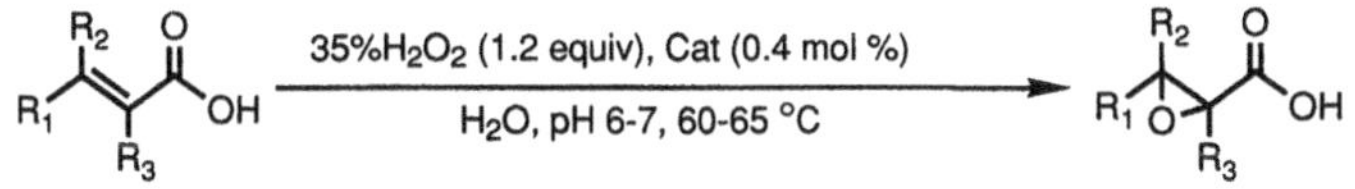

Run	R1	R2	R3	Catal.	Time(h)	Conv.(%)	Selec.(%)
1	CH_3	H	H	CWP[a]	2	90	98
2[b]	CH_3	H	H	CWP	12	8	35
3[c]	CH_3	H	H	CWP	12	no reaction	
4	CH_3	H	H	CMP[d]	12	no reaction	
5[e]	CH_3	H	H	CMP	12	49	45
6	CH_3	H	CH_3	CWP	0.7	89	97
7	CH_3	CH_3	H	CWP	4	70	83
8	*n*-C_3H_7	H	CH_3	CWP	8	93	97
9[f]	*n*-C_6H_{13}	H	H	CWP	8	56	80
10	Ph	H	H	CWP	24	no reaction	

[a] CWP: tris(cetylpyridinium)12-tungstophosphate. [b] pH 4-5. [c] pH 8-9. [d] CMP: tris(cetylpyridinium)12-molybdophosphate. [e] *t*-BuOH was used. [f] EtOH : H_2O (1:1) was used as solvent.

Table 3: Oxidation of Alcohols with 35% aq H_2O_2 by HPA under Various Conditions[a]

Run	Substrate	Catal.	Solv.	Conv.(%)[b]	Product (Selec.%)[b]
1	2-octanol	CWP[c]	*t*-BuOH	90	2-octanone (98)
2	2-octanol	CMP[d]	*t*-BuOH	31	2-octanone (98)
3	2-octanol	CMP	PhH	71	2-octanone (98)
4	2-octanol	WPA/CPC[e]	*t*-BuOH	23	2-octanone (90)
5	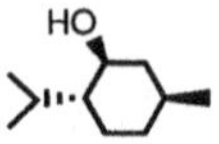	CWP	*t*-BuOH	94	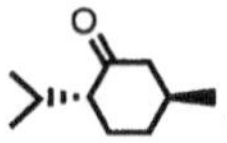 (98)
6	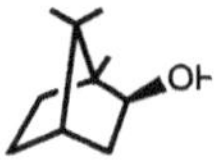	CWP	*t*-BuOH	98	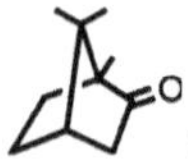 (98)
7	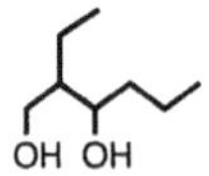	CWP	*t*-BuOH	98	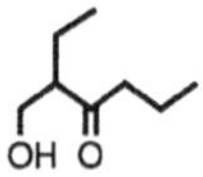 (98)

[a] Alcohol (2 mmol) was allowed to react with 35% H_2O_2 (6 mmol) in the presence of catalyst (2 mol%) in refluxing solvent (15 ml). [b] Determined by gas chromatography. [c] CWP: tris(cetylpyridinium)12-tungstophosphate. [d] CMP: tris(cetylpyridinium)12-molybdophosphate. [e] $H_3PW_{12}O_{40}$ + cetylpyridinium chloride.

Some points need to kept in mind in conjunction with the promising work involving organic oxidations by H_2O_2 catalyzed by d^0 polyoxotungstates. First, complexes and species isolated from catalytic reactions that are derived from the initial form of the catalyst do not necessarily have anything to do with the active catalyst, electronically or structurally. It is probably more common that metal complexes isolated from catalytic reactions are *not* involved than are involved in the main catalytic elementary processes in the mechanism. That is, they represent kinetic cul-de-sacs and do not lie along the principal reaction coordinate for the catalysis. Isolated complexes may or may not be kinetically convertible to reactive species that do lie along the reaction coordinate under the catalytic reaction conditions. Second, there has been little discussion in these systems as to what happens to the d^0 polyoxotungstate species as the reactions terminate. The d^0 metal oxides (one of the forms of WO_3 in these cases) often form and constitute kinetic dead-ends in various chemical and physical processes involving polyoxometalates but they may or may not form in these catalytic reactions. Remember that H_2O_2 is one of the few reagents that can react with and hence dissolve WO_3 and related metal oxides. This includes the reaction between H_2O_2 and tungstic acid, "H_2WO_4"; the latter is usually one form of hydrated WO_3 and is more appropriately written $WO_3 \cdot H_2O$. What is needed in these studies to define what is(are) the catalytically active [poly]tungstate specie(s) is some rigorous kinetic work that would, at a minimum, correlate the time dependence of product distributions and epoxide formation rates with the presence of various [poly]tungstate specie(s) in solution. Until this has been done, the true catalytically active complex(es) in these processes and the detailed mechanism for oxygenation of the substrate will remain enlightened speculation.

5.4. TMSP (IMA) systems

With the goal of selective C-H bond hydroxylation and other difficult reactions in mind, we sought to devise a catalytic system that, in addition to catalyzing electrophilic heterolytic oxidation processes such as those discussed above, could also catalyze the oxo transfer oxidation of organic substrates using H_2O_2 and other oxygen donors. This approach involves the use of transition metal-substituted polyoxometalate (TMSP) complexes, which are inorganic analogues of metalloporphyrins, IMA complexes for short, and hence the title of this section. Figure 3 shows some of the principal processes of interest in oxo transfer catalyzed by transition metal complexes (M) including oxo transfer to substrate (S) from the oxometal form of the catalyst, M=O, (the boxed-in equation) and regeneration of the oxometal form from the resting form of the catalyst.

Three generic routes for oxometal regeneration are indicated. The first, oxo transfer from an oxygen donor (DO), is drawn right under the box. The second, drawn right under this involves metal oxygenation followed by a series of reactions including two-electron reduction (usually in separate steps) and peroxometal intermediates. This type of process is operable in the principal biological system catalyzing oxo transfer and hydrocarbon oxidation in the biosphere, cytochrome P-450, and in other systems [3-17,22,23,25]. The third involves direct oxygenation of metal complexes via a metal-dioxygen complex

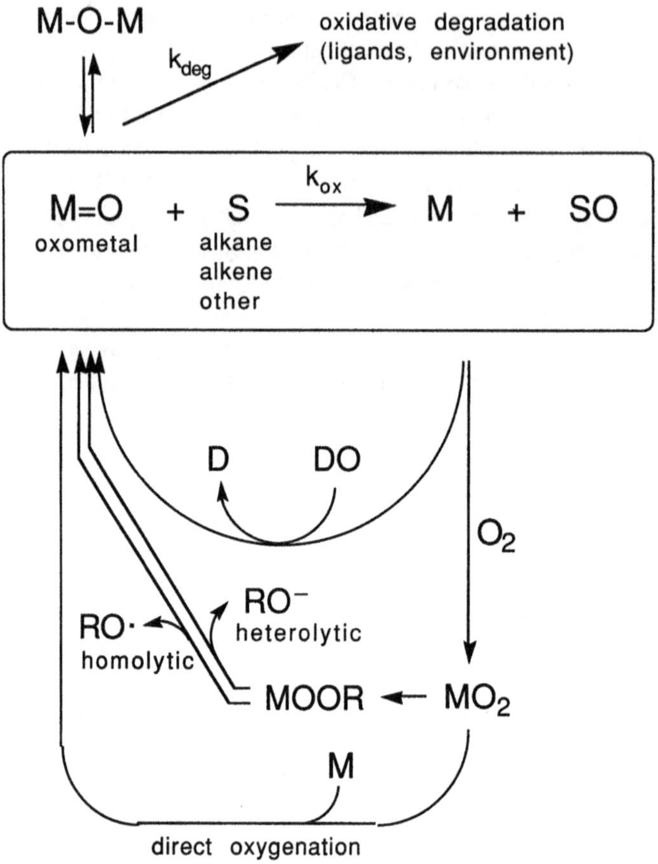

Figure 3. The principal processes of interest in oxo transfer catalyzed by transition metal complexes, M. Organic ligands have been omitted for the various metal species and none of the processes depicted imply specific mechanism. There can be, and in many cases doubtless are, several possible mechanisms for each process. See text.

and other intermediates and generates the kinetically competent oxometal species without consumption of reducing agent. Catalysts that effect this type of process with adequate rates, selectivities, and stabilities are rare and intensely sought by industrial and academic investigators worldwide. The first process, oxo transfer, could involve, in principle, H_2O_2, while the latter two reactions do not involve H_2O_2 as an external agent.

The potential disadvantages of catalytic systems that can effect oxo transfer (boxed reaction in Figure 3) with H_2O_2, is that this process can be more conducive to the generation of freely diffusing radical intermediates than processes involving purely heterolytic peroxo group activation. Many organic substrate oxo transfer oxidation reactions, however, clearly do not involve long lived radicals or radicals in solution. The advantages of catalytic systems that can effect oxo transfer are that a great deal can be done in a logical and systematic way to alter the reactivity of the oxometal intermediate (M=O in Figure 3) and the mechanisms whereby they react with alkenes or other types of organic substrates. Alterable factors will dictate whether or not free radicals will be intermediates [22]. Just as importantly, some oxometal species have reactivities that are simply not seen in H_2O_2-based oxidations (or oxidation by other oxygen donors) that proceed by heterolytic electrophilic activation mechanisms. Included in this category are attack on unactivated saturated C-H bonds and some types of aromatic C-H bonds.

Also shown in Figure 3 is the Achilles' heel of nearly all homogeneous catalysts for the oxidation of organic substrates by reagents such as H_2O_2, namely, the lack of thermodynamic stability to oxidation. Oxidative degradation, as discussed earlier, (k_{deg} in Figure 3) is energetically irreversible and is kinetically inevitable with homogeneous catalysts that contain organic ligands or any component of organic structure. One obvious solution to this problem was to use TMSP/IMA complexes as catalysts. The TMSP/IMA compounds exhibit a lot of the favorable attributes of metalloporphyrins, the catalysts of choice for oxo transfer in the biosphere (cytochrome P-450 has a Fe protoporphyrin IX or heme group at the active site) but unlike metalloporphyrins, TMSP/IMA compounds are not susceptible for oxidative degradation (Figure 4). The molecular properties of the pentadentate O_5 coordination polyhedron of the d^n, $n \neq 0$ transition metal active site in the TMSP/IMA complexes also fulfils another vital role in assuring long catalyst lifetime in oxo transfer processes. It holds the d^n, $n \neq 0$ transition metal ion firmly throughout oxidation state changes preventing deligation and subsequent formation of the insoluble and catalytically inactive (or less active and

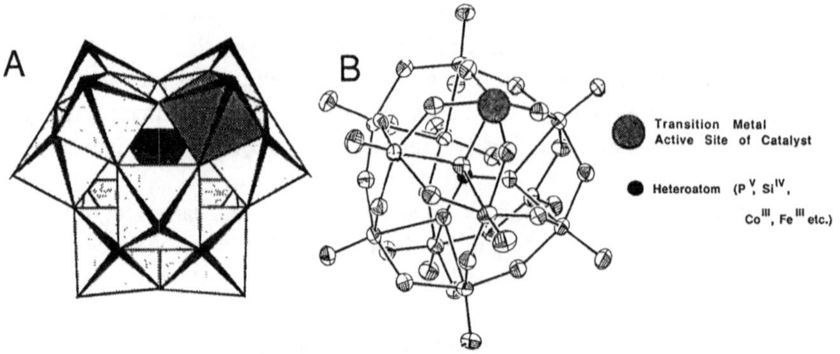

Figure 4. A typical transition metal-substituted polyoxometalate (TMSP) complex which constitutes an oxidatively resistant inorganic analog of the organic metalloporphyrin (inorganic metalloporphyrin analog or IMA for short). This particular TMSP/IMA complex, shown in polyhedral notation in A and atomic notation in B has C_s point group symmetry and is formed by replacing one d^0 W^{VI} ion in the parent Keggin ion, $(\alpha\text{-}X^{n+}W_{12}O_{40})^{(8-n)-}$ with a d^n, $n \neq 0$ transition metal ion. This ion is shaded darker in A and shown as the larger shaded sphere in B. The "heteroatom", X^{n+}, which can be a main group ion, most commonly P^{5+} or Si^{4+}, or a transition metal ion, resides in the middle of the molecule and is shaded darkest in A and in B. Oxidative stability is imparted by the fact that the polyoxometalate ligand environment is constituted only by d^0 transition metal and oxide ions.

selective) hydroxide/oxide of the d^n, $n \neq 0$ transition metal. We [22,50] and others [51-54] have demonstrated that k_{deg} in Figure 3 is effectively zero for these compounds making the key ratio of rates, k_{ox}/k_{deg}, some very large number. Tens of thousands of turnovers have been achieved using alkyl hydroperoxides and iodosylarenes as oxygen donors without catalyst (TMSP/IMA) degradation. Furthermore, these complexes are the only homogeneous catalysts that are stable to oxidative degradation in the absence of the kinetically protecting substrate (i.e. they are not decomposed when k_{ox}[substrate] is 0) [22,50].

Little has been done to date using TMSP/IMA complexes to catalyze H_2O_2 reactions, but work is in progress in our laboratory and elsewhere. As with the H_2O_2 oxidations catalyzed by d^0 systems that proceed by electrophilic heterolytic mechanisms, the structural and electronic factors that minimize disproportionation (eq 5) and other undesirable side reactions and maximize oxygenation (eq 6) must be fully delineated in context with a thorough knowledge of the energetic and mechanistic features of these complex redox reactions.

Acknowledgement

I wish to thank Interox Corporation for supporting our work with H_2O_2 and the National Science Foundation for supporting our basic research dealing with the redox properties of polyoxometalates. I wish further to thank Lucille A. Combs-Walker for preparing the tables.

References

1. R. A. Sheldon and J. K. Kochi, '*Metal-Catalyzed Oxidations of Organic Compounds*', Academic Press, New York 1981, Chapter 3.
2. (a) '*Organic Syntheses by Oxidation with Metal Compounds*', Eds. W. J. Mijs and C. R. H. I. de Jonge, Plenum, New York 1986; (b) G. W. Parshall, '*Homogeneous Catalysis. The Applications and Chemistry of Catalysis by Soluble Transition Metal Complexes*', Wiley-Interscience, New York 1980, and references therein.
3. (a) M. F. Powell, E. F. Pai and T. C. Bruice, *J. Am. Chem. Soc.*, **106**, 3277 (1984); (b) C. M. Dicken, F.-L. Lu, M. W. Nee and T. C. Bruice, *ibid.*, **107**, 5776, (1985); (c) L.-C. Yuan, and T. C. Bruice, *J. Chem. Soc. Chem. Commun.* 868 (1985); (d) T. S. Calderwood, W. A. Lee and T. C. Bruice, *J. Am. Chem. Soc.*, **107**, 5272 (1985); (e) L.-C. Yuan, T. S. Calderwood and T. C. Bruice, *ibid*, **107**, 8273 (1985); (f) C. M. Dicken, T. C. Woon and T. C. Bruice, *ibid.*, **108**, 1636 (1986); (g) L.-C. Yuan, T. C. Bruice, *ibid.*, **108**, 1643 (1986).
4. (a) C. K. Chang and M.-S. Kuo, *J. Am. Chem. Soc.*, **101**, 3413 (1979); (b) C. K. Chang and F. Ebina, *J. Chem. Soc. Chem. Commun.*, 778 (1981).
5. (a) J. P. Collman, T. Kodadek, S. A. Raybuck and B. Meunier, *Proc. Natl. Acad. Sci. USA*, **80**, 7039 (1983); (b) J. P. Collman, J. I. Brauman, B. Meunier, S. A. Raybuck and T. Kodadek, *ibid.*, **81**, 3245 (1984); (c) J. P. Collman, J. I. Brauman, B. Meunier, T. Hayashi, T. Kodadek and S. A. Raybuck, *J. Am. Chem. Soc.*, **107**, 2000 (1985); (d) J. P. Collman, T. Kodadek, S. A. Raybuck, J. I. Brauman and L. M. Papazian, *ibid.*, **107**, 4343 (1985); (e) J. P. Collman, T. Kodadek and J. I. Brauman, *ibid.*, **108**, 2588 (1986).
6. (a) P. S. Traylor, D. Dolphin and T. G. Traylor, *J. Chem. Soc. Chem. Commun.*, 279 (1984); (b) T. Mashiko, D. Dolphin, T. Nakano and T. G. Traylor, *J. Am. Chem. Soc.*, **107**, 3735 (1985); (c) T. G. Traylor, T. Nakano, B. E. Dunlap, P. S. Traylor and D. Dolphin, *ibid.*, **108**, 2782 (1986).
7. (a) J. T. Groves and T. Takahashi, *J. Am. Chem. Soc.*, **105**, 2073 (1983); (b) J. T. Groves, Y. Watanabe and T. J. McMurry, *ibid.*, **105**, 4489 (1983); (c) J. T. Groves and T. E. Nemo, *ibid.*, **105**, 6243 (1983); (d) J. T. Groves and T. E. Nemo, *ibid.*, **105**, 5786 (1983); (e) J. T. Groves and R. S. Myers, *ibid.*, **105**, 5791 (1983); (f) J. T. Groves and D. V. Subramaniam, *ibid.*, **106**, 2177 (1984); (g) J. T. Groves, R. Quinn, T. J. McMurry, M. Nakamura, G. Lang and B. Boso, *ibid.*, **107**, 354 (1985); (h) J. T. Groves and R. Quinn, *ibid.*, **107**, 5790 (1985); (i) J. T. Groves and Y. Watanabe, *ibid.*, **108**, 507 (1986).
8. (a) C. L. Hill and B. C. Schardt, *J. Am. Chem. Soc.*, **102**, 6374 (1980); (b) C. L. Hill and J. A. Smegal, *Nouv. J. Chim.*, **6**, 287 (1982); (c) J.A. Smegal and C. L. Hill, *J. Am. Chem. Soc.*, **105**, 3515 (1983); (d) J. A. Smegal and C. L. Hill, *ibid.*, **105**, 2920 (1983); (e) R. B. Brown, Jr., M. M. Williamson and C. L. Hill, *Inorg. Chem.*, **26**, 1602 (1987); (f) R. B. Brown, Jr. and C. L. Hill, *J. Org. Chem.*, **53**, 5762 (1988).
9. (a) J. R. Lindsay Smith and P. R. Sleath, *J. Chem. Soc. Perkin Trans. II.*, 1009 (1982); (b) J. R. Lindsay Smith, M. W. Nee, J. B. Noar and T. C. Bruice, *ibid.*, 255 (1984); (c) J. R. Lindsay Smith and D. N. Mortimer, *J. Chem. Soc. Chem. Commun.*, 64 (1985).
10. (a) D. Mansuy, M. Fontecave and J.-F. Bartoli, *J. Chem. Soc. Chem. Commun.*, 253 (1983); (b) D. Mansuy, J.-P. Mahy, A. Dureault, F. Bedi and P. Battioni, *ibid.*, 1161 (1984); (c) J. -P. Mahy, P. Battioni, D. Mansuy, J. Fisher, R. Weiss, J. Mispelter, I. Morgenstern-Badarau and P. Gans, *J. Am. Chem. Soc.*, **106**, 1699 (1984); (d) M. Fontcave and D. Mansuy, *J. Chem. Soc. Chem. Commun.*, 879 (1984); (e) D. Mansuy, J. Leclaire, M. Fontecave and P. Dansette, *Tetrahedron*, **40**, 2847 (1984); (f) D. Mansuy,

P. Battioni and J.-P. Renaud, *J. Chem. Soc. Chem. Commun.*, 1255 (1984); (g) D. Mansuy, P. Battioni, J. -P. Renaud and P. Guerin, *ibid.*, 155 (1985); (h) J.-P. Renaud, P. Battioni, J. F. Bartoli and D. Mansuy, *ibid.*, 888 (1985); (i) J.-P. Mahy, P. Battioni and D. Mansuy, *J. Am. Chem. Soc.*, **108**, 1079 (1986); (j) P. Battioni, J.-P. Renaud, J. F. Bartoli and D. Mansuy, *J. Chem. Soc., Chem. Commun.*, 341 (1986); (k) P. Battioni, J. F. Bartoli, P. Leduc, M. Fontecave and D. Mansuy, *ibid.*, 791 (1987).

11. (a) E. Guilmet and B. Meunier, *Tetrahedron Lett.*, **23**, 2449 (1986); (b) O. Bortolini and B. Meunier, *J. Chem. Soc. Chem. Commun.*, 1204 (1983); (c) B. Meunier, E. Guilmet, M.-E. De Carvalho and R. Poilblanc, *J. Am. Chem. Soc.*, **106**, 6668 (1984); (d) B. De Poorter, M. Ricci, B. Meunier, *Tetrahedron Lett.*, **26**, 4459 (1985).
12. (a) M. Nappa and C. A. Tolman, *Inorg. Chem.*, **24**, 4711 (1985); (b) C. A. Tolman, J. D. Druliner, M. J. Nappa and N. Herron in '*Activation and Functionalization of Alkanes* ', Ed. C. L. Hill, Wiley, New York, Chapter 10 (1989).
13. R. J. M. Nolte, J. A. S. J. Razenberg and R. Schuurman, *J. Am. Chem. Soc.*, **108**, 2751 (1986).
14. (a) K. S. Suslick, B. Cook and M. Fox, *J. Chem. Soc. Chem. Commun.*, 580 (1985); (b) B. R. Cook, T. J. Reinert and K. S. Suslick, *J. Amer. Chem. Soc.*, **108**, 7281 (1986); (c) K. S. Suslick, F. V. Acholla, B. R. Cook, *ibid.*, **109**, 2818 (1987),
15. (a) I. Tabushi and K. Morimitsu, *J. Am. Chem. Soc.*, **106**, 6871 (1984); (b) I. Tabushi, M. Kodera and M. Yokoyama, *ibid.*, **107**, 4466 (1985); (c) I. Tabushi and M. Kodera, *ibid.*, **108**, 1101 (1986).
16. (a) T. G. Traylor, J. C. Marsters, T. Nakano and B. E. Dunlap, *J. Am. Chem. Soc.*, **107**, 5537 (1985); (c) T. G. Traylor, Y. Iamamoto and T. Nakano, *ibid.*, **108**, 3529 (1986).
17. E. W. Svastits, J. H. Dawson, R. Breslow and S. H. Gellman, *J. Amer. Chem. Soc.*, **107**, 6427 (1985).
18. F. A. Cotton and G. Wilkinson, '*Advanced Inorganic Chemistry* ', 5th Edition, Wiley, New York 1988, p 456.
19. W. M. Latimer, '*The Oxidation States of the Elements and Their Potentials in Aqueous Solutions* ', 2nd Edition, Prentice-Hall, Englewood Cliffs, New Jersey, 1952.
20. For example see (a) A. F. Ghiron and R. C. Thompson, *Inorg. Chem.*, **28**, 3647 (1989). (b) L. M. Schwane and R. C. Thompson, *ibid.*, **28**, 3938 (1989).
21. R. H. Holm, *Chem. Rev.*, **87**, 1401 (1987).
22. C. L. Hill in '*Activation and Functionalization of Alkanes* ', Ed. C. L. Hill, Wiley, New York 1989, Chapter 8.
23. J. T. Groves in '*Metal Ion Activation of Dioxygen* ', Ed. T. G. Spiro, Wiley, New York 1980, Chapter 3.
24. M. Masarwa, H. Cohen, D. Meyerstein, D. L. Hickman, A. Bakac, and J. H. Espenson, *J. Am. Chem. Soc.*, **110**, 4293 (1988).
25. Recent reviews: (a) J. T. Groves and T. J. McMurry in '*Cytochrome P-450* ', Ed. P. R. Ortiz de Montellano, Plenum, New York 1986, Chapter 1; (b) P. R. Ortiz de Montellano in ref 25a, Chapter 7; (c) C. L. Hill in '*Advances in Oxygenated Processes* ', vol I, Ed. A. L. Baumstark, ? 1988, Chapter 1, p 1; (d) K. S. Suslick, in '*Activation and Functionalization of Alkanes* ', Ed. C. L. Hill, Wiley, New York 1989, Chapter 7.
26. Recent reviews on metalloporphyrin catalyzed oxygenation: (a) B. Meunier, *Bull. Soc. Chim. France*, **4**, 578 (1986); (b) D. Mansuy, *Pure and Appl. Chem.*, **59**, 759 (1987).
27. (a) J. M. Aubry and B. Cazin, *Inorg. Chem.*, **27**, 2013 (1988); (b) J. M. Aubry, *J. Am. Chem. Soc.*, **107**, 5844 (1985).

28. General reviews on polyoxometalates: (a) M. T. Pope '*Heteropoly and Isopoly Oxometalates* ', Springer Verlag, New York 1983; (b) V. W. Day and W. G. Klemperer, *Science*, **228**, 533 (1985); (c) M. T. Pope in '*Comprehensive Coordination Chemistry* ', Eds. G. Wilkinson, R. D. Gillard and J. A. McCleverty, Pergamon Press, New York 1987, Vol. 3, Chapter 38; (d) Y. Jeannin and M. Fournier, *Pure Appl. Chem.*, **59**, 1529 (1987).
29. (a) C. L. Hill, M. Weeks and R. F. Schinazi, *J. Med. Chem.*, **33**, 2767 (1990); (b) C. L. Hill, M. Hartnup, M. Faraj, M. Weeks, C. M. Prosser-McCartha, R. B. Brown, R. F. Schinazi and J.-P. Sommadossi, in '*Advances in Chemotherapy of AIDS, Pharmacology and Therapeutics* ', Eds. R. Diasio and J. -P. Sommadossi, Pergamon, New York 1990, p 33.
30. J. N. Campbell, A. C. Dengel, and W. P. Griffith, *Polyhedron*, **8**, 1379 (1989);
31. N. J. Campbell, A. C. Dengel, C. J. Edwards, and W. P. Griffith, *J. Chem. Soc. Dalton Trans.*, 1203 (1989).
32. C. Venturello, R. D'Aloisio, J. C. J. Bart, and M. Ricci, *J. Mol. Catal.*, **32**, 107 (1985).
33. Structure of $K_6[W_4O_8(O_2)_6(CO_3)]$ $6H_2O$: R. Stomberg, *Acta Chem. Scand.*, **A 39**, 507 (1985).
34. Structure of $[W_2(O)_3(H_2O)_2(O_2)_4]^{2-}$: F. W. B. Einstein and B. R. Penfold, *Acta Cryst.*, **17**, 1127 (1964); see also (b) N. J. Campbell, A. C. Dengel, C. J. Edwards, and W. P. Griffith, *J. Chem. Soc., Dalton Trans.*, 1203 (1989).
35. (a) Polynuclear fluoro peroxo complexes of Zr: B. N. Chernyshow, N. A. Didenko, B. V. Bukvetskii, A. V. Gerasimenko, V. Ya. Kavun, and S. S. Sergienko, *Zh. Neorg. Khim.*, **34**, 2786 (1989); Ti: B. N. Chernyshow, N. A. Didenko, A. V. Gerasimenko, and V. Ya. Kavun, *Zh. Neorg. Khim.*, **34**, 2179 (1989); Mo: C. Djordjevic, J. L. Gunderson, B. A. Jacobs, A. Bradley, and E. Sinn, *Polyhedron*, **8**, 541 (1989). (b) See also polyperoxothorium compounds: C. R. Bhattacharjee, M. K. Chaudhuri, R. N. Dutta Purkayastha, *Inorg. Chem. Acta*, **160**, 147 (1989), and review of peroxometal compounds (in Russian): S. O. Gerasimova and Yu. Ya. Kharitonov, *Koord. Khim.*, **14**, 1443 (1988).
36. In addition to the binary or homoleptic peroxometalate complexes, there are a larger number of polyperoxometalate complexes that contain one or more organic ligands that have been characterized by X-ray crystallography. We will not attempt to review this considerable literature -- one recent example: C. Djordjevic, M. Lee, and E. Sinn, *Inorg. Chem.*, **28**, 719 (1989). Also see older review: C. Djordjevic, *Chem. Britain*, **18**, 554, (1982).
37. For a review see: M. Misono, *Catal. Rev.-Sci. Eng.*, **29**, 269 (1987).
38. (a) J. B. Black, N. J. Clayden, P. L. Gai, J. D. Scott, E. M. Serwicka, and J. B. Goodenough, *J. Catal.*, **106**, 1 (1987); (b) J. B. Black, J. D. Scott, E. M. Serwicka, and J. B. Goodenough, *Ibid.*, **106**, 16 (1987); (c) E. M. Serwicka, J. B. Black, and J. B. Goodenough, *Ibid.*, **106**, 16 (1987).
39. R. Pearce and W. Patterson '*Catalysis and Chemical Processes* ', Hill, London 1981, p 279.
40. For reviews see: (a) I. V. Kozhevnikov and K. I. Matveev, *Russ Chem. Rev. (Engl Transl.)*, **51**, 1075 (1982); (b) I. V. Kozhevnikov and K. I. Matveev, *Appl. Catal.*, **5**, 135 (1983).
41. Representative recent papers on the photooxidation of organic substrates other than alkanes by polyoxometalates: (a) D. Attanasio and L. Suber, *Inorg. Chem.*, **28**, 3781 (1989); (b) R. Akid and J. R. Darwent, *J. Chem. Soc. Dalton Trans.*, 395 (1985); (c) M. A. Fox, R. Cardona and E. Gaillard, *J. Am. Chem. Soc.*, **109**, 6347 (1987); (d) C. L. Hill

and D. A. Bouchard, *ibid.*, **107**, 5148 (1985); (e) K. Nomiya, T. Miyazaki, K. Maeda and M. Miwa, *Inorg. Chim. Acta* , **127**, 65 (1987); (f) E. Papaconstantinou, *Chem Soc. Rev.*, **18**, 1 (1989); (g) E. N. Savinov, S. S. Saidkhanov, V. N. Parmon and K. I. Zamaraev, *Doklady, Phys. Chem. SSSR*, **272**, 741 (1983); (h) M. D. Ward, J. F. Brazdil, S. P. Mehandu and A. B. Anderson, *J. Phys. Chem.*, **91**, 6515 (1987); (i) T. Yamase and R. Watanabe, *J. Chem. Soc. Dalton Trans.*, 1669 (1986); (j) B. Kraut and G. Ferraudi, *Inorg. Chem.*, **28**, 2692 (1989).

42. Unusual transformations of organic substrates catalyzed by polyoxometalates in the presence of light: (a) R. F.. Renncke, and C. L. Hill, *J. Am. Chem. Soc.*, **110**, 5461, (1988); (b) R. C. Chambers and C. L. Hill, *Inorg. Chem.*, **28**, 2509 (1989); (c) C. L. Hill, R. F. Renneke, L. A. Combs, *New J. Chem.*, **13**, 701 (1989); (d) C. M. Prosser-McCartha and C. L. Hill, *J. Am. Chem. Soc.*, **112**, 3671 (1990). (e) D. Sattari and C. Hill, *J. Chem. Soc. Chem. Commum.*, 634 (1990); (f) R. F. Renneke, M. Pasquali, C. L. Hill, *J. Am. Chem. Soc.*, **112**, 6585 (1990); (g) R. C. Chambers and C. L. Hill, *J. Am. Chem. Soc.*, **112**, 8427 (1990).
43. (a) B. M. Trost and Y. Masuyama, *Tetrahedron Lett.*, 25, 173 (1984); (b) B. M. Trost and Y. Masuyama, *Isr. J. Chem.*, **24**, 134 (1984).
44. (a) H. Furukawa, T. Nakamura, H. Inagaki, E. Nishikawa, C. Imai, and M. Misono, *Chem. Lett.*, 877 (1988); (b) H. Inagaki, T. Nakamura, E. Nishikawa, C. Imai, and M. Misono, *Japan. Patent* , 01 190 647, (1989) through *Chem. Abstr.*, 112(11):98357a.
45. M Schwegler, M. Floor, and H. van Bekkum, *Tetrahedron Lett.*, **29**, 823 (1988).
46. F. P. Ballistreri, S. Failla, E. Spina, and G. A. Tomaselli, *J. Org. Chem.*, **54**, 947 (1989).
47. (a) C. Venturello, E. Alneri, and M. Ricci, *J. Org. Chem.*, **48**, 3831 (1983); (b) C. Venturello and R. D'Aloisio, *J. Org. Chem.*, **53**, 1553 (1988); (c) C. Venturello, and M. Ricci, *Ibid.*, **51**, 1599 (1986); (d) C. Venturello, R. D'Aloisio, and M. Ricci, *Eur. Patent*, 109 273 (1984).
48. (a) Y. Matoba, H. Inoue, J. Akagi, T. Okabayashi, Y. Ishii, and M. Ogawa, *Syn. Commun.*, **14**, 865 (1984); (b) Y. Ishii, *Japan. Patent* , 62 234 550 (1987) through*Chem. Abstr.*, 109(10):75696j; (c) Y. Ishii, K. Yamawaki, T. Yoshida, T. Ura, and M. Ogawa, *J. Org. Chem.*, **52**, 1868 (1967); (d) Y. Ishii, K. Yamawaki, T. Ura, H. Yamada, T. Yoshida, and M. Ogawa, *J. Org. Chem.*, **53**, 3587 (1988). (e) Y. Ishii, K. Yamawaki, T. Ura, H. Yamada, H. Nishihara, T. Yoshida, and M. Ogawain '*The Role of Oxygen in Chemistry and Biochemistry* ', Eds. W. Ando and Y. Moro-Oka, Elsevier, New York 1988, p 325.
49. Review: Y. Ishii and M. Ogawa, in '*Reviews on Heteroatom Chemistry* ', vol. 3, Ed. S. Oae, MYU, Tokyo 1990, p 121.
50. (a) C. L. Hill and R. B. Brown, Jr. *J. Am. Chem. Soc.*, **108**, 536 (1986); (b) M. Faraj, C.-H. Lin and C. L. Hill, *New J. Chem.*, **12**, 745 (1988); (c) C. L. Hill, R. B. Brown, Jr. and R. F. Renneke, *Prepr. Am. Chem. Soc. Div. Pet. Chem.*, **32**, [No. 1] 205 (1987); (d) C. L. Hill, R. F. Renneke, M. K. Faraj and R. B. Brown, Jr. in '*The Role of Oxygen in Chemistry and Biochemistry* ', Eds. W. Ando and Y. Moro-Oka, Elsevier, New York 1988, p 185; (e) C. L. Hill., *U.S. Patent* , 4 864 041 (1989).
51. (a) R. Neumann and C. Abu-Gnim, *J. Chem. Soc. Chem. Commun.*, 1324 (1989); (b) R. Neumann and C. Abu-Gnim, *J. Am. Chem. Soc.*, **112**, 6025 (1990).
52. (a) D. E. Katsoulis and M. T. Pope, *J. Chem. Soc. Chem. Commun.*, 1186 (1986); (b) D. Katsoulis and M. T. Pope, *J. Chem. Soc. Dalton Trans.*, 1483 (1989); (c) see also deoxygenation by "heteropoly browns": K. Piepgrass and M. T. Pope, *J. Am. Chem. Soc.*, **111**, 75 (1989)

53. R. G. Finke, D. K. Lyon, K. Nomiya and T. J. R. Weakley, Abstract and talk **INOR 731**, *1989 Int. Chem. Cong. Pac. Basin Societies*, Honolulu 1989.
54. F. Liotta and B. Handwerker, unpublished research.
55. (a) M. Mugdan and D. P. Young, *J. Chem. Soc.*, 2988 (1949); (b) D. P. Young, *Brit. Patent* , 654 764 (1951) to Distillers Co.; (c) C. W. Smith and G. B. Payne, *U.S Patent* , 2 786 854 (1957).
56. J. Prandi, H. B. Kagan, and H. Mimoun, *Tetrahedron Lett.*, **27**, 2617 (1986).
57. S.-I. Murahashi, T. Oda, T. Sugahara, and Y. Masui, *J. Chem. Soc. Chem. Commun.*, 1471 (1987).
58. (a) M. Quenard, V. Bonmarin, and G. Gelbard, *Tetrahedron Lett.*, **28**, 2237 (1987); (b) M. Quenard, V. Bonmarin, G. Gelbard, and L. Krumenacker, *New J. Chem.*, **13**, 183 (1989).
59. For reviews see: (a) H. Mimoun, *Catal. Today* , **1**, 281 (1987); (b) F. Di Furia and G. Modena, *Rev. Chem. Intermed.*, **6**, 51 (1985); (c) H. Mimoun in '*The Chemistry of Peroxides* ', Ed. S. Patai, Wiley, Chichester UK 1983, p 463; (d) O. Bortolini, V. Conte, F. Di Furia, and G. Modena in '*The Role of Oxygen in Chemistry and Biochemistry* ', Eds. W. Ando and Y. Moro-Oka, Elsevier, New York 1988, p 301.

INDEX

Catalysis by Metal Complexes

Series Editors:

R. Ugo, *University of Milan, Milan, Italy*
B. R. James, *University of British Columbia, Vancouver, Canada*

1.* F. J. McQuillin: *Homogeneous Hydrogenation in Organic Chemistry.* 1976
ISBN 90-277-0646-8
2. P. M. Henry: *Palladium Catalyzed Oxidation of Hydrocarbons.* 1980
ISBN 90-277-0986-6
3. R. A. Sheldon: *Chemicals from Synthesis Gas.* Catalytic Reactions of CO and H_2. 1983
ISBN 90-277-1489-4
4. W. Keim (ed.): *Catalysis in C_1 Chemistry.* 1983 ISBN 90-277-1527-0
5. A. E. Shilov: *Activation of Saturated Hydrocarbons by Transition Metal Complexes.* 1984 ISBN 90-277-1628-5
6. F. R. Hartley: *Supported Metal Complexes.* A New Generation of Catalysts. 1985
ISBN 90-277-1855-5
7. Y. Iwasawa (ed.): *Tailored Metal Catalysts.* 1986 ISBN 90-277-1866-0
8. R. S. Dickson: *Homogeneous Catalysis with Compounds of Rhodium and Iridium.* 1985
ISBN 90-277-1880-6
9. G. Strukul (ed.): *Catalytic Oxidations with Hydrogen Peroxide as Oxidant.* 1993
ISBN 0-7923-1771-8
10. A. Mortreux and F. Petit (eds.): *Industrial Applications of Homogeneous Catalysis.* 1988 ISBN 90-277-2520-9
11. N. Farrell: *Transition Metal Complexes as Drugs and Chemotherapeutic Agents.* 1989
ISBN 90-277-2828-3
12. A.F. Noels, M. Graziani and A.J. Hubert (eds.): *Metal Promoted Selectivity in Organic Synthesis.* 1991 ISBN 0-7923-1184-1
13. L.I. Simándi: *Catalytic Activation of Dioxygen by Metal Complexes.* 1992
ISBN 0-7923-1896-X

Kluwer Academic Publishers – Dordrecht / Boston / London

**Volume 1 is previously published under the Series Title:*
Homogeneous Catalysis in Organic and Inorganic Chemistry.

GPSR Compliance

The European Union's (EU) General Product Safety Regulation (GPSR) is a set of rules that requires consumer products to be safe and our obligations to ensure this.

If you have any concerns about our products, you can contact us on ProductSafety@springernature.com

In case Publisher is established outside the EU, the EU authorized representative is:

Springer Nature Customer Service Center GmbH
Europaplatz 3
69115 Heidelberg, Germany

Batch number: 10809441

Printed by Printforce, the Netherlands